THE REMNANTS OF RACE SCIENCE

Race, Inequality, and Health

RACE, INEQUALITY, AND HEALTH
Edited by Samuel Kelton Roberts Jr. and Michael Yudell

The Race, Inequality, and Health series explores how forms of racialization have created a wide range of phenomena, from producing inequities in health and healthcare to inspiring social movements around health. The goal of this series is to publish field-defining works across history, the social sciences, the biological sciences, and public health that deepen our understanding of how claims about race and race difference have affected health and society.

Rob DeSalle and Ian Tattersall, *Troublesome Science: The Misuse of Genetics and Genomics in Understanding Race*
Michael Yudell, *Race Unmasked: Biology and Race in the Twentieth Century*

The Remnants of
Race Science

UNESCO and Economic Development in the Global South

Sebastián Gil-Riaño

Columbia University Press   New York

Columbia University Press
Publishers Since 1893
New York   Chichester, West Sussex
cup.columbia.edu

Library of Congress Cataloging-in-Publication Data
Names: Gil-Riaño, Sebastián, author.
Title: The remnants of race science : UNESCO and economic development in
   the global south / Sebastián Gil-Riaño.
Description: New York : Columbia University Press, 2023. | Includes
   bibliographical references and index.
Identifiers: LCCN 2022061385 | ISBN 9780231194341 (hardback) |
   ISBN 9780231194358 (trade paperback) | ISBN 9780231550772 (ebook)
Subjects: LCSH: Economic development—Developing countries—History. |
   Anti-racism—History | Unesco—History.
Classification: LCC HC59.7 .G489 2023 | DDC 338.9172/4—dc23/eng/20230526
LC record available at https://lccn.loc.gov/2022061385

Printed and bound by CPI Group (UK) Ltd, Croydon, CR0 4YY

Cover design: Milenda Nan Ok Lee
Cover image: Courtesy of UNESCO

To Barry Kenneth Wright (1951–2021)

Contents

Acknowledgments

This book was long in the making, and I owe many people in several different countries and continents thanks for their help along the way. I first learned about the UNESCO Statements on Race when I was an undergraduate at the University of King's College in Halifax, Nova Scotia. When I began my undergraduate honors thesis and told Gordon McOuat, my advisor, I wanted to write about race and science, he told me to have a look at the UNESCO statements. I did have a look and promptly shelved the topic. Though I could not figure out how to write about the UNESCO statements then, I did keep the topic in mind when I went to graduate school at the University of Toronto and began to explore the statements' connections to Latin America. So I must thank Gordon for sending me down this rabbit hole and for serving as an excellent undergraduate mentor at King's along with Stephen Snobelen.

At the University of Toronto, where my interest in UNESCO's race campaign turned into a doctoral dissertation, I was fortunate to have Michelle Murphy, now Murphy, as my supervisor. Murphy was an excellent mentor and has set an incredibly high standard for me to follow as both a scholar and a person. To this day, when I think and write, I often find myself thinking back to conversations I had with Murphy during grad school. In Toronto, I also benefited from the mentorship and feedback of Marga Vicedo and Mark Solovey, who served on my dissertation committee, and from Eric Jennings and Alejandra Bronfman, who were my external examiners. Conversations with Brian Beaton, Anne Emanuelle-Birn, Michael Petit, Sarah Tracy, and Alexandra Widmer were also very helpful during this phase.

My dissertation began to turn into a book during the two years I spent as a postdoctoral fellow at the University of Sydney as part of the Race and Ethnicity in the Global South (REGS) research project supported by Warwick Anderson's Australian Research Council Laureate Fellowship. At the University of Sydney, Warwick helped me to expand the geographic framing of my project as well as my academic networks. Thank you for being a fabulous mentor, Warwick. At the University of Sydney, I also joined a vibrant research team that worked and had many memorable meetings on the top floor of the University of Sydney's sandstone Quadrangle, aka Warwickshire. During this time, I learned immensely from my REGS colleagues and friends including Adrian Atkins, Ann Curthoys, Jamie Dunk, Janet Golden, Vanessa Heggie, Miranda Johnson, Chin Jou, Sam Killmore, Hans Pols, Pamela Maddock, Ricardo Roque, Ben Silverstein, Sarah Walsh, and Christina Winter. From my time in Australia, I must also thank Alison Bashford, David Brophy, Clare Corbould, John Gagné, Michael Goodman, Philippa Hetherington, Marilena Indelicato, Emma Kowal, Michael McDonnell, Dirk Moses, Fiona Paisley, Andres Rodriguez, Hélène Sirantoine, Glenda Sluga, Sophie Loy-Wilson, and Catherine Waldby for their support.

This book followed me to Philadelphia and the University of Pennsylvania where I found an intellectual home in the History and Sociology of Science (HSS) Department. At Penn HSS, I joined an extended and supportive community of scholars who have helped me in countless ways. My HSS colleagues past and present have provided valuable support, both academic and personal. Many thanks to Robby Aronowitz, David Barnes, Etienne Benson, Meggie Crnic, Stephanie Dick, Andi Johnson, Harun Küçük, Susan Lindee, Beth Linker, Amy Lutz, Jessica Martucci, Ramah McKay, Projit Mukharji, John Tresch, Elly Truitt, Beans Velocci, and Adelheid Voskuhl. This book would not have been possible without the research assistance and encouragement of Taylor Dysart who helped with crucial tasks at key moments. Austin Cooper, Nicole Welk-Joerger, Karen Kim, and Felipe Augusto Ribeiro also provided valuable research help. I must also thank Penn HSS's brilliant graduate students and recent alumni for their enthusiasm for my work and for everything I've learned from their projects. A special shout-out goes to Eram Alam, Arnav Bhattacharya, Brigid Prial, Cameron Brinitzer, Claire Conklin Sable, Nikhil Joseph Dharan, Rosanna Dent, Adrien Gau, Ngamlienlal Kipgen, Mary Mitchell, Zachary Loeb, Alexis Rider, Koyna Tomar, and Sam Schirvar.

At Penn and Philadelphia, I have had the good fortune of meeting a wonderful community of Latin Americanists whose insights and support have enriched this project. Many thanks to Daniel Aldana-Cohen, Juan Ignacio Arboleda, Catherine Bartch, Erica Beckman, Marcelo Bucheli, Juan Castrillón, Tulia Faletti, Ann Farnsworth, Tulia Faletti, Kristina Lyons, Marcy Norton, Gabriel Rocha, Melissa Texeira, and Jorge Tellez.

During my time at Penn, I have participated in the Faculty Writing Retreat on multiple occasions, and some of the most crucial parts of this book were written in this venue. A big thanks to the organizers of the retreat, Kayako Ayano and Jennifer Moore, and to the many colleagues whose presence and discussion helped create such a fertile writing environment. During the retreat, I also had the privilege to discuss my work with Christopher Lura who gave terrific editorial advice. At Penn I have also benefited from conversations with Nikhil Anand, David Eng, Dorothy Roberts, and Robert Vitalis.

In Philadelphia, parts of this project were written in writing groups where I benefited from the solidarity of David Amponsah, Judy Kaplan, Brian Kim, Raphael Krut-Landau, Ada Maria Kuskowski, and Jeffrey Womack. Other Penn faculty members who offered crucial bits of career advice that made this book possible are Emily Steiner, Kok-Chor Tan, and Heather Williams. And a special thanks to Cam Gray for some decisive pep talks that helped get this book across the finish line.

This book benefited immensely from a manuscript workshop held by the University of Pennsylvania's Wolf Humanities Institute. I must thank Jamal J. Elias, Pamela Horn, and Sara Varney for their help organizing the workshop. And I must give big thanks to my two excellent readers, Jenny Reardon and Julia Rodriguez, who provided fabulous feedback on an earlier version of the manuscript.

Parts of this book are based on previously published work. A version of Chapter 2 was published as "Relocating Anti-Racist Science: the 1950 UNESCO Statement on Race and Economic Development in the Global South." *The British Journal for the History of Science* 51, no. 2

(2018), 281–303. Parts of Chapter 4 were published in an edited volume as "Becoming an Area Expert During the Cold War: Americanism and Lusotropicalismo in the Transnational Career of Anthropologist Charles Wagley, 1939–1971," in *Cold War Social And Behavioral Sciences in Transnational Contexts*, Mark Solovey and Christian Daye, Eds, (Palgrave, UK: 2021). pp.127-159. A version of chapter 5 was published as "Risky Migrations: Race, Latin Eugenics, and Cold War Development in the International Labor Organization's Puno–Tambopata Project in Peru, 1930–60," *History of Science* 60, no 1 (2021), 41–68. Research for this book was funded by the Social Sciences and Humanities Research Council of Canada, by the Australian Research Council Laureate Fellowship Project "Race and Ethnicity in the Global South" and by the University of Pennsylvania's University Research Foundation.

Beyond Philadelphia, I have incurred many debts to colleagues who expressed enthusiasm for this project and provided crucial feedback or support at different stages. Thank you to Begüm Adalet, Geoff Bil, Eve Buckley, Stephen Casper, Marcos Chor Maio, Alice Conklin, Marcos Cueto, Sarah Dunstan, Vivette García-Deister, Pablo Gomez, Margarita Fajardo, Christopher Heaney, Adrianna Link, Carlos López Beltrán, Terence Keel, Veronika Lipphardt, Yvonne Maggie, Emily Marker, Matt Matsuda, Gisela Mateos, Ian Merkel, Staffan Müller-Wille, Juan Pablo Murillo, Tess Lanzarotta, Antonine Nicoglou, Joanna Radin, Gabriela Soto Laveaga, Mircea Raianu, Marco Ramos, Juno Salazar-Parreñas, Suman Seth, Ashanti Shih, Edna Suárez Díaz, Joel Vargas, Ricardo Ventura Santos, Jaipreet Virdi, Jessica Wang, and Adam Warren. And a special thanks goes to Karin Rosemblatt, who read and provided invaluable feedback on both the proposal for this project and the first draft of the manuscript.

It has been a pleasure to work with the excellent editorial team at Columbia University Press (CUP). I must thank Mike Yudell and Sam Roberts for recruiting me to CUP's Race, Inequality, and Health series and for close reading of the manuscript as well as moral support. Bridget Flannery-McCoy provided help with the initial stages of putting together a proposal, and Stephen Wesley expertly guided me through all the subsequent stages of getting this book to print and provided helpful editing of key parts of the manuscript. Many thanks to my two anonymous reviewers, who provided constructive feedback that made me believe it was possible to get this project to the finish line and who also offered helpful revision strategies. And thanks also to CUP's Faculty Advisory Board, whose members provided timely feedback that has helped sharpen the framing of this book.

Much of this book was written while my family and I lived in Riepe College House at the University of Pennsylvania. Friendship and support from the Riepe community helped my family and I get through some of the more challenging phases of this project, which include the life-altering events triggered by the COVID-19 pandemic. A big thanks to Ralph and Ellen Rosen for their lively conversation, coffee, baked goods, delicious meals, and Trader Joe's runs. Thank you also to Marilynne Diggs-Thompson and Alaina Bailey, who did such a wonderful job of cultivating a sense of belonging among Riepe staff and residents. And thanks also to the other Faculty Fellows who have been an integral part of the extended Riepe family: Carol Muller, Marsha Richardson, and Amy Stornaiuolo. Finally, a big thanks to all the wonderful RAs and GAs who offered camaraderie and, in some cases, child care over the years—Kaylee Arndt, Hector Kilgoe, Aloma Lopez, Natasha Napolitano, Zoe Osborne, Prateekshit Pandey, Aloma Lopez, Janessa Reeves, Josie Wiklund, and Andrew Wu.

This book would not have been possible without the expert child care provided by the teachers and staff at St. Mary's Nursery School. This support became even more crucial during the pandemic, and I must thank Gen Abara, Kevin Adickes, Rosa Brown, Suria Chen, Traci Childress, Fatime Diop, Kai Farrell, Mamatha Naraparaju, Deserae Warren, and Natasha Menn.

Most important, I want to thank the members of my family, who have been a source of steadfast support and patience even as the years to completing this book dragged on. I must thank my mother, Pilar Riaño-Alcala, for her untiring and multifaceted support and also my father, Daniel Gil Pinzon, who has always been a calming influence. And though he regrettably did not live to see this book come to fruition, I must thank my second father, Barry Wright, to whom this book is dedicated, for all his energy and encouragement over the years. My aunt and uncle Yvonne Riaño and Larryn Diamond offered timely advice and support at crucial moments. My mother- and father-in-law, Sandi and Michael Moss, welcomed me into their family and home and have offered timely support over the years. Though they were not able to directly assist in the book writing, my children, Arturo and Iris, have been a huge motivator and source of inspiration. The best part about finishing different phases of this project has been getting to spend more time with them. But above all, I must thank Adriann Moss, my partner and co-parent, who has done more than anyone to make this book possible. Adriann, you have pushed me to continue at crucial moments and gifted me with precious time to write and retreat. Thank you, my dear.

THE REMNANTS OF RACE SCIENCE

Introduction

The Remnants of Race Science

A few months after the United Nations Educational, Scientific and Cultural Organization (UNESCO) released its controversial 1950 Statement on Race, the *UNESCO Courier* published an article titled "An Indian Girl with a Lesson for Humanity." The article celebrated the heights of human potential through the story of Marie-Yvonne Vellard, an "Indian Girl" born into the nomadic "Guayaki" tribe in Paraguay yet raised by the French scientist Jehan Albert Vellard, her adoptive father.[1] With a sense of awe, the article's author explained that Marie-Yvonne came from "one of the most primitive tribes on earth" yet quickly transformed into a civilized French girl who now, at the age of twenty, worked as a laboratory assistant to her adoptive father, a renowned expert in physical and medical anthropology.[2]

In the context of UNESCO's international mission to build peace through science, culture, and education, Marie-Yvonne's story was meant to assure readers that the seeming chasm between so-called primitive and civilized peoples did not stem from inborn natural differences. Her story suggested that UNESCO's actions could bridge this presumed gap between peoples. The article was penned by Alfred Métraux, a Swiss-born and French-trained anthropologist who specialized in Amerindian ethnohistory and served as director of UNESCO's race campaign in the 1950s. To explain how Marie-Yvonne ascended to the privileged world of laboratory science, Métraux appealed to one of the most influential concepts of twentieth-century human science—the cultural environment. Had she not been adopted at the age of two by a French scientist, Métraux conjectured, Marie-Yvonne would have been condemned to grow up in a "primitive and rudimentary culture" in which people wander "at large in the forest" hunting and gathering and whose way of life is "very little different from that of the first bands of men who colonized the empty spaces of South America thousands of years ago."[3] Yet thanks to being "brought up exactly as a white girl," Marie-Yvonne became "an attractive, intelligent girl of twenty and a typical product of the cultural environment in which she has lived for 18 years."[4] For Métraux, Marie-Yvonne's story could be used to convince the "layperson" of the arguments put forward in UNESCO's 1950 Statement on Race. Above all, Métraux argued, Marie-Yvonne's story proved one of the 1950 statement's most important points: that "given similar degrees of cultural opportunity to realize their potentialities, the average achievement of the members of each ethnic group is about the same."[5]

For all its optimism, Métraux's account of Marie-Yvonne's life also reveals how colonial logic informed the attempts of many scientific experts to challenge racism after World War II. His article embraced the liberal premise that cultural achievement does not stem from natural difference. Yet, it also conceptualized non-Western societies as thwarted by cultural environments that cannot stimulate the full development of children's intellectual capacities and thus deprive them of a meaningful future. The anticipatory and possessive logic of the article is thinly veiled—if children like Marie-Yvonne are to thrive, then they must be removed from their stagnant culture and given a proper (Western) education.[6] As it celebrates the biological equality and cultural potentiality of all ethnic groups, Métraux's article simultaneously creates a racialized division between people born into damaged environments (and thus in need of fixing) and those from lands of opportunity, whose duty it is to civilize. Under the guise of opposing racism, the article rationalizes the forced removal and assimilation of children in the name of modernity and progress. Removing children from their families and homes is thus sanctioned by a counterfactual conjecture: if children are not raised by Western standards, then their futures will be bleak. Métraux's article, in other words, reveals that struggles against racism—as they were articulated through international organizations after World War II—were compatible with projects that sought to alter or even destroy the ways of life of those deemed backward in anticipation of a brighter future.

UNESCO AND POSTWAR INTERNATIONALISM

The contradictions in Métraux's account mirror the political complexities at play during the formation of the United Nations (UN) system after World War II. In recent decades, historians have looked back to the founding of the United Nations in an effort to recover the diverse contexts, ideas, and ideologies that shaped its creation. These studies demonstrate that the creation of the United Nations represents a high-water mark of internationalist discourse—what Glenda Sluga has called the "apogee of internationalism"—when intellectuals, politicians, and activists worldwide and from different political orientations responded to World War II by identifying the need to curb aggressive nationalism, to restore peaceful international relations, and to create systems of world governance that would stabilize the international order. Motivated by a sense of political crisis that was rooted in the use of scientific and technological resources for destruction at an unprecedented scale—as in the Nazi Holocaust and the U.S. atomic bombing of Hiroshima and Nagasaki—many postwar internationalists pinned their hopes for future peace and security on the creation of a system of world government. The formation of the United Nations and its specialized agencies such as UNESCO and the International Labor Organization (ILO), World Health Organization (WHO), and United Nations Children's Fund (UNICEF; formerly United Nations International Children's Emergency Fund) was thus spurred by a recognition that the major political issues that arose during World War II were global in scale and required a system of world government that could foster a sense of world citizenship in individuals, establish moral and legal guidelines for international governance, and provide an "international infrastructure" for knowledge making. The imminent threat of nuclear war and genocide coupled with a growing faith in international institutions thus gave this pivotal period an atmosphere that environmental historian Perrin Selcer has aptly described as a "giddy mix of triumph and terror."[7]

As the specialized agency charged with constructing the defenses of peace "in the minds of men" through science, culture, and education, UNESCO epitomized postwar internationalism's loftiest aspirations and its most striking ambiguities. These tensions can be readily observed in UNESCO's international initiatives concerning race and racism. Within the UN system, UNESCO was the specialized agency most often tasked with combating racism, which many internationalists considered one of the main scourges of world conflict. In fact, the international committee that drafted UNESCO's constitution enshrined the fight against racial inequality as one of the institution's core concerns.[8] In this vein, the preamble to UNESCO's constitution identifies "ignorance of each other's ways and lives" as a common cause of war "through the history of mankind" and argues that World War II was made possible by "the denial of the democratic principles of the dignity, equality and mutual respect of men, and by the propagation, in their place, through ignorance and prejudice, of the doctrine of the inequality of men and races."[9] UNESCO's creators thus interpreted racism as a source of international conflict and an intellectual problem whose solution lay in the liberal principles of scientific education, free communication and exchange, and greater "mutual understanding" between groups of people.

During its formative years, educational campaigns against racism occupied a prominent place in UNESCO's activities and generated intense debate and scrutiny. The publication of the 1950 Statement on Race, which prompted Métraux's article about Marie-Yvonne Vellard, serves as the most striking example of these tensions. UNESCO officials hoped the statement would provide the organization, and the UN system more broadly, with a document that could guide its action against racism. The statement began with a request made by the UN's Economic and Social Council (ECOSOC), which tasked UNESCO with creating a "programme of disseminating scientific facts designed to remove . . . racial prejudice." In response, UNESCO's Social Sciences Department (SSD) convened a meeting of international race experts in 1949 for the purpose of drafting a statement that could be widely disseminated and serve as a guide to teaching about race and for taking practical measures

against race prejudice. The resulting statement challenged conceptions of race based on a hierarchy of fixed biological differences. It also
declared that scientists generally agreed on the biological unity of the
human species. By presenting these truths as a matter of scientific consensus, the 1950 statement sought to provide epistemological ballast
for the universalist and humanist ideals upheld in the UN's Universal
Declaration of Human Rights. Yet the statement provoked a strong
backlash among prominent British and French physical anthropologists,
who felt their disciplines were attacked by the statement and argued
that their expert perspective was not reflected in the composition of the
drafting committee.[10] To quell these mounting criticisms and to uphold
UNESCO's professed impartiality and objectivity, Alfred Métraux, who
had recently been appointed to serve as head of a newly created "Race
Division," convened a second committee of experts, composed mainly of
geneticists and physical anthropologists, which produced a revised statement published in 1951. The statements, especially the second one, were
produced through an intricate process that involved circulating drafts
to hundreds of experts and collecting their criticisms and viewpoints
points. The complexity of this process reveals the challenges UNESCO
faced in producing claims about human diversity that would reflect
consensus-based beliefs. Yet despite the controversies that dogged the
statements, they generated widespread attention in the mass media and
went on to exert a significant influence on research agendas in physical
anthropology and human biology in the subsequent decades.[11]

The complexity of the process by which they were drafted and the
controversies they generated make UNESCO's race statements open
to multiple interpretations. Earlier interpretations viewed them as a
decisive moment when experts successfully banished the troublesome
concept of race from scientific inquiry and ushered in a new era of
antiracist approaches to the study of human diversity. Yet more recent
studies have reinterpreted the statements as an attempt not to discard
race altogether but rather to reform it in the technical terms of population genetics and thus narrow its usage to circumscribed scientific discourses. Scholars have also reinterpreted the statements as precursors to
late-twentieth-century projects in human genetic diversity research such
as the Human Genome Diversity Project and National Geographic's Gen-
ographic Project, which touted themselves as antiracist yet garnered

accusations of biopiracy and neocolonialism. From this perspective, the statements can be situated within a genealogy of self-proclaimed anti-racist projects rather than as a bookend to scientific racism. They have thus become invaluable historical sources for what Michelle Brattain has called "historicizing antiracism"—a historical project that requires studying how "movements to dislodge racism are equally contingent, opportunistic, political, and grounded in the same social formations as racism itself."[12]

NARRATIVES OF REDEMPTION

The Remnants of Race Science describes two overlapping histories encapsulated in Métraux's article and in UNESCO's race initiatives in the decades after World War II. The first concerns the history of race science and how scientific experts from Latin America, Oceania, Europe, and North America sought to replace older and essentializing conceptions of human difference with purportedly more objective ones. This familiar story of retreat is encapsulated in the way Métraux presented Marie-Yvonne's story as offering a decisive rejoinder to the rigid hereditary claims of scientific racism. Yet in his attempt to confront racism, Métraux identified European civilization and, ultimately, whiteness as key metrics of achievement and progress. Métraux's article thus points to a more complex history of combating racism—a history in which racism was framed as illogical and irrational and thereby concealed as an operating force in scientific inquiry and humanitarian practice.

Métraux's article also contains another historical discourse not typically present in historiographic treatments of race science. Marie-Yvonne's transformation supposedly offered a lesson not just for UNESCO's readers but also for humanity writ large. If a young girl from one of the world's most primitive tribes could ascend to such heights, why not the rest of the world's "backward" peoples? By exalting Marie-Yvonne's cultural transformation, Métraux's article offered a narrative that vindicated the ambitious projects of "Third World" economic development that were also central to the mission of the UN system. Métraux's article thus rendered Marie-Yvonne an exemplar for the possibilities of the developing, or "third," world. In fact, Métraux explained that there are "hundreds of cases" of similar transformations among "young native children born among what are judged as extremely primitive peoples."[13] The overarching lesson from these cases is that "if a child is removed from its cultural environment before it is old enough to have been influenced, it can be completely assimilated into new surroundings."[14] Métraux's article thus also contains a history of how scientific arguments about human equipotential helped transform the once overtly racial civilizing mission ideologies of European imperialism into the seemingly nonracial discourse of "international development."[15]

To capture these intertwined histories, *The Remnants of Race Science* examines the practical and economic interventions made by human scientists affiliated with UNESCO as part of their efforts to combat racism. Instead of examining what experts like Métraux and institutions like UNESCO opposed, this book interrogates the futures they aimed to bring forth and the versions of the past they sought to leave behind. In raising these questions, *The Remnants of Race Science* takes heed of historian and anthropologist Ann Stoler's searing challenge to historical studies of racism. In her influential essay "Racial Histories and Their Regimes of Truth," Stoler observed that scholarly accounts of contemporary racism tend to be predicated on flattened and reductive histories of what racism once looked like.[16] According to Stoler, these reductive histories of racism have typically assumed that racism once existed in an overt biological form that has since given way to a more complex and culturally coded racism in the present. By taking this contrast between racisms past and present for granted, Stoler argues, antiracist histories of racism have often become scholarly quests for the original moment "in which the die of race was cast."[17] Antiracist histories have thus been guided by the belief that racism can be dismantled by identifying its internal contradictions and by pinpointing its historical origins. And it is this quest for origins, Stoler contends, that explains why antiracist histories have typically appeared as "narratives of redemption"—narratives that seek to identify the moment of original sin and thus absolve the present from the misguided racism of the past.

Marie-Yvonne's story and UNESCO's race campaign were shaped by a similar narrative structure and often sought to render race science a remnant of a less enlightened past. For instance, in his article

Métraux conceded that the "primitive state" of Marie-Yvonne's native tribe could be "logically inscribed to racial inferiority."[18] From this questionable premise, Métraux crafted a redemptive narrative. He extolled Marie-Yvonne's adoption as an experiment that refuted this seemingly logical assumption of her inferiority. Thus, Métraux explained that by adopting Marie-Yvonne, her adoptive father also gained "an opportunity to show that education and cultural background were more powerful than the so-called laws of race alleged to govern the development of the individual."[19] Métraux's reasoning thus interpreted Marie-Yvonne's story as one that would render such archaic racial thought a relic from the past. In Métraux's telling, Marie-Yvonne became an object lesson for the errors and sins of an outmoded race science and a portent of a more enlightened future.

By examining the underlying assumptions of UNESCO's race campaign and its geographic scope, *The Remnants of Race Science* demonstrates how mid-twentieth-century race experts unwittingly concealed how race and racism have structured the modern world. The enduring influence of UNESCO's approach can be seen in political theories of multiculturalism, which often identify cultural difference as an ostensibly nonracial way of describing and managing human diversity in contrast with more overtly racist pasts.[20] Such framing can be observed in the work of the influential political theorist Will Kymlicka, who has appealed to a redemptive history of racism in order to situate political theories of multiculturalism within their historical context. Respect for cultural diversity can be observed throughout human history, Kymlicka has argued, yet struggles for multiculturalism and minority rights in "Western democracies" emerged in response to the decolonization and civil rights movements of the 1950s and 1960s, which sought "to contest the lingering presence or enduring effects of older hierarchies."[21] In Kymlicka's narrative, the horrors of World War II serve as a crucial turning point and create the conditions of possibility for multicultural politics to emerge in the West:

> Prior to World War II, ethnocultural and religious diversity in the West was characterized by a range of illiberal and undemocratic relationships of hierarchy, justified by racialist ideologies that

explicitly propounded the superiority of some peoples and cultures and their right to rule over others. These ideologies were widely accepted throughout the Western world and underpinned both domestic laws (e.g., racially biased immigration and citizenship policies) and foreign policies (e.g., in relation to overseas colonies). After World War II, however, the world recoiled against Hitler's fanatical and murderous use of such ideologies, and the United Nations decisively repudiated them in favor of a new ideology of the equality of races and peoples.[22]

Kymlicka's narrative is a classic example of the "narratives of redemption" that often structure scholarly accounts of racism. His account creates a stark contrast between a dark past where an overt and pristine form of racism prevailed and a more enlightened contemporary where racism is submerged and exists only as a "lingering presence."[23] Insofar as it assumes that "the West" is the key site for this transition from illiberal to more egalitarian relationships, this narrative also points to a geographic dimension often implicit in redemptive accounts of racism. In other words, like Marie-Yvonne's story and many of the modernization theories that emerged after World War II, Kymlicka's redemptive narrative privileges "the West" as an aspirational model of democracy and civilization for the rest of the world.

THE PERSISTENCE OF RACE

For a long time, historical accounts of race science favored a similarly redemptive narrative.[24] According to traditional narratives the period after 1945 marked a sharp break with the preceding century and a half, where racism thrived through scientific research programs that presumed fixed racial hierarchies rooted in anatomical and hereditary differences. This conventional story tended to champion the post–World War II period as one where enlightened experts decisively repudiated scientific racism after decades of struggle. It thus assumed a narrative structure that emphasized the period prior to 1945 as typified by overt scientific racism and the period after World War II as a redemptive break from this unenlightened past. In this canonical account, UNESCO's 1950 and 1951 Statements on Race loomed large as the culmination of a campaign to cleanse scientific studies of human diversity of their ideological baggage.[25]

Yet this standard narrative of retreat also assumed that "the race concept" is a singular, and ultimately flawed, scientific object whose coming into being and passing away can be precisely charted—like the rise and fall of phlogiston or the luminiferous aether.[26] On this view, the race concept is something that crystallizes in Western Europe and North America as a taxonomic device from the late-eighteenth-century science of comparative anatomy. Through comparative studies of skulls and other skeletal remains, racist physical anthropologists then imbued the race concept with all sorts of explanatory import during the first half of the nineteenth century, thereby justifying the institution of slavery in the United States and European imperialism worldwide. After the introduction of Darwin's theory of evolution by natural and sexual selection and the rediscovery of Mendel's laws of inheritance, the race concept was eventually tied to a eugenic agenda in the first half of the twentieth century that culminated in the policies of forced sterilization in North America and Scandinavia and the genocidal policies of racial hygiene of the Nazi regime in Germany. After the race concept had been taken to these extremes, cultural anthropologists and population geneticists—especially those from minority backgrounds in the United States and the United Kingdom—began to recoil in the mid-1920s and initiated an intense trans-Atlantic campaign to repudiate scientific racism once and for all.[27] Although there was an outpouring of scientific statements against racism during the 1930s and 1940s, the repudiation was not complete until after World War II when the UNESCO Statements on Race ushered in a new scientific consensus that hammered the final nails into the coffin of the now archaic "race concept." Once relegated to an unfortunate and regrettable past, it was superseded by more objective conceptions of human diversity that were free of the essentialist and typological thinking of the now discarded regimes of truth. Or so the story went.

Instead of charting the rise and fall of a singular race concept, more recent histories of human biology and physical anthropology have unsettled the discrete periodization of the retreat narrative. They instead chart a persistence and reconfiguration of race concepts in science after 1945.[28]

These studies have focused on developments in the life sciences where human biologists began to tout population-based conceptions of human diversity as a more neutral alternative to fixed racial typologies. Yet instead of interpreting the work of human biologists as constituting a definitive repudiation of race science, these recent histories interpret it as an attempt not to lay the race concept to rest but to render it politically innocuous. As exemplified by the work of the U.S. geneticists Theodosius Dobzhansky and L. C. Dunn and the physical anthropologist Sherwood Washburn, this project involved redefining conceptions of race through highly technical discourses that would serve to differentiate between objective and scientifically sanctioned conceptions of race and folk concepts mired in ideology.[29] As a result of this embrace of population thinking during the Cold War, human biology was given an image as socially progressive and antiracist. Yet by taking up this redemptive conception of their disciplinary identity, human biologists were poorly prepared to respond to criticisms of their research from indigenous groups who criticized the Human Genome Diversity Project of the late 1990s by labeling it a "vampire project" that perpetuated colonial relations of power.[30] Critical studies of human genomics, in turn, have focused their attention on the ways recent genomic diversity studies and genetic ancestry tests unwittingly reinforce the archaic conceptions of race they were meant to have displaced.[31] In these accounts, human biology has emerged as a discipline burdened by many of the same ethical issues concerning race and colonialism that mid-century scientists believed they had escaped.[32] In other words, instead of assuming the political neutrality of population concepts, such studies have shown their non-innocence as well as the ways "race is the grammar and ghost of population."[33]

By reexamining the intellectual debates surrounding the UNESCO Statements on Race and the projects of social change they made possible, *The Remnants of Race Science* raises fresh questions about the persistence of race after 1945. While histories of human biology have effectively dislodged facile narratives of the decline of race science, they have typically left the social sciences unexamined. And yet, as Alfred Métraux insisted at the height of the controversies surrounding the race statements, UNESCO's interest in the "race question" was "situated within realms of social anthropology, sociology and social psychology."[34] What stories can we tell about the persistence of race in the social sciences after World War II? And how are these related to those told in human biology? Studies concerning the persistence of race in science after 1945 have tended to assume a separation between the realms of social science and biology. For instance, according to sociologist Rogers Brubaker, the retreat of scientific racism after 1945 resulted in a "tacit and largely uncontested division of jurisdiction between social scientists and biologists."[35] Whereas social scientists agreed to have "exclusive jurisdiction over the phenomenon of race as it was experienced and enacted in social, cultural, economic, and political life," biologists and life scientists were happy to cede jurisdictional authority over "race" to the social sciences and to restrict themselves to the more reliable and politically untainted object of "populations."[36] As we've seen, historians of human biology have contested the political neutrality and nonracial status of population concepts. But what are the politics shaping the conceptions of race that social scientists claimed as their jurisdiction? In other words, how might we also historicize conceptions of race as a social construct and the forms of scientific inquiry they have underwritten? And to what end?

Instead of narrating a definitive retreat of racial thought, *The Remnants of Race Science* describes a more complex transnational project—often led by scientists from the Global South—of aligning conceptions of race and human difference with post-colonial projects of economic development and social integration. Although this transformation in racial thought has conventionally been described as marking the demise of static racial typologies, *The Remnants of Race Science* reveals how scientific experts simultaneously revived assumptions about race from the Southern Hemisphere that emphasized practices such as racial mixing and acculturation to European norms as strategies of post-colonial nation building.

The Remnants of Race Science examines this shift by tracing the history of UNESCO's race campaign after 1945. Through an international campaign that sought to educate people worldwide about the differences between conceptions of race anchored in science and those mired in ideology, UNESCO became a key player in the post-1945 shift in racial thought. Instead of describing UNESCO's campaign as signifying a major rupture point, *The Remnants of Race Science* illustrates how it drew from conceptions of race that were common to the development schemes of independent nations and European colonies in the Southern Hemisphere. These southern conceptions of race emphasized the plasticity and alterability of racial groups and thus stood in stark contrast to the rigid conceptualizations of Mendelian eugenicists. Yet this emphasis on plasticity and—above all—racial improvement was also conveniently aligned with the UN system's emphasis on economic development and international health in the newly coined "Third World." Thus, what *The Remnants of Race Science* makes visible are the continuities between colonial and post-colonial projects of racial improvement and the international development and global health industry that flourished through the United Nations and its allied agencies after World War II. *The Remnants of Race Science* situates the history of race science, and its alleged decline, within the history of applied social science projects of modernization, acculturation, social hygiene, and mestizaje ("race-mixing") that proliferated in the Southern Hemisphere during the twentieth century.

Historical studies of modernization and development theory have also favored a narrative arc that presumes the disappearance of race thinking after World War II in favor of technical discourses for improving the poor. This literature has described how development discourse and modernization theory created the idea of the "Third World" as a social problem to be solved by technical experts from the Global North and how these domains emerged out of the bureaucratic and economic infrastructure created by European empires during the interwar period.[37] Yet despite its attention to power imbalances at multiple scales, most of this literature has not directly engaged with the history of race science and has instead assumed that development is a product of social science—or "high modernist" discourse, to use James Scott's phrase— and thus removed from overt racial thought.[38]

The few studies that have directly examined the conceptual and institutional links between development discourse and race science identify important genealogical affinities. For instance, Helen Tilley's excellent work on colonial science in British Africa from 1870 to 1950 demonstrates how imperial scientists and local experts from this period advanced ambitious research agendas that offered forceful critiques of race science and eugenics while adopting cultural, ecological, and interdisciplinary approaches for studying local realities and for managing and developing "native" populations. Tilley's work provocatively argues that in the context of British Africa, science and development offered tools for imperial management and control and also began to "decolonize Africa by challenging stereotypes, destabilizing Eurocentric perspectives, and considering African topics on their own terms."[39] According to Tilley, race science and eugenics proved to be imperfect tools for imperial management and were thus eclipsed by social anthropological and ecological approaches that embraced "modernization" as an aspirational horizon. Tilley's account thus suggests the need to treat critiques of race science and the formation of modernization discourse as allied projects that gained impetus in imperial contexts during the interwar period.

Though arriving at a different set of conclusions, Michelle Murphy has similarly examined how overt racial thinking gave way to an ensemble of social scientific metrics for assigning economic value to human populations, a phenomenon that Murphy names "the economization of life." Examining quantitative practices that grappled with "economy and population as objects of governance and intervention" during the Cold War, Murphy argues that eugenics and race science gave way to a set of practices including family planning, development projects, and global health, which increasingly valued human populations according to their ability to contribute to the gross domestic product of a nation-state. Murphy observes how these practices eschewed race as an object of study, yet were "replete with methods for governing brown, black, poor, and female bodies that recast racial difference in terms of economic futures." "Race," Murphy contends, "did not have to be named in order to enact racist practices."[40]

Tilley's and Murphy's analyses offer important reference points for the history of UNESCO's race campaign that this book describes. Many of the applied social science projects this book tracks shared the indifference toward racial science and concern with interdisciplinary knowledge that Tilley identifies as a defining feature of colonial science in British Africa. As we will see in chapter 1, the work of British social anthropologists that Tilley describes was warmly received by the Brazilian anthropologist Arthur Ramos, who similarly challenged Eurocentric approaches to knowledge production. *The Remnants of Race Science* thus embraces Tilley's call to not rush to judgment when considering science produced in colonial circumstances and to consider its anti-colonial potential. Yet with Murphy's account of the economization of life in mind, this book is also attentive to how projects concerned with challenging racism can also enact racist practices and perpetuate asymmetries of power.

RACE IN THE SOUTHERN HEMISPHERE

The standard retreat of scientific racism narrative centers on a North Atlantic geography. It foregrounds a tight-knit group of actors located in a handful of elite institutions in the Northeast corridor of the United States or prominent universities in the United Kingdom. It then culminates in Paris with the drafting of the 1950 and 1951 UNESCO Statements on Race. From the vantage of this North Atlantic geography, historians crafted narratives describing a decisive shift from racism to antiracism in science. Yet Marie-Yvonne's story points to a much more expansive geography and less decisive change. Marie-Yvonne originally hailed from the Alto Parana region of Paraguay, where the Aché—her native tribe—still live today. Her story is told by Alfred Métraux, who trained at the Institut d'Ethnologie in Paris, spent the first five years of his career at the University of Tucuman in Nothern Argentina, was renowned for his encyclopedic knowledge of indigenous groups in South America—especially those of the Gran Chaco region—and also conducted ethnographic research in Rapa Nui, Haiti, and Benin.[41] Marie-Yvonne's story hence conveys the need to *relocate* the standard retreat narrative to the Southern Hemisphere and to activate Latin America and other regions as crucial sites within the history of mid-twentieth-century race science. Yet, as this book demonstrates, once conventional narratives are rerouted through the Southern Hemisphere, the period after World War II does not so readily appear as a stark break with the past.

By rerouting the retreat of scientific racism through the Southern Hemisphere, this book unsettles the often assumed centrality of the North Atlantic region and thereby contributes to the more global picture of racial thought and practice that scholars have been crafting in the past two decades.[42] In so doing, it heeds Warwick Anderson's urging "to recognize that North Atlantic scientific debates did not occur in isolation from the rest of the world, especially the Southern Hemisphere."[43] By rerouting the familiar story of scientific racism, the book also offers a different vantage from which to examine questions concerning the history of racial thought after World War II. The canonical (northern) history of race science emphasizes fixed racial typologies, rigid Mendelian eugenics, and an aversion to race mixing as defining features of scientific racism. As a consequence of defining race science in this way—that is, as a project concerned with purity, essences, and rigidly demarcated groups—the conventional history ends up adopting a narrative form that charts a mid-nineteenth-century zenith of racial thought followed by a mid-twentieth-century nadir and retreat. Yet when viewed from a southern standpoint, this supposed waxing and waning of race science is far less clear. Scientific studies of race in many parts of the Southern Hemisphere rarely emphasized racial fixity (with the exceptions of South Africa and parts of Australia) and instead demonstrated a keener interest in climatic adaptation to the environment, the plasticity of racial kinds, race mixing as a nation-building strategy, and sociocultural evolution and progress.

Latin America is an exemplar of these Southern Hemispheric conceptions of race. From the perspective of Latin America, UNESCO's appeals to racial plasticity—couched in the language of culture—are continuous with the intellectual debates that emerged in the period between decolonization in the early nineteenth century and the nation-building projects of the late nineteenth century and first half of the twentieth century. In this period, the question of how to create a modern republican citizenry and labor force in newly formed nations marked by severe social divisions and extensive race mixing vexed the fair-skinned elites who occupied positions of power.[44] After the independence wars of the late eighteenth century and early nineteenth century, Creole elites, often inspired by Enlightenment philosophies, overhauled the racial divisions of the colonial era through an ostensibly liberal discourse that associated the traits of citizenship with the privileges of whiteness and masculinity—namely literacy, property ownership, and individual autonomy. By contrast, ruling elites deemed those who did not possess these traits—slaves, Indians, women, and the propertyless—to be lacking the requisite levels of individual autonomy for full citizenship.[45] Yet precisely because the populations of these newly created nations were made up by large numbers of those deemed unsuited for citizenship, the independence period posed fundamental questions that would continue to worry elite Latin Americans well into the twentieth century: Should democratic representation and citizenship be extended universally in nations composed primarily of mixed-race and non-European populations? Can those unsuited for citizenship acquire the requisite traits through civic education and tutelage? Will Latin American societies be irredeemably condemned to backwardness by their mixed-race and non-European populations?

By the late nineteenth century, slavery had been abolished throughout Latin America, and most countries were emerging out of the social turmoil of the early republics and attempting to forge distinctive national identities and to develop apace with Europe and North America. In this period of intense nation building, ruling elites turned to scientific and medical discourses including positivism, criminal anthropology, fingerprinting technologies, tropical medicine, and bacteriology, which promised new insights on how to solve the dilemmas raised by Latin America's racial diversity as well as new tools for categorizing and controlling their populations.[46] As they sought to propel Latin American countries toward modernity, liberal elites also embraced Comtean and Spencerian conceptions of evolution and progress that emphasized improvements to the environment and social milieu as mechanisms of racial improvement.[47] Through these paradigms of social change, which assumed neo-Lamarckian conceptions of heredity, liberal scientists, politicians, and intellectuals throughout Latin America found ways to counter the pessimistic conclusions that were drawn from the nineteenth-century racial sciences of the North Atlantic. Rather than viewing their black, indigenous, and mixed-race populations as an inevitable source of racial degeneration, Latin American intellectuals instead embraced the view that their nations could be progressively whitened (and thus modernized) through a combination of educational reforms, sanitary improvements, and selective immigration from white European nations. As Latin American nations sought to establish their political and economic independence, intellectual elites repudiated nineteenth-century race science yet also associated slavery, blackness, and indigeneity with backward social conditions that had to be left behind in order to advance social progress.

In the interwar period, prominent human scientists such as Gilberto Freyre, Manuel Gamio, and José Vasconcelos celebrated Latin America's unique pattern of racial blending as reflecting a history of racial tolerance.[48] They also repudiated Mendelian-inspired concerns about race mixing by praising the mestizo populations of Latin America as marking the emergence of distinct and potentially superior or even "cosmic" races.[49] In this same period, medical authorities, intellectuals, and politicians in various Latin American countries embraced eugenic measures as mechanisms for racial improvement. Yet the majority of these self-described eugenicists distinguished their "Latin" approach—which included hygiene and sanitation campaigns, teaching puericulture (the science of child-rearing) to new or expectant mothers, and increased physical education classes in schools—from the unnecessarily severe methods of their "Nordic" or "Anglo-Saxon" counterparts.[50] By the time war was declared in Europe in 1939, Latin America represented a region where experts of various kinds had been debating and producing distinctive racial theories for almost two centuries, if not longer.

Similar conceptions of race emphasizing plasticity and oriented toward uplift and progress can be found beyond Latin America. Indeed, similar racial ideas can be found in the narratives of civilization, development, and progress that abounded in turn-of-the-century social science and liberal theory and served as justifications for European and U.S. imperialism.[51] In Africa, Asia, and the South Pacific, late colonial administrators from Britain and France crafted policies of "indirect rule" and "mise-en-valeur" that were informed by the human sciences and sought to economically develop "native" societies along their own lines.[52] In the formal and informal colonies of the United States, human scientists mobilized modernization theories that upheld U.S. society as a model for the so-called third world and became ubiquitous during the Cold War era.[53] From the perspective of these racialized development schemes, the rejection of biological determinism that was such a prominent part of UNESCO's race campaigns is not a discursive rupture so much as a reformulation and amplification of existing currents of thought and practice.[54]

The Remnants of Race Science examines how UNESCO experts efforts to dismantle scientific racism sanctioned interventions in the

"Third World" that bore a striking resemblance to earlier colonial civilizing missions, which drew on overtly racist ideas. It demonstrates how the development regimes assembled during the first decades of the UN system framed the traditional customs of indigenous peoples as incompatible with productive industrial economies. As a result, indigenous peoples were conceptualized as "backward" and in desperate need of assimilation and modernization. By approaching this history from the Southern Hemisphere, *The Remnants of Race Science* demonstrates how postwar antiracist science also functioned as a colonial governmentality concerned with dismantling ostensibly primitive ways of life and with obliging new and modern forms of life to come into being.[55] It demonstrates how—despite the often good intentions of those involved—scientific campaigns against racism legitimated economic development projects that, paradoxically, sought to improve indigenous and other non-European peoples by destroying their ways of life. By describing how human scientists moved effortlessly from denouncing Mendelian race science to projects of indigenous modernization in the "Third World," *The Remnants of Race Science* shows how internationalist antiracism paved the way for the expansion of modern capitalism.

RECALLING ANTIRACISM AND THE REMNANTS OF RACE

Though UNESCO's race campaign was concerned with combating racism, I will mostly refrain from using the term "antiracist" to describe its actions. I do so for several reasons. In the period this book covers, UNESCO experts did not typically use the term "antiracism" to describe their actions and instead preferred terms such as reducing "race prejudice" or combating "racial discrimination." In fact, "antiracism" does not appear in UNESCO's race campaign until 1960 and 1961 where it is used, in both instances, in translation from English to French and from Spanish to English, respectively.[56] Even beyond the timeframe covered by this book, the United Nations and its allied organizations have not consistently used "antiracism" to describe their major initiatives against racism. Examples include the UN's 1971 international year for action to combat racism and racial discrimination, UNESCO's 1978 declaration on race and racial prejudice, and the UN's 2001 world conference against racism, racial discrimination, xenophobia, and related intolerance in Durban, South Africa. In strict historical terms, then, "antiracism" is not used by the actors in this story and is thus anachronistic.

Beyond the question of historical accuracy, another reason for not using the term "antiracism" is that the projects described in this book were often demonstrably racist. Métraux's account of Marie-Yvonne's assimilation serves as a striking example and illustrates one of the central tensions this book explores: how earnest and well-intentioned initiatives to combat racism and improve the lives of others can seamlessly coexist with and reproduce the forms of racism they oppose. A striking feature of UNESCO's race campaign is that, in its early stages, experts often conceptualized racism as a mistaken view of the natural world, one that presumed the immutable biological inferiority of non-Europeans. Yet when opposing this form of racism, UNESCO experts often invoked another form of racism—what David Theo Goldberg has aptly called a "historicist" racism, which avoided biological conceptions of race yet infantilized non-Europeans on the basis of their supposed lack of progress and backwardness. This "historicist" form of racism is most visible, this book argues, when we examine the links between UNESCO's race campaign and the UN system's economic development initiatives, especially in regions of the Southern Hemisphere like South America and Polynesia.

Although this book refrains from using the term "antiracism" to describe UNESCO's initiatives, it nonetheless makes the case that UNESCO's race campaign must be situated within historical genealogies of antiracism. The arguments made against the presumptive biological and psychological inferiority of non-European races in UNESCO's documents and booklets played an important role in buttressing legal and political movements against racial segregation in the United States and apartheid in South Africa. UNESCO's reformulation of race as population also served as an important stepping stone toward late-twentieth-century genomic projects with more clearly articulated antiracist ends.[57] UNESCO's race campaign thus laid the foundations for later developments that are more readily identifiable as part of the history of antiracism, a history that has been little told.

By situating UNESCO's race campaign within a genealogy of antiracism, this book offers readers critical insights for understanding how antiracism has historically been constituted and for thinking about how it might be reconfigured. In doing so, it takes inspiration from anthropologist Ghassan Hage's idea of "recalling antiracism." Hage argues that, as a current of thought and social movement, antiracism has a long history encompassing the abolition of slavery, anticolonial struggles, the civil rights movement in the United States, and the anti-apartheid movement in South Africa. Yet despite its long history and noticeable victories, Hage claims that antiracism has lost its force in recent decades, which have been marked by a sharp surge of racist movements. The enduring and increasing force of racism, Hage argues, has made it necessary to reconfigure antiracism through the process of "recalling it."

In proposing a "recalling of antiracism," Hage invokes multiple meanings of the verb "to recall." In one sense, recalling antiracism means remembering its founding principles and functions through historical study. Yet Hage also suggests recalling antiracism like a company "recalls" a defective product to restore public trust by showing "the care it takes with the quality control of its goods and the safety of their users."[58] This is necessary because antiracism has "often failed to perform and rise to the situations it is confronting" and thus stands in contrast with racism, which Hage argues has been through several "recalls" that have made it "operationally suitable for a variety of socioeconomic and cultural environments." Antiracism, argues Hage, has become "conceptually rather ossified" and is thus "always trying to catch up with the racists' fluid modes of classification."[59]

As part of Hage's effort to recall antiracism, this book examines UNESCO's race campaigns. I interpret UNESCO's race campaigns as projects that offer both striking examples of the ossification of antiracism that Hage describes and openings and possibilities for ways of doing antiracism otherwise.[60] Although UNESCO's race campaign does not perhaps warrant the antiracist label, the tactics and lines of argumentation that UNESCO experts deployed left behind a strategic template that has endured into the twenty-first century. Through their insistence on combating racism through logical and empirical argumentation, UNESCO's race experts offer an early example of what Hage calls "academic antiracism": a variety of antiracism that has made significant contributions to antiracism's conceptual ossification. This can be most clearly observed in the way academic antiracists typically respond to contemporary manifestations of racism. Hage observes that while racists care little about logical contradictions and inconsistencies, academic antiracists have insisted on criticizing racists as if they are "fellow academics with whom they are having disagreements in a tutorial room about how to interpret reality." For instance, academic antiracists accuse racists of being bad thinkers—"essentialists"—or of "making false statements about reality that [they] can empirically correct by highlighting a lot of statistical data that proves them incorrect."[61]

The overarching ambition of UNESCO's race experts was to eliminate race prejudice by empirically refuting beliefs about immutable racial hierarchies. In the buoyant spirit that defined liberal internationalist projects after World War II, UNESCO race experts wagered that their educational initiatives would relegate typological and essentialist conceptions of race to the historical dustbin. In this sense, UNESCO race experts believed themselves to be grappling with the remnants of race and racism—with objects that would soon figure as a mere trace from the past. They thus adopted an antiracist tactic that quickly ossified. Yet, as this book demonstrates, UNESCO and UN officials were not unaware of this ossification, as demonstrated by the recurring statements, declarations, and conferences from the 1960s and 1970s, which introduced a different repertoire of tactics for challenging racism. By using the remnants of race to describe UNESCO's race campaign during the 1950s, this book thus identifies a notable instance of the ossification of antiracism that Hage describes.

Yet by shifting away from the presumed centrality of the North Atlantic world, which has prevailed in existing accounts of UNESCO's race campaign, this book also tracks how antiracism has been configured otherwise, in some cases even as a project that recognizes the seemingly intractable nature of race and racism in the modern world.

STRUCTURE OF THE BOOK

This book tells a series of interconnected storylines that unfold primarily between the 1920s and the 1960s. To tell the history of scientific campaigns against racism from the perspective of the Southern Hemisphere, it tracks the intellectual and professional itineraries of the experts who played key roles in the race campaign organized by UNESCO's Social

Sciences Department. Most of these figures trained in the human sciences during the 1920s and were well into their scientific careers when they came to UNESCO after World War II. Though they hailed from different countries and regions, held different cultural backgrounds, and trained at different institutions, they were all connected in different ways to the political and intellectual life of countries and regions in the Global South, primarily South America and the South Pacific. During the formative phases of their careers in the late 1940s and early 1950s, these experts crafted conceptual frameworks and political rationalities informed by human science methods that emerged in response to the social and political dilemmas of late colonial and post-colonial nations. And they brought these ways of seeing with them to UNESCO's race campaign. Although this book introduces the readers to many actors and institutions, the activities and writings of Alfred Métraux, who was at the heart of UNESCO's race initiatives during the 1950s, serve as a focal point for many stories. Métraux thus serves as a protagonist for this book, and his career arc and writing exemplify many of the complexities it seeks to describe.

The book follows a rough chronological order focused primarily on the period between 1920 and 1960. The book is transnational in scope and links together scholarly networks from South America, North America, Europe, and the South Pacific. Yet it identifies significant continuities across this period and geography, which are exemplified by projects that sought to replace static typological conceptions of race with processual and relational conceptions of social life such as acculturation, cultural change, and race relations that were better aligned with late colonial and post-colonial modernization projects. In order to trace the Southern Hemispheric influence of UNESCO's race campaign, part I, "Confronting Racism in the Southern Hemisphere, 1890–1951," follows the epistemic itineraries of key figures in UNESCO's race campaign and shows how contexts and institutions such as the Brazilian state of Bahia, the indigenismo movement in Central and South America, Polynesian anthropology, and the Bishop Museum in Hawai'i exerted a major influence on key figures of UNESCO's race campaign, such as the Brazilian anthropologist Arthur Ramos, who was the head of UNESCO's Social Sciences Department at the start of the race campaign; the 1949 committee of race experts who drafted the 1950

Statement on Race; and Alfred Métraux, who was appointed as head of UNESCO's Race Division after Arthur Ramos's unexpected death. Rather than focus on individual actors as units of analysis, part II, "Race in the Tropics and Highlands and the Quest for Economic Development, 1945–1962," examines two of the UN system's most ambitious development projects from the postwar era: the International Institute of the Hylean Amazon (IIHA) from the late 1940s and the Andean Indian Program from the mid-1950s. By examining the conceptions of race that informed these two economic development projects and the pivotal contributions of figures from UNESCO's race campaign, notably Alfred Métraux, Charles Wagley, and Ernest Beaglehole, the two chapters in part II demonstrate how Cold War development discourse contained active remnants of race science and eugenic discourse. Part III, "Engineering Racial Harmony and Decolonization, 1952–1961," examines UNESCO's mixed results in the development of "race relations" as a field of inquiry during the 1950s. The two chapters in this part examine two major race-relations projects of the postwar era: the cycle of race-relations studies that UNESCO sponsored in Brazil in the early 1950s and a UNESCO-sponsored attempt to create an international society for race relations, which stemmed from a monthlong conference held in Honolulu, Hawai'i, on "Race Relations in World Perspective." At stake in both of these projects was UNESCO's institutional ambition to identify portable sociological models of "racial harmony." Yet as these chapters demonstrate, this ambition proved to be an illusory ideal that researchers in Brazil and in other nonaligned nations increasingly identified as an ideological construct of North American liberalism.

PART ONE

Confronting Racism in the Southern Hemisphere, 1890–1951

CHAPTER ONE

Substituting Race

Arthur Ramos, Bahia, and the "Nina Rodrigues School"

> If, in the works of Nina Rodrigues, we substitute the terms *race*
> for *culture* and *miscegenation,* for *acculturation,* for example,
> his concepts become completely and perfectly contemporary.[1]
>
> ARTHUR RAMOS, *A ACULTURAÇAO NEGRA NO BRASIL*

In October of 1949, just three months after arriving in Paris to serve
as director of UNESCO's Social Sciences Department (SSD), Arthur
Ramos, the influential Brazilian psychiatrist and anthropologist, died
from a heart attack. During his spell as SSD director, Ramos organized
the meeting of race experts that produced the 1950 UNESCO Statement
on Race and also sketched out a research program centered on race
relations in the Southern Hemisphere. With this tragic death, Ramos
suffered the same fate as Nina Rodrigues—the Bahian doctor of legal
medicine whose own life ended unexpectedly in Paris in 1906.[2] By the
time of the unfortunate end of his life, Ramos had fulfilled his career-
long ambition to become Rodrigues's intellectual successor. Yet despite
the intellectual affinity shared by Ramos and Rodrigues, the work of
Ramos and the work of Rodrigues have been interpreted as representing
opposing poles in Brazilian racial thought. For instance, the eminent
historian Thomas Skidmore has described Nina Rodrigues as "the most
prestigious doctrinaire Brazilian racist of his era."[3] According to Skidmore, Rodrigues's research on Afro-Brazilians from the 1890s, which
became foundational for the field of Afro-Brazilian studies, was tainted

by the scientific racism of his time. Rodrigues's research, argued Skidmore, displayed an attitude of "racial pessimism," in which Rodrigues augured a dim future for Brazil because of the high percentage of blacks and *mestiços* ("mestizos") within its population.[4] By contrast, Skidmore situated Arthur Ramos within a new wave of optimistic interwar researchers such as Gilberto Freyre who embraced a pure "environmentalist hypothesis," enthusiastically celebrated Brazil's African heritage, and offered a positive reinterpretation of Brazil's history of race mixing. Skidmore also pointed out that Ramos wrote a manifesto, with Freyre and other researchers, that denounced the arrival of racist ideas from Nazi Germany.[5] Within Skidmore's narrative, Rodrigues and Ramos represent the diametrically opposed poles of "racial determinism" and social "environmentalism" and presumably racism and antiracism.

Instead of contrasting their work, this chapter situates Arthur Ramos and Nina Rodrigues in a shared intellectual tradition that emerged in the northeastern province of Bahia and took the study of Afro-Brazilian culture and history as one its central objects. In doing so, it seeks to recover the Bahian and Southern Hemispheric outlook that Ramos brought to UNESCO but also to demonstrate the continuities between late-nineteenth-century race science and UNESCO's race campaign. Although his time at UNESCO was short-lived, Ramos crafted a framework that shaped the organization's race projects for the following decade. Ramos not only organized the 1949 meeting of international race experts who drafted UNESCO's 1950 Statement on Race but also corresponded intensely with his Brazilian colleagues and used their input to outline an ambitious research program that aimed to reorient the SSD program toward the Southern Hemisphere through comparative studies of race relations and "backward" groups in Brazil, the Americas, and Africa. Instead of interpreting the race-focused and lofty comparative vision that Ramos brought to the SSD as a post-WWII product, this chapter traces Ramos's research agenda back to late-nineteenth-century debates about Brazil's multiracial population and the country's capacity for modernization. The conceptual associations that emerged in UNESCO between antiracism and economic development can thus be seen as the remnants of conceptual debates from the late nineteenth century in which the southern nation's place within the modern world was at stake.

The ambitious research program that Ramos brought to UNESCO represents a career-long endeavor to redeem Nina Rodrigues's methods. Ramos first encountered Rodrigues's work as a young medical student in Salvador, Bahia, in the late 1920s and quickly became one of its most vigorous champions. As this chapter demonstrates, Ramos recognized that Rodrigues's evolutionist and hereditarian orientation led to racist conclusions yet believed that his ethnographic insights on Afro-Brazilian religion made it foundational for the field of Afro-Brazilian studies, which began to coalesce during the 1930s. He thus sought to preserve Rodrigues's empirical insights while reformulating its bleak outlook with a reformist optimism. For Ramos, this comparatively optimistic outlook for Brazilian society was anchored by a strong belief that race was not an impediment to progress and that people from all sectors of Brazilian society could be molded into modern citizens through institutional and infrastructural reforms. While Rodrigues's work suggested that Brazil's prospects for modernization were fraught, Ramos believed that applied human sciences—like mental hygiene and applied anthropology—could function as tools for modernizing the most "backward" sectors of Brazilian society. Like many of the experts who participated in UNESCO's race campaigns, Ramos endeavored to render race an epistemic relic of the past yet continued to uphold a hierarchical view of human diversity.

BAHIAN SCHOOLS

Arthur Ramos was a product of the Bahian school of medicine. In the late nineteenth century, this school of medicine established its reputation on the strength of groundbreaking research on tropical medicine conducted by the Tropicalistas—a group of Bahian physicians who countered prevailing European interpretations of the tropics as an inherently inferior environment. Instead of holding Brazil's tropical environment responsible for the supposed biological degeneration of its occupants, the Tropicalistas discerned social and microbial determinants of disease and postulated that Brazil's social and health problems could be remedied through improvements to the nation's medical, sanitary, and education infrastructure. In this sense, the Tropicalistas anticipated the optimistic discourses of modernization and improvement that figured prominently in UNESCO's race campaign after World War II.

Arthur Ramos was a product of Bahia through and through. He was born in 1903 to a comfortable upper-middle-class white family in the municipality of Pilar, Alagoas. Like many of Brazil's northeastern states, Alagoas was at the forefront of Brazil's slave trade and sugar plantation system during the nineteenth century, and a large portion of its population was made up of formerly enslaved people brought from Africa and their descendants. As a teenager, Ramos wrote vivid descriptions of the festivals, costumes, tales, and songs of Alagoas's Afro-descended inhabitants.[6] Like his father, Ramos trained as a physician in Salvador, Bahia, at the Faculty of Medicine of the Federal University of Bahia in 1921. Through his medical training, Ramos acquired a pragmatic and even therapeutic approach to scientific practice and its relation to societal change. Like many physicians trained in the Bahian medical school, Ramos viewed improving the health and well-being of the Brazilian population through government intervention and public health advocacy as a means of unlocking the nation's path toward modernity.[7]

Since the 1860s, medical experts from the Bahian school of medicine had been at the forefront of tropical medicine and had played a crucial role in challenging European conceptions of Brazilian racial inferiority. The medical school in Salvador, Bahia, that Ramos attended from 1921 to 1926 is one of two medical schools established by the Portuguese Crown in 1808 (the other was established in Rio de Janeiro). During the first half of the nineteenth century, medical instruction at these institutions was largely humanistic and rigidly adhered to the study of Western and especially French medical literature. Yet in Salvador, Bahia, during the 1860s, a group of physicians trained in German and Scottish medical schools and inspired by the German physician Rudolf Virchow began meeting regularly and began to question Brazil's medical status quo. This group of physicians—the "Escola Tropicalista de Bahiana"—did so by adapting European bacteriological innovations and the German conception of "social medicine" to the Bahian context and by calling for an approach to science and medicine that prioritized the study of local realities as opposed to rote adherence to European medical wisdom. The Tropicalistas also challenged prevalent and often deterministic European and North American assumptions about the debilitating effects of tropical environments and the stereotypes of Brazilian and Latin American racial inferiority that these assumptions legitimated.[8] The founding members of the *escola* ("school")—the German émigré Otto E. H. Wurcher, John L. Paterson, and José Francisco da Silva Lima—also shared a progressivist faith in the power of science and technology and argued that the obstacles to health in Brazil were primarily due to social inequality and thus no more daunting than those faced by Western Europe or any other region of the world. Generally speaking, the Tropicalistas professed that neither Brazil's climate nor its racial makeup posed insurmountable barriers to its ability to progress into a modern nation. As such, the Tropicalistas maintained that many of the dangers typically associated with Brazil's tropical climate could be reversed through enhanced sanitation infrastructure and educational measures to improve the personal hygiene of the most "backwards" sectors of Bahian society.

Yet when Ramos attended Bahia's medical school, the optimistic ideas of the early Tropicalistas had been eclipsed by a less sanguine approach typified by Nina Rodrigues and inspired by the Italian school of criminal anthropology led by Cesare Lombroso. Adopting evolutionary,

FIGURE 1.1 Arthur Ramos (second from the left) during his medical school days. Faculdade de Medicina da Bahia, ca. 1920s (photograph, Bahia, Acervo da Funação Biblioteca Nacional, Brazil).

positivist, and statistical frameworks, Lombroso sought to understand the causes of race degeneration, which led him to argue that criminals were born as such. He proposed that criminality was an inherited trait and theorized that criminals represented atavistic throwbacks to a primitive evolutionary state who possessed distinct physical traits that could be identified through anthropometric measurement. Following in Lombroso's footsteps, Rodrigues and other Bahian researchers turned away from their predecessors' concern with tropical disease and instead introduced a psychiatric concern with studying the observable links between criminality and insanity. The rise of Rodrigues and his criminal anthropological approach coincided with the abolition of slavery in 1888 and the establishment of a Brazilian republic.

Through his medical training in Bahia, Arthur Ramos absorbed many of criminal anthropology's key tenets. At Bahia, Ramos completed a dissertation titled "Primitivo e Loucura" (Primitive and madness) in 1926 and wrote a series of essays on madness and crime that were later published as a collection in 1937. As Brad Lange has argued, Ramos's work during this period bore the clear imprint of Rodrigues and criminal anthropology yet also signaled some new directions.[9] While Lombroso and Rodrigues viewed crime and mental illness as hereditary phenomenon that could be detected through racial phenotype, Ramos studied patterns between crime and mental illness but did not find a correlation with race. Instead, Ramos conducted clinical observations of patients who had committed crimes and, borrowing from the psychoanalytic and ethnographic theories of Sigmund Freud and Lucien Lévy-Bruhl, theorized that mental illness and criminal behavior stemmed from "deficits" in "psychic functions" or the survival of a "primitive mentality." During the 1930s, Arthur Ramos moved to Rio de Janeiro where he worked for five years as director of the Orthophrenology and Mental Hygiene section of the Institute for Education Research (IPE). As Jerry Dávila has shown, during this period, the IPE advanced a eugenic agenda supported by the Getulio Vargas government and which sought to counter Brazil's supposed racial degeneration through a policy of social and cultural whitening.[10] As director of the Orthophrenology unit, Ramos kept files on individual children in Rio's school system where he tracked their development using anthropometric and psychometric testing and observations. In 1939, he used this data to publish a book called *A criança problema* (the Problem Child) where he argued that children classified as "abnormal" based on psychological tests were often better described as "problem children" who acted out because of "maladjustments" to their environment.[11] The lessons that Ramos learned from criminal anthropology thus stayed with him throughout his career.

Although he had strayed from many of Rodrigues's conclusions, by the 1930s Ramos took to vigorously defending Rodrigues's work and took every opportunity to describe himself as an adherent of the "Nina Rodrigues school." What was it that Ramos found so appealing and enduring in Rodrigues's work? Rodrigues was born in 1862 to a slave-owning family in the rural northeastern state of Maranhão. He spent his childhood on a large family estate where he played with enslaved children, learned to read from an enslaved woman, and heard stories about a nearby *quilombo*—a community of runaway slaves.[12] He began his medical studies in Salvador, Bahia, in 1884 and then transferred to the medical school in Rio de Janeiro where he graduated in 1887. After graduating, Rodrigues returned to Salvador and became professor of clinical medicine at the Federal University of Bahia in 1887, which he held until he became professor of legal medicine in 1891.[13]

When Nina Rodrigues began teaching at Bahia's medical school, Brazil had abolished slavery (1888) and become a republic (1889), which provoked renewed and intense debates about the future of the country and its racial makeup. Rodrigues admired the Tropicalista school's emphasis on direct study of the sick (as opposed to narrow book knowledge) and embraced its dual commitment to empirical rigor and political advocacy. Yet Rodrigues departed from the school's optimistic assessment that changes to Bahia's environment through sanitation and hygiene reforms would necessarily uplift the general health and well-being of the state's population. Unlike the Tropicalistas, who attached little significance to heredity as a cause of disease and social disorder, Rodrigues viewed the hereditary makeup of populations as crucial to understanding disease patterns. Rodrigues argued that Brazil's social and medical problems could not be solved without careful study of environmental forces coupled with consideration of the hereditary contributions made by the distinct "races" that combined to make up Brazil's population.

Yet unlike the Tropicalistas and abolitionists who believed that Bahia's predominantly black and mixed-race population could be progressively whitened through educational and sanitary reforms, Rodrigues viewed race mixing as a complex and open-ended process that could also lead to the blackening of Brazil's population.[14] As historian Anadelia Romo has argued, Rodrigues was a complex and multivalent thinker who attempted to balance concerns about the heredity makeup of Brazil's population with advocacy for social reform that stemmed from a belief in the formative influence of tropical environments and social institutions.[15] Rather than pure racial types, Brazil's population was characterized by the emergence of various *mestiço* types, which natural selection would ultimately mold into a "future national type."[16]

Rodrigues's open-ended conceptualization of race mixing and the influence of criminal anthropology is evident in his 1899 article titled "Métissage, Dégénérescence, et Crime," which claimed to offer evidence of racial degeneration through an intimate ethnography of a mixed-race population.[17] In this article, Rodrigues challenged the views of European race scientists such as Gobineau, Quatrefages, Agassiz, and Gustave Le Bon and argued that their views concerning race mixing in Latin America were questionable because they were not based on direct observation. In fact, throughout his career, Rodrigues positioned himself as a rigorous positivist who eschewed metaphysical speculation and grounded his theoretical claims on rigorous observations and evidence.[18] A proper understanding of race mixing, insisted Rodrigues, should be based on "small localities where it is easier to identify the different causes of degeneration." It must also stem from "studying the social capacity of a population by examining its biological capacity as calibrated through its medical history."[19]

For his study of race mixing, degeneration, and crime, Rodrigues chose Serrinha—a small Bahian community with a "proverbially salubrious" environment that many Bahians considered a "sanitarium of the first order for tuberculosis."[20] Rodrigues reasoned that Serrinha's salubrity made it an ideal location to observe how the forces of nature and nurture shape populations and to test whether Brazil's mestizos were degenerating. If a place existed where "Brazilian métis" might prosper and offer hope for the future, then Serrinha would be it, reasoned Rodrigues. Yet his clinical observations purportedly demonstrated the contrary. Instead of thriving, Rodrigues claimed that Serrinha's various mixed-raced types, such as "pardos" (the predominant mestizo type, which he defined as a cross between "white, black, and yellow" races) displayed indolence and laziness and were plagued by "hereditary" deficiencies. In fact, Rodrigues claimed to observe that Serrinha's population was marked by a "propensity to mental illness, to serious afflictions of the nervous system, and to physical and psyche [psychic] degeneration of the most accentuated kind."[21]

Rodrigues explanation for Serrinha's degeneration is striking and reveals his distance from the enviromentalist framework of his Tropicalista colleagues. While he considered five potential causes for Serrinha's degeneration—local conditions, climate, hygienic conditions, sanitation, and "consanguinity"—he insisted that "consanguinity" was the most "efficient cause" for the prevalence of mental and physical degeneration in Serrinha's population. Indeed, given the salubrity of Serrinha's environment, Rodrigues concluded that the degeneration of Serrinha's "pardos" stemmed from an ill-suited racial mixing. "The crossings of races as different, anthropologically, as are the white, black and red races," Rodrigues explained "produced a poorly balanced product of low physical and moral resistance."[22] It was due to this "poorly balanced product" that Serrinha's *mestiços* struggled to thrive in the Bahian climate and to compete in "the social struggle of the higher races." Serrinha's population, Rodrigues warned, offered cautionary tales about the pitfalls of race mixing. At its worst, argued Rodrigues, race mixing produced "rigorously average types" that brutally revealed a conflict between the " . . . psychic qualities and very unequal physical and physiological conditions of two deeply dissimilar races whose characters heredity has melted in a combination, in a defective amalgam seen in the Métis product resulting from the crossing of two races."[23] Thus, although he grounded his study on intimate insights of a local population, he ultimately drew abstract moral lessons about the dangers of uncontrolled race mixing.

Yet Rodrigues offered a faint glimmer of hope for Bahia's racial future. Although he observed many examples of "defective amalgams" in Bahia's population, Rodrigues also referenced studies by other Brazilian scientists, which purportedly showed that the best *mestiços* are the product of "white stock combined with at least one-fifth of indigenous blood." He also referenced a study suggesting that in "mulatto" families, "children with the most accentuated characteristics of the black race are sometimes the most intelligent."[24] What was notable about these purportedly successful examples of race mixing, Rodrigues argued, was that the outcome of fusion had "more or less returned to the equilibrium of one of the pure races."[25] Rodrigues's arguments thus suggest that he did not condemn race mixing per se nor categorically reject blackness. Rather his anxieties about race mixing seem to be based on an abiding concern with order and equilibrium, which, like many positivists, he viewed as being achievable only through scientific management.

Rodrigues's scientism and commitment to empirical rigor also informed his ethnographic work on Afro-Brazilians. Rodrigues's major work on Afro-Brazilian religions came in the form of an 1896 book titled *O animismo fetichista dos negros baianos (L'animisme fétichiste des Nègres de Bahia.* In this study, Rodrigues combined careful historical and ethnographic study of Afro-Brazilian religious practices and beliefs with a theoretical framework that presumed a racial hierarchy and distinguished between "primitive," "polytheistic," and "monotheistic" religions. *O animismo fetichista* rigorously documented the animistic beliefs and practices of Afro-Brazilian cults in Bahia and the numerous religious streams that they drew from including African, such as Yoruba religions and Islamic beliefs, Catholic, and indigenous. It also situated these beliefs within the context of the African diaspora and compared them to those of other Afro-descendant groups in the Americas. Rodrigues was concerned with thoroughly describing these "cults" in order to counter a persistent and dogmatic belief that the Bahian population was entirely Christian—a belief that Rodrigues described as implying either "systematic contempt" for the "black Africans and their mestizos who make up the vast majority of the [Bahian] population," or a "naivety of vulgar nescience."[26]

In contrast to these mainstream beliefs about the prevalence of Catholicism in Bahia, Rodrigues argued that "African fetishist animism" served as the "underground subsoil" for Bahia's Afro-Brazilian cults like Candomblé, which had been diluted by the superstitions of "the white race" and by "the incipient animism of the American aborigines."[27] He also provocatively argued that during this process of syncretization, Catholicism was being shaped by fetishism and not the other way around. "Here, in Bahia," Rodrigues wrote, ". . . far from the black converting to Catholicism, it is Catholicism that receives the influence of fetishism [and] adapts to the rudimentary animism of the black . . . [which] gives body and objective representation to all monotheistic mysteries and abstractions."[28] Like his studies of race mixing, Rodrigues's study of Afro-Brazilian religions was concerned with documenting and making blackness visible and with understanding the way it was shaping Brazilian society.

Yet despite his efforts to shine a spotlight on Afro-Brazilians and their contributions to Brazil, Rodrigues clung to an evolutionary framework and often assumed black inferiority. For instance, Rodrigues argued that the prevalence of fetishism in the Americas stemmed from Afro-descendants' racial inferiority. Certainly, Rodrigues made an effort to consider other potential causes for the persistence of African fetishistic beliefs such as the number of enslaved people brought from the Yoruba region to Bahia, the early freedom and financial success enjoyed by some of Bahia's enslaved peoples, and the continued commercial ties and exchange of people between Salvador and Lagos. Yet he concluded that once Afro-Brazilian cults were organized, "blacks" of other nations and origins preferred them because they were "more within the reach of their rudimentary intelligence, and more in line with [their] way of feeling" compared to the "Catholic cult that they could understand nothing or little of."[29] Similarly, Rodrigues described religious fusion in Bahia as a process that "tends to adapt the understanding of Catholic monotheistic conceptions to the weak mental capacity of the black."[30]

Yet for all of his racialist excess, through his commitment to ethnographic rigor and occasionally sanguine views of evolution, Rodrigues anticipated the arguments that Ramos and other experts would use during UNESCO's race campaign. As we've seen, like many race experts from the Southern Hemisphere, Rodrigues viewed racial groups as plastic and subject to environmental and evolutionary forces. And in the context of his religious studies, Rodrigues at times suggested that shifts in religious practice tended to follow a logic of evolutionary progress. For instance, when describing the process of religious syncretism he observed in Bahia, Rodrigues argued that it mirrored a process that had taken place in Europe at the beginning of Christianity when "polytheistic Europe" converted to "Christian monotheism." According to Rodrigues, this parallelism suggested that "the laws of psychological evolution are fundamentally the same in all races." Introducing a theme that would become prominent in the work of the twentieth-century experts who participated in UNESCO's race campaign, Rodrigues thus intimated that racial differences were not fixed but rather a reflection of different stages of an evolutionary process, one that encompassed mental capacities. For all of its unabashed racialism, Rodrigues's work exemplifies the tendency to conceptualize race as a remnant from a distant past, an old world, something that was being lost or buried as a result of contemporary processes and movements.

AFRO-BRAZILIAN STUDIES AND THE "NINA RODRIGUES SCHOOL"

In the 1930s and 1940s, when he turned his attention to institutionalizing anthropology in Brazil, Arthur Ramos anointed Nina Rodrigues as the founding figure of Afro-Brazilian studies. Throughout his major publications from the interwar period, which included monographs on Afro-Brazilian culture and religion and introductory anthropology textbooks, Ramos touted the formative influence of Nina Rodrigues and his disciples—what he called the "Nina Rodrigues school"—on the nascent field of Afro-Brazilian studies. For instance, in his textbook *Introdução à antropologia brasileira* (Introduction to Brazilian anthropology), Ramos described how Rodrigues "revolutionized studies of the Negro" by taking them out of their "pre-scientific" phase and articulating a foundational methodology that was adopted by subsequent anthropologists: "the method of comparative study of African cultures and their 'survivals' in Brazil."[31] The transformative influence he ascribed to Rodrigues also led Ramos to describe the trajectory of black studies in Brazil as composed of three phases: (1) the pre–Nina Rodrigues phase; (2) the Nina Rodrigues phase; and (3) the post–Nina Rodrigues phase. As the self-appointed torch bearer of this third phase, Ramos discerned dense connective tissue between his and Rodrigues's work.

Despite these palpable links, existing histories from North America have tended to situate Rodrigues and Ramos at opposing ends of racial thought. In her incisive critique of comparative studies of race relations, Micol Seigel situates Arthur Ramos within a tradition of radical trans-Atlantic scholarship on the African diaspora that emerged during the interwar period and centered on questions of "race consciousness" and "black nationalism." According to Siegel, Ramos's work typifies the "cohort of innovative Afro-Brazilianists" who emerged in the 1930s and formed part of "transnational networks of antiracist intellectual production and struggle" and ultimately effected "a sea change in academic views of race." Seigel situates Ramos in sharp contrast to Nina Rodrigues, whose work she describes as incredibly "derogatory to Brazil." By the 1930s, argues Seigel, Rodrigues's derogatory approach was replaced by the anthropological and historical approach of Ramos's cohort, "who exalted Brazilian culture in general and venerated Afro-Brazilians in particular as worthy and valuable subjects of study."[32] Like Skidmore and Lange, Seigel thus interprets Rodrigues and Ramos as representing opposing poles in Brazilian racial thought.

If their work represents such polar opposites, how do we make sense of Ramos's enthusiastic identification with Rodrigues's work during the interwar period and beyond? One reason is epistemic. Like many mid-twentieth-century scholars who confronted scientific racism, Ramos prized empirical insights over abstract theory. From this pragmatic perspective, Ramos seemed willing to look past the racist themes in Rodrigues's work because it was also rich with observational data on the customs and practices of Afro-Brazilian communities. When discussing Rodrigues's racist views, Ramos explained that he believed "that the fairest method [in science] is the one that leads to the most fruitful results [and] that leads to the collection of rich data and the scientific systematization of these data." By placing an emphasis on results and data, Ramos implied that even racist theories could be salvaged as long as they could generate useful facts. "A theory is only a working hypothesis," Ramos insisted, " [and] what counts is the fecundity of the results. What matters are the facts." From this perspective, Ramos reasoned that Nina Rodrigues had been a "pure evolutionist" yet had also been "very fruitful during his time." Rather than discard Rodrigues's contributions, Ramos argued that contemporary scholars could simply "adjust his theories to ones from our period."[33]

Yet Ramos did not just attempt to salvage Rodrigues as someone working with outdated theories. He went a step further and suggested that Rodrigues had actually anticipated the subfield of "acculturation studies," which had taken off among U.S. anthropologists during the

1930s. "It was Nina Rodrigues," Ramos asserted, "who gave us the first descriptions of that general mechanism that modern anthropologists today prefer to call acculturation."[34] According to Ramos, this aspect of Rodrigues's work could be most clearly seen in his observations of Afro-Bahian religious beliefs. Ramos thus pointed out how Rodrigues had used terms such as "hybrid associations" and "mixed beliefs" to describe Afro-Bahian religions and had interpreted them as a product of the "contact between African religions and Catholicism." To further emphasize Rodrigues's prescience, Ramos argued that while "scholars of the Negro in Brazil" now commonly interpreted Bahian religions as a "fusion" between Yoruban, Catholic, and Amerindian beliefs, when Rodrigues had done so in the 1890s it "sounded weird" and was not warmly received. By portraying Rodrigues as out of sync with his time, Ramos thus implied that his views were more modern than what their evolutionary and racial framing suggested.

Ramos's intense identification with Rodrigues can also be explained by the disciplinary and regional tensions that shaped the nascent field of Afro-Brazilian studies in the 1930s. Ramos's ardent methodological proposals spoke not only to his own epistemological sensibilities but also to debates about the disciplinary and political orientation of the burgeoning field. At stake were major questions about the methods and content of the field as well as questions about whether Bahia or Recife would serve as the field's institutional home. These debates occurred during a dramatic period in Brazilian history that was marked by political polarization and gave rise to the authoritarian and populist regime of Getulio Vargas, who ruled with military support and through the suppression of free speech and democratic opposition from 1930 to 1945. As part of his effort to affirm a distinctively Brazilian national idenity, Vargas championed race mixing as a tool of whitening and also sought to strengthen the nation's sense of Brazilianness through celebration of folk, popular, and Afro-Brazilian cultural traditions. In this context, Afro-Brazilian cultural and artistic traditions such as samba and capoeira flourished while European traditions lost ground.[35]

In this context of heightened national sensitivity, Bahian elites and Afro-Brazilian leaders crafted a cultural identity for Bahia that was rooted in Africa and marked the birthplace of Brazilian identity writ large. Yet in doing so they ran afoul of the historian Gilberto Freyre, who had proposed a differing origin story that also recognized the African influence on Brazilian society but ultimately championed the Portuguese roots of Brazil's identity. Tensions between these competing visions of Brazilian identity came to a head during the two Afro-Brazilian congresses that marked the maturation of the field. Freyre played a leading role in organizing the first of these congresses, which took place in Recife in 1934 and consisted of an academic program that featured historical studies of slavery and studies of race mixing that relied on anthropometric measurements. During this same year, Freyre published his best-known book *Casa Grande e Senzala* (The masters and the slaves), which argued that Portuguese colonization was relatively benign and that Brazil was not doomed by its racial mixture and tropical climate. In 1937 a second Afro-Brazilian congress was held in Salvador, Bahia. The Afro-Brazilian anthropologist Edison Carneiro organized this second congress with close input from Arthur Ramos and from Afro-Brazilian leaders of the Candomblé religion. Yet before the second congress took place, Gilberto Freyre attempted to sabotage the event out of fear that Afro-Brazilian studies would develop beyond his geographic realm of control.[36] In public interviews and personal correspondence with prominent scholars such as the American anthropologist Melville Herskovits, Freyre asserted that the idea of holding Afro-Brazilian congresses was his and that he would not participate in the Bahian event due to concerns over its content and politics. Thus, in an interview for the Bahian press, Freyre admonished the organizers of the Bahian congress for straying from the strictly "scientific" focus of the first congress and devoting too much attention to "picturesque" topics such as capoeira, samba, and the drumbeats of Candomblé.[37] Similarly calling into question the scientific rigor of the congress in his correspondence with Herskovits, Freyre complained that the second congress had been taken over by "political progandists."[38] Freyre also criticized the participation of black scholars and community leaders in the second congress and couched this criticism in a knock on Nina Rodrigues's methodology. In the previously mentioned interview, Freyre lauded the Recife congress's scientific contributions by contrasting it with Nina Rodrigues's work. While Nina Rodrigues had faced "the black and black mestiço" as a problem of "biological pathology," experts at the Recife congress, argued Freyre, had made a major breakthrough in Brazilian social science by approaching these issues as a problem of "social disadjustment." Yet Freyre cautioned that this achievement was under threat by the second congress, which he insinuated was slipping toward the political and "demagogic discourse" of "people of color." Freyre's attempts to discredit the Bahian congress suggest that he felt threatened not only by its geography but also by the assertion of race and blackness that it represented for him.

Ramos's persistent, if awkward, championing of Nina Rodrigues's importance to Afro-Brazilian studies functioned as a retort to Freyre's posturing. This combative sensibility is evident in Arthur Ramos's writings in the period between the two congresses. For instance, in 1935, Ramos published an incendiary preface to his book *O Folclore Negro*, which described the emerging field of Afro-Brazilian studies as striking a "protest" against the state of historical blindness that preceded it. Without naming specific works or authors, Ramos's preface lambasted studies of Afro-Brazilians with an exclusively historical focus and described them as flawed and based on "artificial unilateralism." Freyre was one of Brazil's most prominent historians at the time, so it is quite possible that Ramos had him in mind here. In Brazil, Ramos argued, proper historical analysis was next to impossible because the Brazilian government had destroyed most historical documents pertaining to the slave trade when it abolished slavery. As a result, historical studies of Afro-Brazilians lacked relevant primary sources and thus produced "flawed" studies with "often imaginary" claims. Ramos further argued that these flawed historical studies had led to the rise of "false scientists" who now called for an "aryanizing whitening" of Brazil's population and sought to erase what they saw as the "black stains" from Brazil's historical record. In the post-abolitionist period that Ramos admonished, Nina Rodrigues's work stood as the sole exception to the prevailing state of ignorance and "blindness" with respect to the "problem of the black race in Brazil." Ramos thus spiritedly asserted that in his own studies, he would "not tire" from bringing attention to Nina Rodrigues's Bahian studies and considered them an "indispensable starting point for the continuation of a systematic and serious study [of Afro-Brazilians]."[39]

Although he did not name any specific figures in this preface, Ramos did make it clear that his dissatisfaction with the state of Afro-Brazilian studies stemmed from the First Afro-Brazilian Congress organized by

Gilberto Freyre in Recife. Ramos described how the congress committed the "injustice" of not paying homage to Nina Rodrigues, "the great Bahian master." Though he admitted that the Recife congress signaled the start of a new phase of Afro-Brazilian studies, Ramos also insisted on keeping Bahia at the center of this new direction. "It would be unfair to ignore the great merit of the school of Nina Rodrigues, which revealed to today's generations the first scientific studies on the issue, now continued by his disciples," Ramos wrote.[40]

FIGURE 1.2 Nina Rodrigues, o Mestre bahiano, precursor dos estudos sobre o problema do Negro no Brasil (portrait, in *Novos Estudos Afro-brasileiros*, vol. 2, ed. Gilberto Freyre et al. Rio de Janeiro: Civilização Brasileira, 1937).

Ramos continued this line of argumentation the following year (1936) when Freyre invited him to contribute a preface to the Recife conference proceedings. In Freyre's summary of the proceedings, which was also included in the publication and which Ramos read before writing his preface, Freyre acknowledged that Nina Rodrigues was of some importance to the field of Afro-Brazilian studies yet ultimately downplayed his continued relevance. In a caption of a portrait of Nina Rodrigues (figure 1.2) that the congress participants voted to include in the proceedings and which Freyre described as a homage to the Bahian professor, Freyre also diminished Rodrigues's influence by describing him as a "precusor" to Afro-Brazilian studies and thus as someone whose work ultimately lay outside of the intellectual core of the field.[41] In his preface to the proceedings, Ramos responded to these provocations, again insisting on the indispensability of Rodrigues's work and by calling for the second congress to go further than the first by not merely paying "homage" to Rodrigues's work but also "definitively consecrating" the Bahian scholar's memory.[42] Instead of a fleeting photo tribute, Ramos thus insisted on keeping Nina Rodrigues's methodology—which he interpreted as ethnographic and comparative—at the center of Afro-Brazilian studies and as a baseline that future scholars could continue to correct and improve.

FROM SOCIAL PSYCHOLOGY TO APPLIED ANTHROPOLOGY

During the late 1930s and 1940s, Ramos increasingly moved away from the psychological and psychiatric focus of his previous work and pivoted to institutionalizing anthropology in Brazil and positioning himself—the self-appointed torchbearer of the "Nina Rodrigues school"—as its figurehead. As Getulio Vargas's control over Brazilian society tightened and war and fascism intensified in Europe, Ramos also strengthened his international connections and spent time teaching and lecturing in the United States. With the help of Melville Herskovits, who introduced him to key figures in U.S. cultural anthropology, Ramos spent several months at the end of 1940 and the start of 1941 touring the United States, including a three-month stay at Louisiana State University where he gave a course on race relations in Brazil in the sociology department.

During this stint in the United States, he consolidated his reputation as a professional anthropologist by joining the American Anthropological Association, publishing articles in North American journals, and collaborating with leading anthropologists like Herskovits and Ralph Linton. By establishing these professional connections, Ramos bolstered his own anthropological bona fides and international reputation to the degree that Herskovits referred to him in his letters of introduction as "the only

FIGURE 1.3 An illustration titled "Three Girls from the Same Street" from Arthur Ramos's introductory anthropology textbook. Dimitri Ismailovitch, *Relações de raça no Brasil: Três meninas da mesma rua* (original illustration, in Arthur Ramos, *Introdução à antropologia brasileira*, vol. 2, 1947. Acervo da Fundação Biblioteca Nacional, Brazil).

full-time professor of anthropology in Brazil."[43] And when he returned to Brazil, he worked toward institutionalizing anthropology in Brazil by creating a national society and by creating an anthropology department within the Federal University. As his authority and reputation grew, he also began to exert intense control over the intellectual orientation of Brazilian anthropology during this time period and to fend off what he viewed as scholars and ideas that were not in line with his vision.[44] This included besmirching the reputation of the U.S. anthropologist Ruth Landes, whose work on the prominent role played by women and homosexuals in Bahia's Afro-Brazilian religious cults he attacked by suggesting that Landes had paid her informants with sexual services.[45]

During this period of turmoil, Ramos also began to directly confront racism and to tout Brazil's distinctive standing as a "laboratory of civilization" and "racial democracy." For instance, in 1935 Arthur Ramos published a declaration along with several prominent Brazilian social scientists (including Gilberto Freyre) titled "Manifesto dos Intelectuais Brasileiros contra o Preconceito Racial" (Manifesto of Brazilian intellectuals against racial prejudice), and in 1942 he served as a signatory to a similar manifesto issued by the Brazilian Society for Anthropology and Ethnology, an organization he had helped to create.[46] In English-language articles on Afro-Brazilians, which he published in the *Journal of Negro Education*, Ramos also diligently replaced notions of a "primitive" or "pre-logical" cultural mentality with analyses of Afro-Brazilian religions that foregrounded the lasting effects of slavery and the precarious economic conditions it had created. He also began to celebrate what he viewed as a distinctly Brazilian approach to race relations—what he called "one of the purest racial democracies of the Western Hemisphere"—and suggested that Brazil's most urgent task ahead was to raise "the economic standards of all races" and specifically the need to raise these standards much higher than the one "prevailing throughout the vast territory in which a large colored population is to be found."[47]

As he reckoned with racism, Ramos also began to trace its roots back to European conquest. Unlike Gilberto Freyre, whose work would align with the neo-imperial and Lusocentric policies of Getulio Vargas and the Portuguese dictator António de Oliveira Salazar, Ramos's writings during World War II identified European imperialism as the root cause of racism. In a 1944 lecture on the "Social Sciences and the

Problems of the Postwar," which he gave at the Casa do Estudante in Rio de Janeiro, Ramos offered a fiery rumination on the historical origins of the social sciences and their future after World War II. In this lecture, Ramos argued that the imperial context that gave rise to the social sciences ultimately led to their dehumanization. Social scientists' task after the war would be to figure out how to "re-humanize" their fields of study. To make this point, Ramos repeatedly invoked the idea of "cultural compulsives" proposed by the U.S. literary critic and Marxist, V. F. Calverton. In his own trenchant critique of evolutionism in anthropology, Calverton had argued that evolution represents a type of idea—a "cultural compulsive"—that persists in social science and social thought not because of its "abstract accuracy or inaccuracy" but because of the "strengths of the interests [it] represent[s]."[48] Calverton thus questioned the political neutrality of social science and argued that its presumptive objectivity was in fact a "defense mechanism" and unconscious attempt to "cover up the presence of compulsive factors and convictions."[49] Following Calverton's lead, Ramos argued that the social sciences are inextricably tied to and reflect the dominant aspirations and ideologies of their historical context—what Ramos referred to as their historical *Zeitgeist*. From this standpoint, Ramos argued that the social sciences and especially sociology and anthropology had been indelibly shaped by a process of "Europeanization," which he defined as "the implantation, by force or economic corruption, of European techniques or philosophies to the remotest peoples on Earth."[50] Ramos described this *Zeitgeist* of Europeanization as one that emerged with European voyages of discovery and led to several lamentable consequences: "the slave trade, the retaliation of the continents, the violent exploitation of human labor, the struggle between races, the 'barbarization' of peoples, [and] the dehumanization of man."[51] He also described Europeanization as having important epistemic consequences, notably the formation of an ideological division of labor—one that Ramos labeled a "cultural compulsive"—that led to the classic distinction between sociology as the study of "civilized" European societies and anthropology as the study of "backward" African and other non-European societies. The alignment between the social sciences and this process of Europeanization reached its peak with British imperialism during the late nineteenth century when the social sciences became infused with what Ramos described as a "deterministic" conception of "progress." At this point, argued Ramos, "entire groups of humans were destroyed while others were detribalized and demoralized." According to Ramos this process intensified during the twentieth century as "economic competition gave rise to the concept of inferior groups" and eventually "Europeanization gave rise to racism, imperialism, and slavery whose modern names are fascism and nazism." As he considered the future of the social sciences after World War II, looked back on the twentieth century as one of great wars and global transformations wrought by a "machine civilization" in which "technological advancements served the interests of capitalist and imperialist domination."[52] By his reckoning, the process of Europeanization in the nineteenth century thus culminated in a process of intellectual and material dehumanization that figured prominently in the major wars of the twentieth century.

Yet despite this bleak outlook, Ramos saw new possibilities emerging from the rubble of Europeanization—a process that he suggested was now in decline. Evolutionary conceptions of progress were now on the wane, and in their stead Ramos discerned the emergence of a new unified science of "human relations," which he described as the "crowning achievement of all the others." As an exemplar of this new science, Ramos cited the introductory anthropology textbook written by Harvard anthropologist Eliot Chapple and the notorious Carleton Coon. For Ramos, Chapple and Coon offered a new methodology that blurred the boundaries between sociology and anthropology through a new focus: "the study of the relations between men; between man and his environment; between man and nature; between man and society; [and] between man and culture."[53] As another example of what a reformed social science might look like after World War II, Ramos cited a collection of papers from a Carnegie-Mellon symposium on technology and society and described them as a offering a "re-humanized" approach to technical knowledge focused on studying its social and cultural dimensions for the benefit of society. From the vantage of this reformulated approach to social science, Ramos augured, the distinction between pure and applied social science would collapse and give rise to an "applied sociology" that would serve not just the colonial interests of "indirect rule" but the "social improvement of all humanity without distinction of race, creed, or social class."[54]

Foreshadowing the optimistic rhetoric that would later be embraced by UNESCO experts, Ramos insisted that he did not want to describe his postwar vision in political terms and argued that what was most important is "harmonious relations between men, in social life . . .; fundamental rights guaranteed by a social contract . . .; and the benefits of science for human improvement."

UNESCO AND BACKWARDNESS IN THE SOUTHERN HEMISPHERE

> This is the new humanism we all hope will triumph in the post-war world. It involves, not the imposition of the European way of life and culture, but the pooling of different experiences, as part of a generous and democratic view of existence.
>
> —ARTHUR RAMOS, "THE QUESTION OF RACE AND THE DEMOCRATIC WORLD," P. 13.

Though it meant leaving his cherished position as chair of anthropology at the University of Brazil in Rio, Ramos could not resist the opportunity to extend Nina Rodrigues's legacy abroad when he was recruited to serve as director of UNESCO's Social Sciences Department in 1949. The opportunity arose thanks to the efforts of Ramos's Brazilian colleague Paulo Carneiro, who was a prominent biochemist and member of Brazil's positivist church and represented Brazil on UNESCO's Executive Board. Carneiro played an important role when UNESCO was first created and helped to define its core policies. As such, Carneiro was in close contact with some of the most prominent figures of UNESCO's early years, notably the British biologist Julian Huxley, who was UNESCO's first director-general, and the Mexican poet and intellectual Jaime Torres Bodet, who succeeded Huxley as director-general.[55] When UNESCO began searching for someone to direct the SSD in 1949, Carneiro wrote glowing recommendations of Ramos's work to Torres Bodet, who was ultimately persuaded by Carneiro's praise.[56] The Canadian social psychologist Otto Klineberg, who was then acting head of the SSD, also expressed enthusiasm for Ramos's candidacy, and Ramos was appointed to the department in September of 1949.[57]

When Arthur Ramos began his SSD directorship in the northern autumn of 1949, he promptly initiated a plan to reorient research at the

FIGURE 1.4 Portrait of Arthur Ramos. Dimitri Ismailovitch, *Busto masculino três quatos a esquerda, usando um óculos de lentes octogonais* (drawing, 1940. Acervo da Fundação Biblioteca Nacional—Brazil).

department away from the study of international tensions and toward issues concerning the assimilation and acculturation of "backwards" peoples in the Southern Hemisphere. At the time, the SSD program was strongly marked by the efforts of Ramos's predecessor, Klineberg, who had made the Tensions Affecting International Understanding

Project a focal point for the SSD. The purpose of the Tensions Project was to identify the central factors leading to tensions between nations and to develop "techniques which might be applied to the reduction of such tensions."[58] The language UNESCO officials used to frame the Tensions Project drew heavily from social psychology and the behavioral sciences—disciplines that were on the rise in the United States thanks to major financial support from government, military, and philanthropic organizations. Given this framing, Otto Klineberg—a pioneering social psychologist and anthropologist of race differences—was an ideal person to serve as director of the Tensions Project, which he did from 1947 to 1949. Under the rubric of the Tensions Project, Klineberg and the SSD sponsored a bevy of studies including the production of a series of monographs on the "Way of Life" of different nations, a series of community studies in different countries (India, Australia, France), studies using "public opinion surveys" in order to collect data about the stereotypes held about other nations, and studies of the formation of attitudes and stereotypes in children. Broadly speaking, the Tensions Project adopted nation-states and individuals as its central units of analysis and used methods from social psychology and the behavioral sciences to investigate the psychological dimensions of intra and international conflict.

When he arrived in Paris to begin work at the SSD, Arthur Ramos offered a muted assessment of the ongoing research program. In his first written impressions, Ramos observed that the department was diligently carrying out plans and programs for which "a whole team of specialists and other qualified people, devoting their best energies and abilities to the work, is required." He also noted that the ongoing research concerned three main areas of social science: sociology, political science, and social psychology. Yet despite the very specialized research taking place, Ramos lamented that "cultural anthropology has not yet found its place in the Department's programme." Although the Tensions Project touched on issues concerning "races and minorities," Ramos was astonished that "no regular attention has yet been given to the special problems of Man himself."[59] More puzzling still, Ramos could not understand why the "comparative study of cultures"—and particularly the study of native African cultures—had been relegated to the Department of Philosophy and Human Sciences and was not something that concerned the SSD. And it was here that Ramos saw an opportunity to make his mark on the department's research.

To correct these oversights, Ramos proposed a new research program with "backward peoples" as the main object of study. In a sketch of this research, Ramos wrote that the problem of "backward peoples" should be "among the most important of the activities of the Department of Social Sciences." To begin, Ramos suggested a general study "of the living conditions and the indigenous civilization of the backward peoples in their native places throughout the world." Yet, reflecting the focus on syncretism and acculturation prominent in his previous work on Afro-Brazilians, Ramos also emphasized the importance of considering "backward peoples" in "comparison with the dominant cultures." "In other words," stressed Ramos "a study of the life and culture of non-European peoples considered in relation to the decisive factors of European civilization." More concretely, Ramos urged UNESCO's SSD to take up the "question of the assimilation of the Indian and Negro peoples of the New World and of their introduction to our culture."[60] And here Ramos saw an opportunity for UNESCO to link its research to work being done by Latin American institutions, notably the Instituto Indigenista Interamericano in Mexico City and the Afro-American Institute in Cuba.

In fact, one of Ramos's central ambitions when he came to UNESCO was to turn Latin America into a laboratory for the study of human relations. According to the Brazilian sociologist Luiz Aguiar Costa-Pinto—one of Ramos's protégés—the day Ramos left Rio for Paris he convened all his colleagues from the Social Sciences Department of the National Faculty of Philosophy of the University of Brazil to generate ideas and plans for his tenure at UNESCO. In the exchange of ideas that took place, Ramos's colleagues suggested that UNESCO "seriously consider turning Latin America, and Brazil especially, into a laboratory for the study of human relations."[61] According to Costa-Pinto, Ramos and his colleagues argued that Brazil and other Latin American nations offered particularly fruitful research sites because they had social structures that raised problems of "universal scientific interest" and could not be found anywhere else in the world. The meeting participants also interpreted Latin American nations as going through a "long and painful transition phase" characterized by the "co-existence" of "problems common to developed capitalist societies, alongside problems typical to backwards agrarian structures." Accordingly, Ramos's Rio colleagues insisted that Latin America and especially Brazil offered a "dramatic" opportunity for social scientists to observe "organisms that . . . participate in two epochs, in two historical styles, we might even say, in two worlds."[62]

In the three months following this meeting in Rio, Ramos continued to collect ideas from Latin American colleagues and assured them that he was steering the SSD program toward their interests. He also diligently maintained connections with prominent Brazilian politicians who could offer support for a Nina Rodrigues inspired research program focused on Afro-Brazilians. For instance, shortly before leaving Rio, Ramos wrote to his "eminent friend," the Bahian lawyer and politician Clemente Mariani, who was then serving as the Brazilian minister of education. In a pattern that would continue in his correspondence from this period, Ramos described himself as facing "precarious conditions" at UNESCO given that he was subject to a "3 month probation period." Yet he also vowed that he was going to make every effort to "find favorable conditions in my country and abroad so I can extend my post to the maximum of two years." Ramos thus assured Mariani that he was committed to working for Brazil and especially the issues they had previously discussed, namely " . . . the study of negro groups within their habitat of origin and their destiny within the New World and especially in Brazil."[63] Ramos also explained that Mariani's name was becoming increasingly known in UNESCO circles thanks to his fundamental education campaign in Brazil and that Ramos would be honored to cooperate with him in this field. As a point of reference, Ramos explained that he was preparing for further work on "the assimilation and acculturation, with the end of education in its long sense, of backward, maladjusted, or marginalized groups in our country and especially of its black population." In order to bring this plan to fruition, Ramos explained that he would need to be in permanent contact with the person chosen to replace him as chair of anthropology at the University of Brazil and declared that this chair serves as "a point of reference for my studies and concerns" as well as an embodiment of "our Bahian school of Nina Rodrigues," and that he would therefore like to have it "converge with the practical results that might be obtained through UNESCO."[64]

Once his appointment to UNESCO became known, other Afro-Brazilian studies scholars encouraged Ramos to promote their growing field. Shortly before his departure for Paris, Ramos received a letter from Roger Bastide—the French sociologist and scholar of African religions who succeeded Claude Lévi-Strauss as chair of sociology of the University of São Paulo and taught there from 1938 to 1957. In this letter, Bastide praised Ramos for his appointment and described it as a fitting recognition of "a life dedicated to scientific work and of the great value of your anthropological oeuvre." Bastide also took the liberty to add, as a "half-Brazilian," that his nomination marked "the importance of Brazil in international scientific work." Bastide counseled Ramos to participate in some of the ongoing work at the SSD—notably Otto Klineberg and Georges Gurvitch's efforts to create an international institute of sociology—yet also implored Ramos to not forget their "dear Brazilian negroes." Bastide suggested that Ramos might even instigate a workplan focused on "negroes in the Americas" from the point of view of "physical (and medical) anthropology, cultural anthropology and sociology/social (economic) conditions." He also suggested that Ramos loop the American anthropologist Melville Herskovits into this project and seek financial support from the United Nations' "aid program."[65]

A month and a half into his UNESCO tenure, Ramos wrote a letter addressed to the economist Jorge Kingston of the University of Brazil and the "other colleagues" he convened in his final meeting in Rio. The purpose of the letter was to bring his colleagues up to speed concerning his first activities at the SSD. Ramos struck an apologetic tone and lamented that his possibilities for "autonomous work" were limited given that the research agenda for the year had been set well before his arrival. Yet Ramos assured his colleagues that there were several noteworthy activities that he "would like to bring to the attention of [his] dear colleagues from the Faculty." The first was the creation of the International Sociological Association (ISA) at a meeting that Ramos attended in Oslo as UNESCO's representative. Ramos rued the fact that only two Latin Americans were present at the meeting and that Brazil was not represented (according to him, he was unable to represent both UNESCO and Brazil simultaneously). Yet he also reassured his Rio colleagues that there was plenty of time for a Brazilian society to join the organization and that requests to join should be directed to the newly elected president—the Chicago sociologist Louis Wirth—and to UNESCO's SSD. Ramos also noted that during the following week, UNESCO would host a meeting for the purpose of creating an International Association for Political Science and promised to send further instructions on membership to their colleague Victor Nunes Leal.[66]

A couple of weeks later, Ramos wrote a similar letter to Luiz Aguiar Costa-Pinto in which he bemoaned the fact that "possibilities for cooperation with Brazil are very limited for the rest of the year" and described the various obstacles he faced. Chief among these were work processes that were "too slow and cautious" and "excessive control mechanisms." Indeed, Ramos grumbled that "we cannot veer a single line away from this rather rigid structure" and that "things are completely different from a university job." Yet Ramos maintained his optimism and explained that his opportunity would come once he presented a new program at the General Conference in Florence scheduled for May 1950, which, if approved, would be executed in 1951. In drafting this new program, Ramos assured Costa-Pinto that he had not forgotten the suggestions he collected during their last departmental meeting in Rio and especially his suggestions for "teaching social sciences in primary school and on field research in Latin America." In the interim, Ramos explained that one of the plans for next year was to create a new division within the department that would be devoted to the study of "race questions." Although this new division had been planned by Otto Klineberg, Ramos assured Costa-Pinto that he intended to give this new initiative "a personal orientation" and that he would "count on his full participation" in this project. To this end, Ramos further explained that he was beginning to plan for a "preliminary meeting" on race questions for the upcoming month of December and that once the details were worked out, he would surely invite Costa-Pinto to serve as a "representative of Brazilian sociology."[67]

A couple of weeks before his death, Arthur Ramos wrote a letter to the Brazilian historian Pedro Calmon, then dean of the Social Sciences Department at the University of Brazil in Rio. Ramos again struck a note of disappointment. "I have found little at the Department," Ramos wrote, "in relation to the countries of Latin America." Not only did current programs fail to focus on Latin America, but also the department itself had almost no contact with "institutions and persons of interest." As such, Ramos explained that one of his most pressing tasks was to organize "a thorough list of contacts" of interested individuals and institutions in social science. Ramos also reiterated that he was hampered by the current program but was nonetheless finding ways to collaborate with Brazilian scholars and mentioned ongoing contacts with the Brazilian National Commission for the purpose of initiating a study on the influence of ethnic minorities in Brazil and their influence in international relations. Yet Ramos also assured Calmon that he had "big plans" in the works for the following year, which he would soon present at UNESCO's General Conference in Florence. These plans included a series of studies on "non-mechanized" groups and the problems they raise for integrating into the modern world. If the plans are approved, Ramos assured Calmon, they would present an enormous opportunity to study Brazil's "negro and indigenous groups" and their contacts with "the dominant cultures, from the points of view that I have defended on so many occasions in my courses and written work." Ramos also explained that he was seeking the necessary political support for such a plan and explained that he had sent his research plan to "our friend"—the Brazilian education minister, Clemente Mariani—and was planning to write to other personalities in Latin America to raise support for this program.[68]

Ramos suffered an unexpected heart attack that ended his life a few months into his tenure and was thus not able to see his research plans come to fruition. Yet shortly before his death, Ramos returned to some of the themes he had lectured on during World War II and published an article titled "The Question of Races and the Democratic World" that reveals the redemptive possibilities he envisioned for UNESCO and the postwar era. In this article—a précis of the lecture titled "The Social Sciences and the Problems of the Post-War" that he gave at the Brazilian Ministry of Foreign Relations in 1944—Ramos argued that the world was in need of a "new humanism" because anthropology—presumably the bearer of the "old humanism"—had been so thoroughly "deflected from its true ends" during the war. In the name of anthropology, Ramos wrote, "whole nations have resorted to conflict, to defend the false ideal of racial or ethnic supremacy."[69] Given this lamentable history, Ramos urged that it was thus "entirely natural that anthropology, restored to its proper place and stripped of the myths in which it had been veiled, should now deliver its scientific message to the world." As an example of what a reformed anthropology might look like, Ramos offered the gendered explanation that anthropology is no longer a "mere descriptive or illustrative study of the so-called 'primitive' peoples" and has instead become a proper science of "human relations" concerned with understanding the relations between "man and his environment, man and nature, man and society, man and civilization, man and man." According to Ramos, the rise of a "normative attitude" exemplified by applied anthropology was one of the consequences of anthropology's more objective reorientation toward this proper science of human relations. Although applied anthropology first emerged as "merely a technical means of securing better treatment for colonial peoples," developments during the interwar period had transformed it such that it was "now concerned with the more far-reaching task of analysing and adjusting human relations."[70]

Yet in describing the redemptive possibilities of this new humanism, Ramos looked well beyond the history of twentieth-century warfare and once again invoked a sweeping history of "Europeanization." In the truncated history he offered of this process, Ramos identified the European conquest of the "New World" in the sixteenth century and the debates about the relative humanity of new world inhabitants it sparked as an origin point. Ramos described the conquest of the new world as a process by which "whole peoples were massacred, their civilizations destroyed, their riches looted . . ." and argued that ideas concerning "the 'savagery' of the Indian, his non-Christian origin, and the need to teach him the Catechism served to justify . . . the large-scale pillaging in which Spaniards, Portuguese, Dutch, English, and French vied for pride of place." Ramos further described "Indians" as besieged by a "process of tribal and cultural destruction" conducted by means of the corrupting influence of alcohol and "the imposition of 'superior' culture by gunpowder." In the face of these weapons, "Indians" were forced to initiate a "retreat" that "has not yet come to an end" and is exemplified by the "Indians" in Brazil "who have been driven from their homes" and continue a utopian search for "the land where one does not die."[71]

Although he painted a portrait of deplorable cruelty in the Americas, Ramos argued the continent of Africa offered the scene of "colonial exploitation *par excellence*." It was in Africa that colonizers developed the "idea of the barbarous and savage negro," which they wielded as a "device of power" to justify "all the brutalities of Europeanization." And it was in Africa, more than anywhere else in the world, argued Ramos, that Europeans developed the institution of slavery and refined it into a method to establish "the most unbelievable type of domination." In two centuries, Europe promptly destroyed the Sudanese and Bantu empires, annihilated entire kingdoms, imprisoned entire tribes, and demolished important works of art. Europe thus violently installed herself, argued Ramos, "in the Dark Continent and set about imposing on it her own methods and outlook, through the slow destruction of a way of life and philosophy replete with unsuspected humanism, which it is now being attempted to rebuild."[72]

It was from the vantage of this centuries-long process of Europeanization that Ramos offered an analysis of World War II. In the twentieth century, Ramos explained, the "weapon" of Europeanization was "eventually turned against those who forged it." According to Ramos, it was in the twentieth century that "racialists" began resorting to "methods characteristic of colonial policy and slavery" in their relations with communities within their nations—notably "the cases of the Jews and Negroes," which offered "living examples of the 'superiority myth' in action."[73] In this present century, the "racial" technique gave rise to "one of the greatest states of disequilibrium that exist, namely war." In his assessment of the causes of "the European nations' Second Great War," Ramos argued that, although multiple, "one cause was undoubtedly the philosophy of racial domination espoused by the racialists of our time, that is to say the Germans."[74]

To restore equilibrium in human relations, Ramos argued that the first task would be to scientifically correct the "odious frame of mind" that was one of its root causes—the "false philosophy of racial supremacy." For Ramos, correcting this empty philosophy involved laying bare the workings of the "historical rationalization" that reached its "zenith in European racialism." This "ideological battle" was just as important as the "military battle fought by armies," argued Ramos, and entailed a "disarmament of the mind" that "can only be achieved through a reasonable, humane and scientific policy, designed to bring about harmonious contact between the different races of the world."[75]

Yet despite this impassioned plea for a new, and ostensibly less European, way of doing things and anticipation of a brighter future, the solutions Ramos invoked were drawn from a familiar imperial toolkit. Indeed, the two measures that Ramos recommended "for the purpose of ordering the relationships between different racial and cultural groups living in the same country" were assimilation (in the metropole) and indirect rule (in the colonies). Regarding assimilation, Ramos argued that it is advisable in occasions where the "environment, the liberal atmosphere and relative economic equality render it opportune." Although he did not offer an explicit definition of assimilation, Ramos did put forward a lengthy summary of anthropological arguments against the existence of pure races. "Inter-breeding of races (*miscegenation*) has," Ramos explained, "from prehistoric times, been the rule" and "modern means of communication are making the world ever smaller" meaning "exchanges between the peoples are now the rule rather than the exception."[76] Contact between races and presumably assimilation were now inescapable according to Ramos, and therefore racial discrimination betrayed an "ethnocentricity" without biological justification. Although he spoke favorably of this process of cultural and biological absorption and implied that it was inevitable, Ramos also suggested that in "special circumstances" groups of "a particular race or culture wish to preserve their cultural originality and to form racial minorities in an environment other than their country of origin." As examples of these "special circumstances," Ramos cited "Europeans or Orientals who have emigrated to America, and Europeans who have settled in the colonies." As such, though Ramos's conception of assimilation generally implied a process of cultural and biological fusion, Ramos made an exception when it came to preserving the distinctiveness of European (and in some cases "Oriental") culture. In other words, it was not so much the Europeanization of the world that Ramos objected to but rather the violent methods through which it had taken place.

In the postwar world, Ramos's arguments implied, equilibrium would be restored and Europeanization could continue as long as people were left to freely adopt European standards. Given this framing, it is not surprising that Ramos wrote glowingly of the policies of indirect rule implemented by the British colonial administration during the interwar period. After World War I, explained Ramos, British colonial officials were struck by the degree to which their colonized subjects had become demoralized due to the "breaking up of the tribal and cultural units, the flight of the natives, the loosening of social and political ties, the loss of the characteristic marks of the culture." It was in response to the natives' "disillusioned melancholy" that the British Empire began making progress, according to Ramos, and introduced "indirect rule" thereby recalling "many native chiefs who had been banished from their territories and reinstat[ing] them as leaders of their former subjects."[77] In addition to shifting to an approach that involved ruling through "the native chiefs," British administrators also "tried to reconstruct, so far as they could, the aboriginal cultural framework which was in process of disappearance or dissolution." In Ramos's reckoning, indirect rule was thus "designed to repair the serious damage previously caused to the natives' traditional civilizations" and found its greatest expression in the creation of institutions that "set about trying to give concrete effect to the liberal principles of applied social anthropology."[78] For the postwar future, Ramos recommended perfecting the policy of indirect rule until "natives have a sufficient modicum of material culture of European origin" yet insisted that this stance did not entail an affirmation of European cultural superiority. Instead, Ramos argued that colonial administrators had shown that "many Africans assimilate European methods not in order to 'become whites,' but to further their own tribal ideals." Anticipating themes that would become central to UNESCO's philosophy in the 1950s, Ramos suggested that "for the rest [of the world]" what must be done is "to bring about a fruitful exchange of experiences" and argued that "it will only be possible for different groups of human beings to live together in harmony if there are exchanges of cultural experience, which will, in any case, make life more interesting."[79] It was against this backdrop of a reformed process of Europeanization yielding to multiculturalism that Ramos saw an opportunity for a new humanism that "we all hope will triumph in the post-war world." In contrast to the old humanism, whose roots lay in a violent historical process where Europeans forcefully imposed their civilization on those they colonized, Ramos described new humanism in terms that echoed the language of *unity in diversity*, which became a mantra during UNESCO's early years. Indeed, Ramos described this new humanism as a process involving "the pooling of different experiences, as part of a generous and democratic view of existence."[80]

CONCLUSION

In histories of UNESCO the cosmopolitan ideal of 'unity in diversity' is often attributed to the French anthropologist Claude Lévi-Strauss and his bestselling UNESCO booklet *Race and History* (1952).[81] In this booklet, Lévi-Strauss offered a purportedly non-biological explanation for the material and technological differences between the civilizations of white and those of "colored" peoples. Though he did not cite Arthur Ramos, Lévi-Strauss crafted an explanation that appealed to themes that were central to the work of the Bahian physician and the "Nina Rodrigues School" he claimed to represent. Like Ramos, Lévi-Strauss minimized the explanatory value of race and instead appealed to the importance of cultural contacts and exchanges in attempts to understand human diversity. In *Race and History*, Lévi-Strauss theorized that the differences in scientific and technological achievement between civilizations were not the result of innate racial aptitudes but rather the outcome of historical processes that hinged on the degree to which a society establishes and maintains mutual exchanges with other societies. Societies that prioritize mutual exchange, Lévi-Strauss argued, become "cumulative cultures" and stand a better chance of pooling knowledge and techniques that allow for technological and cultural progress. By contrast, societies that remain isolated and hostile towards others—what he called "stationary cultures"—diminish their chances of accumulating innovations and doom themselves to stagnation.[82] Though he did not necessarily use these terms, Lévi-Strauss's *Race and History* thus championed processes not unlike the 'syncretism' and 'acculturation' that often served as a focal point for Arthur Ramos and his generation of Afro-Brazilianists. While histories of UNESCO have tended to ascribe this attention to cultural exchange to the influence of

French anthropology, Arthur Ramos trajectory suggests that Brazilian and Bahian intellectual traditions also exerted a significant influence.

What is clear from the brief period that Ramos spent at UNESCO is that he viewed his appointment as an opportunity to expand Nina Rodrigues's studies of religious syncretism and race mixing in Bahia to the international realm. In doing so, Ramos hoped to complete his career-long project of redeeming the Bahian scholar's work by sanitizing its racist overtones. As the epitaph to this chapter suggests, Ramos viewed Rodrigues's work as epistemically sound yet dated in its terminology. Accordingly, Ramos believed that the topics Rodrigues framed as questions of race mixing and heredity could be reframed as ones of acculturation and education and thus rendered inoffensive. Yet as his language of "backward" peoples suggests, Ramos was not compelled to challenge the hierarchies at play in Rodrigues's work. Indeed, in Ramos's own intellectual trajectory, the boundaries between racism and antiracism cannot be tidily drawn.

In fact, throughout the 1930s, the anthropological and psychiatric research that Ramos conducted in Rio's school system fit seamlessly within the eugenic programs of racial whitening put forward by the authoritarian regime of Getulio Vargas. Given that so much of the historiography of race science interprets UNESCO's postwar antiracist activities as a repudiation of eugenics and scientific racism, Ramos's proximity to eugenic projects seems perplexing.[83] Yet, as historians of eugenics in Latin America have shown, medical experts and scientists in Brazil (as in many other parts of Latin America) embraced a capacious conception of eugenics that brought together disparate strands of racial thought in the human sciences.[84] More than anything, these disparate strands were concerned with the racial redemption of Brazil through accelerated economic development. The very malleability of the eugenic label made it useful for consolidating numerous discourses concerned with modernizing Brazil's predominantly non-European population. Yet the indeterminacy of eugenic discourse meant that Brazilian experts could just as easily distance themselves from it in line with shifting geopolitical winds. Thus, this chapter showed that though his studies of Afro-Brazilian religious practices and problem children fit within eugenic agendas during the 1930s, by the 1940s Ramos grew disillusioned with the Vargas regime and instead embraced applied anthropology and internationalism as tools for social reform. Though his research agenda drew from multiple disciplines and advanced eugenic agendas, one of Ramos's central and abiding goals was to invert and replace the racialist doctrines of his mentor Nina Rodrigues—rooted in the anatomical and phrenological preoccupations of criminal anthropology—with a conceptual framework emphasizing the social and cultural patterning of the psyche. Ramos's work can thus be seen as an antiracist project that did not challenge the social structure of Brazilian society or the international world order. Rather it was an antiracist vision that paternalistically anticipated a future in which Afro-Brazilians and other supposedly "backward" groups would find better ways to adjust to a modernity that Ramos imagined as primarily white and European. As we will see in the next chapter, he was not alone in sharing this vision.

CHAPTER TWO

Relocating Race Science After World War II

Situating the 1950 UNESCO Statement on Race in the Southern Hemisphere

The minutes to UNESCO's 1949 Meeting of Experts on Race Problems in Paris begin with an affectionate tribute to the recently deceased Arthur Ramos, who was described by Luiz Aguiar Costa-Pinto as the meeting's "guiding spirit." Among the renowned social scientists who listened to Costa-Pinto's heartfelt tribute and participated in this 1949 meeting, at which the 1950 UNESCO Statement on Race was drafted, were Claude Lévi-Strauss, Ashley Montagu, Juan Comas, and E. Franklin Frazier. "One of the outstanding characteristics of Dr. Ramos' work," Costa-Pinto explained to the committee of experts before him, "was his deep sympathy with all the backwards peoples and oppressed races."[1] According to Costa-Pinto, Ramos's noble sentiments stemmed from a remarkable career spanning several decades during which he achieved fame as "an expert on African problems and on integrating Negroes in the culture of the New World." For Costa-Pinto, Ramos's tireless efforts on behalf of backward peoples made him "one of the greatest representatives of the new scientific humanism."[2]

As we saw in the last chapter, the scientific humanism that Costa-Pinto cited in his tribute was one that Ramos built on the foundation of Nina Rodrigues's studies of race mixing and religious syncretism and updated through a wide engagement with anthropological and social psychological theories of cultural change. By the time Ramos landed in

Paris, his intellectual outlook had become deeply shaped by the tradition of applied anthropology that arose in British colonial Africa, and he was convinced that a reformed discipline of anthropology—redeemed from its racialist excess—offered a conceptual toolkit for the integration of "backward people" into the modern world. Ramos tellingly described this process as Europeanization and believed that the postwar period offered an opportunity to transform this process of Europeanization from one that was forcefully imposed upon non-Europeans to one they would freely adopt. Although Ramos developed this framework in Brazil, it was one that translated easily into the international context of UNESCO's Social Sciences Department (SSD). Indeed, as this chapter demonstrates, the SSD and the social scientists who drafted the 1950 Statement on Race were steeped in similar traditions of imperial social science inquiry that took the comparative development of societies as their central research object and are thus forerunners to the field of international development that crystallized with the help of international organizations in the period after 1945.[3]

By foregrounding the committee members' engagement with questions of social change, this chapter offers an alternative interpretation of the 1950 UNESCO Statement on Race. The 1950 statement and its companion statement from 1951 have typically been interpreted as landmark documents reflecting scientists' efforts to distance the study of human biological diversity from its association with scientific racism. According to these interpretations, the UNESCO statements reflect a significant conceptual shift away from the static racial typologies of nineteenth-century physical anthropology that were propelled by the modern evolutionary synthesis in biology and the consolidation of cultural anthropology.[4] Although there is no denying that disputes about the biological reality of race were central to the controversies surrounding the UNESCO statements, this chapter demonstrates how these now well-rehearsed debates were secondary within the agenda of UNESCO's race campaign. The SSD was more concerned with identifying tools that could be used to combat racial prejudice worldwide than it was with resolving contentious questions concerning the reality and biological basis of racial differences. As such, UNESCO officials envisioned the race campaign as one that was firmly situated within the social sciences and which was primarily concerned with the study and management of

interpersonal and group relations. The makeup of the 1949 committee—
which included sociologists, social psychologists, cultural anthropologists, and educators—reflects this concern with the social dimensions
of racism. By examining the participants of the 1949 meeting's intellectual itineraries, this chapter shows that, like Arthur Ramos, many of the
tools these experts relied on to study comparative societal development
were also put to work for the purpose of combating racial prejudice.
Instead of signifying the demise of scientific racism, this chapter argues
that the 1950 Statement on Race can and should be situated within histories of international development and social science.

PRACTICAL MEASURES AGAINST RACIAL PREJUDICE

The social scientists who drafted the 1950 Statement on Race were
influenced by diverse currents in the human sciences of which Boasian cultural anthropology was but one among many. These currents
included purposive conceptions of human biology and social change
that drew from Spencerian and Lamarckian conceptions of evolution
and racial progress rather than from the historical contingency of the
evolutionary synthesis. Many of the participants of the 1949 meeting
were in fact situated within traditions of human science concerned with
the improvability of so-called backward races and with the relations
between consciousness, the environment, and social change. For the
1949 committee members, the problem of combating racial prejudice
was also deeply intertwined with the issue of how to uplift the lives of
populations deemed culturally and psychologically maladjusted to the
economic strains of emerging urban-industrial societies in the Southern Hemisphere. It was in the context of this concern with adjusting
non-European bodies and minds to the rigors of modern, technological
life that rigid racial typologies often became an object of critique for the
participants in the 1949 meeting.

The 1950 and 1951 UNESCO Statements on Race have typically been
examined in relation to their significance for debates concerning the
existence of biological races. Although questions concerning the biological basis of race were prominent in the published statements, the UN
officials who proposed the idea for the 1949 meeting and the reports
and documents that UNESCO officials circulated in anticipation of the

meeting suggest a different set of concerns. Instead of the theoretical issue of how to define race in biological terms, UNESCO and UN officials expressed greater interest in pragmatic questions concerning the socioeconomic relations between racial groups and, more important, in the development of practical tools for combating racial prejudice. For instance, during the first round of discussions of the 1949 meeting, Edward Lawson—a Black diplomat who served as the meeting's representative from the United Nations—offered an important clarification for why the United Nations had tasked UNESCO with addressing the issue of race. According to Lawson, the idea of asking UNESCO to address the question of race arose as a result of discussions held during an Economic and Social Council (ECOSOC) meeting at Lake Success, New York, which led to the creation of a "Sub-Committee on the Prevention of Discrimination and the Protection of Minorities." During this meeting, the subcommittee discussed the question of "a definition of race" at some length and, in collaboration with representatives from the Division of Human Rights of the Department of Social Affairs, arrived at the conclusion that "the concept of a definition of race was scientifically illegitimate, and that there was no way of defining race in any generally acceptable sense." As a result, ECOSOC specifically requested that UNESCO not "concern itself with theoretical considerations" and instead "take practical measures to solve the problem of racial prejudices." During the opening discussions of the 1949 meeting, Lawson thus counseled the committee to keep the issue of establishing a definition of race on its agenda "if only to arrive at the conclusion that it was not a scientifically [sic] illegitimate concept" and that their most important task was "not to define race" but to make "a clear scientific statement of facts which could not be challenged by anyone in the world."[5]

Given this emphasis on practical measures to confront racial prejudice, it is not surprising that the committee the SSD assembled consisted mostly of social scientists (even though they did attempt to include some biologists). The literature that SSD officials asked the meeting participants to read in preparation for the meeting also drew heavily on theories and concepts from the social sciences. A few weeks before the meetings took place, SSD officials provided the committee members with a considerable list of documents and reading materials in preparation for the meeting. The committee's reading list offered a daunting combination of policy documents, organizational manifestos, and academic literature. For instance, the list of documents included: internal UN reports describing how the meeting stemmed from a concern for the "prevention of discrimination and the protection of minorities," which was a focal point during the fifth session of ECOSOC's Commission on Human Rights; excerpts from the manifestos and resolutions issued by scientific and nongovernmental organizations such as the American Anthropological Association, the American Psychological Association, the Union International Contre le Racisme, and the International Genetics Conference; and excerpts from thirty-seven different scientific articles on race written by authors organized into different disciplines (anthropologists, physical anthropologists, social anthropologists, sociologists, psychologists, psychiatrists, and biologists).[6]

Many of these preparatory documents emphasized the importance of taking practical measures against racial prejudice as opposed to arriving at a precise definition of race. In orienting the committee toward practical action, the preparatory documents emphasized the use of social science techniques for managing social relations and combating social tension. One of the most notable preparatory documents was a report penned by the Chicago-trained sociologist and pioneering urbanist Louis Wirth. Wirth's report titled *Implementation of the Resolution of the Economic and Social Council on the Prevention of Discrimination and the Protection of Minorities* offered an overview of the specific actions requested by ECOSOC as well as commentary on how to operationalize key concepts from ECOSOC's resolution through social science research. For instance, Wirth's report praised ECOSOC's report on the prevention of discrimination for offering a "valuable account of various definitions of discrimination and prejudice and the distinctions and interrelations between these phenomena."[7] Yet Wirth found the report to be wanting in its discussion of one of its central concepts—the concept of a minority. Wirth argued that although the report recognized the existence of different kinds of minorities (i.e., race, ethnic, class, political opinion, and sex), it failed to conceptualize many of the specific forces that shape the specificity of minority groups. The report failed to recognize, argued Wirth, that minority groups differ "according to the nature and degree of their group consciousness and their relationships to the dominant group" and that their political ambitions are often quite different.[8] Indeed, Wirth argued that when it comes to the protection of minorities, it makes a big difference whether the group in question seeks a separate existence, assimilation with the dominant group, to establish a separate nation-state, or dominance over another "minority or the dominant group itself."[9] If UNESCO was to take effective action toward the prevention of discrimination and protection of minorities, Wirth stressed, it would need to rely on "thorough-going knowledge" of the societal, historical, and institutional forces shaping prejudice and "the attitudes which are held by the people involved."[10]

As an example of the kind of thorough-going knowledge UNESCO could draw from, Wirth pointed to the subfield of race relations research that was central to much of the work done by sociologists trained in the Chicago tradition. In describing the promise of race relations research, Wirth describe it as part of a broader domain of knowledge concerning "man in his social relations" that built on knowledge from all of the social science disciplines. Indeed, Wirth stressed that race relations—aside from the "complicating factor of biological or cultural differences"—were no different from other intergroup relations. As such, race relations research could learn much from studies concerning other fields of social relations, notably "family relations, play groups, labour relations and international relations."[11] Because they were embedded in other social relations and "manifest themselves in the ordinary routines of living," Wirth argued that the attitudes that underpin race relations would have to be studied in everyday sites: "in the school, the local community, the church . . . the army, the civil service . . . in connexion with health, welfare and housing problems."[12] In addition to these kinds of studies targeting local manifestations of racial sentiments, Wirth also proposed comparative studies of "race relations and minority problems on a world-wide scale" with an eye to "the range of possibilities and hence alterability of minority status." Wirth also suggested that racial and cultural problems were related to the historical processes of "migration, conquest, colonization, the rise of cities and the development of an industrial civilization."[13] The effect of these historical processes, Wirth argued, is that "peoples of various characteristics have mingled and sometimes blended both biologically and culturally," thus minimizing "isolation."[14]

The pragmatic outlook reflected in Wirth's reports and the meeting's preparatory documents suggests that determining the precise biological basis of race was not the most important agenda item for UNESCO's race campaign. Instead, the wording of the ECOSOC resolution that prompted the 1949 meetings described the role of scientific knowledge in rather instrumentalist terms. One of the key clauses of the resolution requested UNESCO "to consider the desirability of initiating and recommending the general adoption of a programme of disseminating scientific facts designed to remove what is commonly known as racial prejudice."[15] Although this certainly offered plenty of scope for a program focusing on disseminating facts concerning the biological basis of race, the discussions in the preparatory documents and, as we will see, the conversations held during the 1949 meetings slanted toward sociological questions concerning the political and economic forces that give rise to and condition race prejudice and techniques for minimizing tensions and changing harmful attitudes. The issue of how social scientific knowledge could be translated into effective antiracist action was also reflected in the intellectual trajectories of those invited to participate in the 1949 meeting. As it turned out, the meeting participants were steeped in traditions of social science inquiry that had pragmatic issues concerning the modification of individual attitudes for the purpose of harnessing social change at their core.

COMMITTEE OF RACE EXPERTS
AND SOUTHERN RACIAL REGIMES

Given the organizational emphasis on identifying practical measures against racial prejudice, it makes sense that SSD officials turned primarily to social scientists when they convened the 1949 Meeting of Experts on Race Problems. When social science disciplines like sociology, anthropology, psychology, and political science became distinct professions in research universities in Europe and North America in the late nineteenth century, questions concerning race and comparative societal development emerged as core concerns. The conceptual links that social scientists established between the study of racial anatomy and the comparative development or evolution of societies reflects the imperial context in which the social sciences emerged. In the case of sociology, it was the global expansion of European empires and the United States that furnished the social sciences with their "main conceptual framework and much of [their] data."[16] In this imperial context, the social sciences developed expansive frameworks for making sense of a wide spectrum of human experience. For instance, classic sociologists such as Émile Durkheim and his collaborators studied modern industrial societies as much or even less than non-modern societies. According to Raewyn Connell, only 28 percent of the reviews published in *L'Année Sociologique*—the international survey of sociological research published by Durkheim and his collaborators between 1898 and 1913—concerned contemporary societies in Europe or North America. Instead, the bulk of sociological research material in this period concerned "ancient and medieval societies, colonial or remote societies, or global surveys of human history." This could include, among others, studies of "holy war in Israel, Malay Magic, Buddhist India, technical points of Roman law, medieval vengeance, Aboriginal kinship in central Australia, and the legal systems of primitive societies."[17] To make sense of such a broad swath of human history, sociologists adopted a central framing device to organize their studies—the comparative study of the difference between the metropole and "an Other whose main feature was its primitiveness."[18] And it was in this imperial context where grand comparative theorizing became a central feature of sociology and other disciplines that the study of the conditions in which "progress" from the primitive to the advanced became pervasive.

As described in the previous chapter, scholars in the Southern Hemisphere such as Nina Rodrigues and Arthur Ramos appropriated the comparative and evolutionist methodologies of turn-of-the-century social science. But instead of crafting narratives of progress that doomed Brazil and its *mestiço* population, Rodrigues and especially Ramos conceptualized the environment and the cultural milieu as entities that could be reformed in an effort to kickstart social progress. We can see similar themes and concerns in the research programs of the participants of the 1949 meeting. Yet we can also see attempts to break from this frame. For instance, in his UNESCO-sponsored study *O Negro no Rio de Janeiro* (1952), Luiz Aguiar Costa-Pinto insisted that the study of race relations needs to move beyond the ethnographic approach of his mentors (Ramos and Rodrigues), which he characterized as narrowly focused on documenting the bizarre and exotic aspects of Afro-Brazilian culture and attempting to trace African survivals. Costa-Pinto argued that this narrow ethnographic approach was ill suited to the rapidly shifting economic structure of Brazilian society and proposed that instead of the ethnographers' concern with the "integration of the African into Brazil," what was needed was "serious sociological study" of the "integration of the black Brazilian into Brazilian society."[19] Indeed, Costa-Pinto argued that a sociological approach that took the economic and political interests of Afro-Brazilians seriously was required by the rapid industrialization and urbanization taking place in Brazil's major cities, which was transforming the economic structure of Brazilian society and increasing racial tension. Thus, in contrast to his mentors, who viewed race through the prism of culture and folklore, Luiz Aguiar Costa-Pinto conceptualized race as something emergent from and embedded in the interplay between economic structures, interpersonal relations, and political ideologies.[20]

The Brazilian representatives at the 1949 meeting were primarily concerned with race as it pertained to the future of Afro-descendant populations in the Americas. By contrast, Juan Comas, the Spanish-Mexican anthropologist at the 1949 meeting, was primarily concerned with race insofar as it related to the assimilation of indigenous populations in Mexico and the Americas. A Spanish exile from the Franco regime, Comas trained as a physical anthropologist with the Swiss anthropologist Eugène Pittard during the 1930s then fled to Mexico in 1940 where he taught physical anthropology and became an important figure in the *indigenismo* movement through his work for the Instituto Indigenista Interamericano (III) in Mexico City.[21] *Indigenismo* was a diverse political, economic, and cultural movement from the early twentieth century that was led by intellectual and cultural elites who sought to make the indigenous heritage of South and Central America nations a central part of their national identity. Although it had distinct national variants, historians of *indigenismo* have noted that it was characterized by contradictory impulses in which eugenic conceptions of acculturation coexisted with a politics that called for the celebration of indigenous heritage as a nation-building strategy.[22]

Juan Comas's writings on race perfectly illustrate the ambivalent trajectory of *indigenismo*. In work published by UNESCO, Comas offered potent critiques of biological conceptions of race while calling for the rapid assimilation of indigenous groups in the Americas. For instance, in the UNESCO booklet *Racial Myths* (1951), Comas argued that the concept of race is one that implies the "existence of groups presenting certain similarities in somatic characteristics which are perpetuated according to the laws of biological inheritance."[23] On the basis of this hereditary conception of race, Comas argued that "racial prejudice" did not exist before the Iberian conquest of the Americas in the fifteenth century. While there were certainly many examples of cultural antagonism toward other groups before European colonization, Comas insisted that these were not made on the basis of a "division of mankind into antagonistic races" or prejudice based on "implacable laws of heredity." The origins of modern racism, Comas argued, could thus be found in "the beginning of African colonization and the discovery of America and of the trans-Pacific sea route to India" when "there was a considerable increase in race and colour prejudice" fueled by "economic self-interest" and a need to justify the practice of slavery.

A year after the publication of *Racial Myths*, Comas drew upon *indigenista* conceptions of acculturation in an article describing the importance of cultural anthropology for UNESCO's fundamental education projects in the Americas. Although he did not speak directly about race in this article, his assessment of the indigenous peoples of the Americas emphasized how they were held back by entrenched cultural differences and patterns of oppression stemming from colonial conquest. The Americas, Comas argued, were home to more than 30 million "Indians" or "Natives" with "specific cultural characteristics" that differed from what was known as "white" or "western" civilization.[24] Because of "three centuries of conquest and colonization," this huge population now existed at an "exceedingly low social level, expressed in the poorest economic conditions and standards of living, which prevents its rapid assimilation as an active factor of national production and consumption."[25] Comas further argued that UNESCO could play a crucial role in lifting the indigenous populations out of this position of chronic backwardness through culturally sensitive fundamental education projects emphasizing basic literacy and hygiene and cited the work of leading *indigenistas* such as the Mexican anthropologist Manuel Gamio as models for how this work could be done. Gamio argued that Mexico's indigenous populations were biologically and racially deficient due to centuries of socioeconomic oppression, and he vigorously promoted applied anthropological studies of the culture and material conditions of indigenous populations that would inform state-backed initiatives in the realms of education and public health.[26] From this developmentalist framework, Comas and Gamio thus conceptualized indigenous groups as racially and culturally inferior due to centuries of European oppression but anticipated a prosperous future in which indigenous groups would be freed from their degraded status through absorption into the modern apparatus of the postrevolutionary Mexican state.

This interest in social evolution and racial improvement was shared by the other committee members who drafted the 1950 UNESCO Statement on Race. But whereas the South American participants often blurred distinctions between culture and biology in their work, the other participants tended to create sharp distinctions between these domains. For instance, Morris Ginsberg and Ernest Beaglehole were both products of a British sociological tradition that embraced comparative studies of the social institutions of different races as a means of studying evolution in a strictly social sense. This sociological approach was one that emerged out of the sociology department at the London School of Economics (LSE) and was strongly shaped by the writings of the liberal philosopher Leonard Trelawney Hobhouse. Hobhouse was an admirer of Herbert Spencer and sought to reformulate Spencer's conception of evolution by further placing human psychology and human agency at the center. In contrast to Malthusian and eugenic conceptions of race that saw little hope in social reform, Hobhouse articulated a purposive conception of evolution that recognized the formation of social institutions such as moral codes, cultural traditions, and systems of mutual aid as a domain where evolutionary process occurred that could not be explained in biological terms.[27]

This conception of social evolution was also shared by Morris Ginsberg, who was born into a Jewish family from Lithuania and later trained with Hobhouse and eventually succeeded him as chair of LSE's sociology department. Ginsberg's approach to sociology was also deeply indebted to the classic works of nineteenth-century liberalism and especially the work of Herbert Spencer. Ginsberg argued that one of the central insights of Spencer's philosophy was the importance of "comparing rudimentary societies with one another and with societies in different stages of progress."[28] Through such comparison, Ginsberg suggested, sociologists could discover "common traits of structure and function as well as certain common traits of development." Indeed, Ginsberg suggested, "to discover the conditions of social growth, arrest, and decay, is one of the principal tasks of sociology."[29]

This Spencerian interest in discovering the conditions of social growth and decay permeates a landmark book by Hobhouse and Ginsberg titled *The Material Culture and Social Institutions of the Simpler Peoples: An Essay in Correlation* (1915). In this book, Hobhouse and Ginsberg sought to establish a "social morphology" of the social and political institutions of "those less fortunate races which range from the lowest known *Naturmenschen* to the confines of the historic civilization."[30] To craft this social morphology, Hobhouse and Ginsberg combed over a vast body of colonial ethnographic data on "simpler peoples" from Australia and Oceania, Latin America, Asia, Africa, and North America. They then used these data to test whether "the advance of human knowledge" as it concerned the "understanding and control of natural forces" carried "any distinct movement in morals, law, religion, the general organisation of society,"[31] and to classify the social institutions of the "less fortunate races" according to their varying stages of economic advancement.[32] Their methods also involved crafting tables and charts that established statistical correlations between economic development and social institutions such as methods of justice, family and marriage practices, and property arrangements. In this classic treatise in imperial sociology, Hobshouse and Ginsberg thus sought to offer a nonbiological account of racial progress and social evolution that was rigorously grounded in the thick data of a colonial ethnographic archive.[33]

This concern with the growth and decay of the social institutions of simpler peoples was also central to the work of the New Zealander Ernest Beaglehole, who conducted his doctoral studies in LSE's sociology department with Ginsberg and also served on the 1949 committee of race experts. In his first book, *Property; a Study in Social Psychology* (1932), Beaglehole conceptualized the psychological basis of property as stemming from a "sentiment of possession" or "grouping of various emotional tendencies about the property object itself" as opposed to an innate biological instinct.[34] This affective conception of property,

Beaglehole reasoned, required taking into account both "psychological fact" and "social data" and investigating the manner in which "aggregated sentiment" is "moulded by, and in turn moulds, the economic culture patterns of a group."[35] To understand property sentiments in their most basic form and observe how they are molded by culture and built up into social institutions, Beaglehole assumed the unity "of all animal life" and compared the behaviors and practices of animals (insects, birds, and mammals), "simpler peoples," and children in civilized societies. Among animals, Beaglehole identified a "primitive property value" that was the result of an "organic striving which appropriates from the environment material for self-provision, self-development and racial perpetuation."[36] Among simpler peoples, Beaglehole described a "cultural patterning of property values" where a set of cultural *mores* "moulds the raw stuff of human nature into conformity with its own demands and practices."[37] And in his discussion of groups with "higher levels of intelligence," Beaglehole argued that this cultural patterning of property values molded the individual personality and "adapt[ed] one child to a communist organization of society and another to a Western capitalist society."[38] Reflecting the Spencerian leanings of his mentors, Beaglehole thus crafted a social psychological taxonomy of property sentiments and probed the ways in which social institutions emerge and develop out of the raw biological material of human nature.

After his doctoral work, Beaglehole's interest in cross-cultural comparison led him to the anthropology department at Yale, which was then the leading center for the culture and personality movement led by the linguistic anthropologist Edward Sapir. During his time at Yale, Beaglehole conducted extensive fieldwork in Hawai'i, New Zealand, and Polynesia with the support of the Bernice P. Bishop Museum in Honolulu. In Honolulu, Beaglehole collaborated closely with the Māori physician-turned-anthropologist Te Rangi Hīroa (Peter Buck) and the physical anthropologist Harry Shapiro, both of whom studied race mixing enthusiastically and conducted extensive anthropometric studies of Polynesians.[39] Te Rangi Hīroa and Shapiro's influence is particularly evident in a 1943 article where Beaglehole argued that the concept of race is a valuable tool for "the physical anthropologist and the human biologist" but has no value for the social scientist trying to understand "contact between peoples."[40] Though tainted by popularizers of the "race superiority" doctrine such as Hitler and Gobineau, the concept of race was, at Beaglehole's insistence, nevertheless necessary for studying human biological variation. Careful to avoid conceptualizing human variation as static, Beaglehole described how the human species was characterized by a terrific "variability increased by an incredible amount of inter-mixing, blending, and hybridization" and by "an occasional stabilization of physical type in the past due to the isolation of peoples behind impenetrable geographical barriers."[41] Given this continual process of racial formation, Beaglehole argued that the concept of race had some value for human biology. Yet for the social sciences, Beaglehole argued that race generally implies the notion that social and psychological phenomena are "determined by heredity and therefore fixed and unalterable" and thus had no value for "understanding the contact between peoples, whether this contact is cooperative or competitive."[42] Thus, a common thread in Beaglehole's work before the 1949 meeting is the contention that social institutions and processes must be understood separately from the biological basis of human nature and human variation. However, his work also insisted that society, culture, and biology are interrelated and that cultural patterns and choices thus play an important role in shaping human biological variation.

Although he shared a similar trajectory to Beaglehole, Ashley Montagu arrived at the conclusion that race has no legitimacy or value in the study of human variation. Before commencing his PhD work at Columbia with Franz Boas, Montagu studied social anthropology with Bronislaw Malinowski at LSE and wrote a paper on a debate that was central to settler colonial anthropology: whether Aboriginal Australians were "nescient," or rather ignorant, of the link between the act of coitus and biological reproduction.[43] Montagu developed this work into his first book, *Coming into Being Among the Australian Aborigines* (1937), which sought to dispel the idea that aboriginal "nescience" about the link between coitus and pregnancy stemmed from a primitive mentality.[44] Yet instead of conducting fieldwork with indigenous Australians, Montagu reviewed settler colonial ethnographies and concluded that "Aborigines'" beliefs about pregnancy were in fact consistent with their "particular conceptual system." In other words, Montagu reasoned, the "mind of the native functions in exactly the same way as our own" but is differentiated by the "premises upon which that functioning is based" and by "categories and forms of judgement" which are "different though quite as rigorously organized as our own."[45] Thus, Montagu explained, Aboriginal Australians could not be said to be ignorant of the physiological relationships between mother, father, and child but rather operated within a conceptual framework wherein "the concept of blood relationship or consanguinity is impossible to them."[46] By adopting this relativist perspective, Montagu sought to redeem Aboriginals by rendering their beliefs the product of culture rather than a sign of innate biological primitivism. His analysis, in other words, left an opening for improvement and reform.

Though *Coming into Being* sought to position so-called primitive peoples within the same cognitive and biological plane as "civilized" Westerners, Montagu's later antiracist texts mobilized Malinowski's concept of "cultural contacts" to describe Aboriginals as occupying a culturally and temporally distant space. For instance, in his acclaimed antiracist book *Man's Most Dangerous Myth* (1942), which argued that the concept of race was so ridden with fallacies that it should no longer be used, Montagu described Aboriginal Australians as peoples stuck in an anterior time, existing anachronistically within the modern world.[47] What truly separated Aboriginals from "modern civilization," Montagu argued, was not racial type or genetic endowment but rather a difference in "experience" and "variety of cultural contacts." Indeed, Montagu mused, "we of the Western world have packed more experience into the past two thousand years than have the Australian aborigines, during their 60,000 years of continuous settlement of Australia."[48] Thus, although he is best known for his critiques of racial typologies, Montagu's interpretations of Aboriginal primitivism reveal the latent influence of settler colonial anthropology and racialized development.

While Montagu's work explained civilizational differences by evoking anachronistic tropes, E. Franklin Frazier—a black U.S. sociologist and elected chairman for UNESCO's 1949 meeting—sought to understand the "adjustment" and assimilation of African Americans to "modern civilization" in the historical context of slavery and the transition to freedom.[49] When the 1949 meeting took place, Frazier's career was on the rise. He was head of Howard University's sociology department and had just finished serving a term as the first black president of the American Sociological Association. From 1951 to 1953, he acted as chief of the Division of Applied Social Sciences at UNESCO and used the opportunity to conduct research in the West Indies on race and culture contacts. Frazier trained with the Chicago sociologist Robert Park in the 1920s and retained much of his ecological approach to studying the assimilation of racial minorities to the modern industrial culture of U.S. cities. During the 1930s, Frazier published two influential books on the sociology of black families—*The Negro Family in Chicago* (1932) and *The Negro Family in the US* (1939)—which established him as a leading expert on the effects of slavery on black families in the United States. In these studies, Frazier sought to understand the "Negro Family" as both a "natural human association and as a social institution subjected to the severest stress and strains of social change."[50] Unlike the anthropologist Melville Herskovits, whose work sought to document the survival of African cultural practices in the Americas, Frazier argued that Afro-descendants in the Americas lost their cultural heritage during slavery and were forced to develop new cultural patterns and family structures dictated by the desires of their masters.[51] His research thus sought to understand the ongoing social adaptations of black families and how they were shaped by processes of acculturation and assimilation. In tracing the social history of black families, Frazier's work also sought to understand how they could be more effectively assimilated in the future. For instance, *The Negro Family* concluded that although African American assimilation had been near impossible in the past, the shift to a highly mobile and urbanized society where "caste prescriptions lose their force" was creating the conditions for ever-closer association between "Negroes and whites in the same occupational classes." These emerging associations held the potential for a more comprehensive social change. "Intermarriage in the future," Frazier augured, "will bring about a fundamental type of assimilation" and "the gains in civilization which result from participation in the white world will in the future be transmitted to future generations through the family."[52] Like many of the participants in the 1949 meeting, Frazier thus placed great faith in racial mixing, cultural exchange, and intermarriage as practical measures for integrating African Americans—and other racial minorities—into U.S. society, thus improving their standing in the civilizational hierarchy presumed by the social science of this period.

This practical concern for understanding the mechanisms of social change was also shared by perhaps the meeting's best-known participant—the French anthropologist Claude Lévi-Strauss. Lévi-Strauss's participation in the 1949 meeting was his first of several contributions to UNESCO's antiracist efforts and marks the beginning of a long and at times strained relationship with the organization.[53] In his best-selling UNESCO booklet *Race et Histoire* (1952), Lévi-Strauss sought to refute "the original sin of anthropology," which he described as the confusion between the idea of race "in a biological sense" and the psychological and sociological "productions of human civilizations." More narrowly, Lévi-Strauss sought to explain the seeming historical advance of the "white man's civilization" relative to the "civilisations of the coloured peoples" without positing the existence of "innate racial aptitudes."[54] Instead of innate racial aptitudes, Lévi-Strauss posited that the differences in scientific and technological achievement between "civilizations" were shaped by "geographical, historical, and sociological circumstances" and especially by the degree to which a society establishes and maintains mutual exchanges with other societies.[55] Societies that prioritize mutual exchange and remain open to others, Lévi-Strauss argued, become what he called "cumulative cultures" and stand a better chance of pooling together knowledge and techniques that allow for cultural progress. By contrast, societies that remain isolated and hostile toward others—which he called "stationary cultures"—diminish their chances of accumulating innovations and thus doom themselves to stagnation.[56] In offering this nonracial interpretation of the mechanisms that underpin societal development, Lévi-Strauss presented a redemptive narrative that glossed over European colonialism and positioned "the West" as an unquestioned pinnacle of civilizational achievement. Similarly, by rejecting explanations appealing to biological difference, Lévi-Strauss offered hope to those "stationary cultures" trapped by their self-imposed isolation and suggested they might join the fold of modernity through greater exchange with advanced cumulative cultures. By emphasizing the capacity of all civilizations to advance and the importance of exchange and cooperation, Lévi-Strauss's message was tailor-made to the liberal ideals of UNESCO and the UN system.[57]

The research programs of the 1949 "race experts" thus reveal common themes and modes of analysis drawn from social science traditions

FALLACIES OF RACISM EXPOSED

UNESCO PUBLISHES
DECLARATION BY
WORLD'S SCIENTISTS

FIGURE 2.1 "Fallacies of Racism Exposed: UNESCO Publishes Declaration by World's Scientists," in the *UNESCO Courier*, vol. 3, no. 6–7, July–August 1950, p. 1 (UNESCO Digital Library).

that emerged in tandem with or in response to European colonialism and U.S. expansionism. Though the participants shared an aversion to fixed racial typologies, this did not mean they abandoned racial conceptions altogether. In fact, their research relied upon conceptions of race that theorized human biological variation as subject to continual change through cultural and environmental forces and focused on

FIGURE 2.2 "Racial Harmony—New Zealand," in the *UNESCO Courier*, vol. 3, no. 6–7, July–August 1950, p. 8 (UNESCO Digital Library).

immaterial objects such as social institutions, cultural patterns, family structures, and personality types as crucial sites for studying the historical formation and evolution of racial consciousness. Within this antiracist racial regime, the apparent alterability of social institutions, cultural patterns, and the physical body was upheld as evidence that "simpler peoples" could and should escape their hapless state of cultural and even biological inferiority. For these antiracist scientists, typological conceptions of race thus stood in direct opposition to the purposive conceptions of social change in which they were so invested and became mechanisms for denying backward "races" their rightful incorporation into an imagined industrial and urban modernity. In this regard, the outlook of the 1949 committee was well aligned with Louis Wirth's proposal for comparative studies of race relations and racial prejudice on a worldwide scale. And it was precisely this grand comparative outlook and concern for understanding the socioeconomic dimensions of race development and decay that reveal the committee's indebtedness to imperial social science.[58] In short, for the 1949 committee, critiques of racial typologies paved the way for an increased investment in studying processes of social, cultural, and historical change and for developing frameworks to guide projects of racialized development.

DRAFTING THE 1950 STATEMENT ON RACE AND LOOKING TO THE FUTURE

The final statement was both a collective artifact and a document that reflected the influence of Montagu and his ambition to undermine the scientific credibility of "the race concept." Because of Montagu's influence, the 1950 Statement on Race rehearsed a series of well-known arguments from North Atlantic debates concerning the differences between typological and population approaches to human biology. Yet a close look at the discussions that took place during the 1949 meetings in Paris reveals the influence of the southern streams of racial thought described in the previous section. For instance, when the committee attempted to define race, E. Franklin Frazier reminded the committee that definitions of "race" varied significantly across countries. Frazier drew attention to the fact "that his own 'race' would vary widely according to the country where he found himself: it would be defined very differently in the United States, Brazil, Porto Rico [sic], or Jamaica."[59] Accordingly, Frazier suggested that the committee should consider a statement "outlining the development of the concept of race." But instead of looking to genetics, Frazier proposed a statement including physical anthropologists' criteria for the "separation of groups," consideration of "the confusion of race and culture," and a review of "acculturation and changes in the culture of the so-called backward peoples of the world." Frazier's envisioned statement also gestured toward social psychology and sociology and proposed data on "racial psychology" and its relation to "racial and cultural differences such as emotional, temperamental and intellectual ones" and a discussion of how "race prejudice was influenced by physical and cultural differences between groups." In addition, Frazier proposed sociological discussions of how children acquire "racial consciousness," as well as discussions of "the types of interaction and contacts which led to a development of race consciousness or 'feeling.'" Keeping with this sociological emphasis, Frazier also proposed a discussion of "how different types of social organization (caste and 'class' relationships)" influence "'race' relations," and of how "the various definitions of race involved valuations of physical and cultural differences."[60] Unlike Montagu's statement, Frazier's proposal thus foregrounded a sociological and psychological framework not concerned with the formation of genetically distinct populations but with the formation of racial consciousness amidst social change.

Frazier's emphasis on racial consciousness, geographic difference, and social change carried over to discussion of the second item on the meeting's agenda: "imagining future studies of race." Montagu played a marginal role in these discussions, which had little to do with biology or physical anthropology and instead focused on the sociological and psychological underpinnings of "race prejudice" and the geographic distribution of racial attitudes. This round of discussions began with a provocation from the Brazilian sociologist Costa-Pinto, who pointed out that all scientific disciplines could agree "that no pure races existed" and that relations between groups were "based on ideology and not on any scientifically definable differences." Proposals for future studies of race should therefore "begin by recognizing that race prejudice had its roots in social and political differences not in physiological or mental ones."[61] Like his critique of Afro-Brazilian ethnography, Costa-Pinto's comments thus reflect his ambition to shift studies of race away from not only biology but also narrowly construed conceptions of culture.

Costa-Pinto's comments sparked a quick response from Ginsberg, who proposed that UNESCO initiate a series of studies with the aim of determining the extent to which prejudice was really "racial." Keeping with the comparative emphasis of his sociological training, Ginsberg proposed the following series of research questions:

> [(1)] was it really the case that there was a fundamental difference between Latins and Anglo-Saxons in their attitude toward race and colour; (2) what were the variations in the attitude toward "race" (South Africa, Indonesia, Brazil); (3) what historical changes in that attitude had occurred in some parts of the world (disappearance of prejudice towards Indians and growth of it with regard to Negroes in the United States); (4) determination of the circumstances coinciding with a demand for segregation of races.[62]

In addition to these questions, Ginsberg also observed that there were important differences in the colonial policies of "Anglo-Saxon" and "Latin peoples" in relation to "the assimilation of indigenous peoples." While assimilation was encouraged in French colonies, argued Ginsberg, it was discouraged in British ones.[63]

By pointing toward this expansive imperial terrain, Ginsberg's comments dramatically shifted debate away from genetics and instead brought questions concerning the administration of racial groups within colonial bureaucracies and the formation of racial attitudes to the fore. Ernest Beaglehole echoed Ginsberg's comments by pointing out that these contrasting imperial attitudes toward assimilation could be observed in colonial census practices in the South Pacific. Whereas "census forms in the French possessions in the South Sea Islands made no mention of half-breeds," Beaglehole explained ". . . they were specifically referred to in censuses in the British and New Zealand possessions." According to Beaglehole, this provided compelling evidence that assimilation was officially encouraged in the French territories and opposed in the others. Beaglehole also noted that "racial prejudice was particularly strong in areas previously under German control and, exceptionally, in the Islands of Samoa, where the New Zealand authorities had followed the same policy."[64]

E. Franklin Frazier responded to Beaglehole's comments with enthusiasm and noted that "the position of half-breeds in various parts of the world" could be a fruitful area of future study. During his own research on the "attitude of Europeans and North Americans towards indigenous people in Brazil," Frazier explained, he realized that differences in attitude "were due not only to psychological but also to political, economic, religious and even demographic factors."[65] Lévi-Strauss also endorsed this emphasis on comparative studies of racial attitudes and suggested that there was real promise in "studying the different attitudes adopted by any particular civilization towards various cultural minorities." For example, though the French displayed a "liberal attitude" toward "negroes," the same could not be said of their attitudes toward "other minorities." Further, Lévi-Strauss suggested that "in France, as in Mexico," racial prejudice against blacks was being introduced from the United States, which meant that there "could be no question of an unvarying attitude on the part of any particular group toward any other group."[66]

Juan Comas was similarly enthused by comparative studies of racial consciousness and pointed out that there were important differences between the attitudes of Anglo-Saxon and Spanish and Portuguese colonizers. In particular, Comas suggested that Portuguese colonialism was comparatively benign as evidenced by the fact that Portuguese settlers "behaved towards the native peoples in North Africa in the same way as they did towards the Indians in South America." Comas also agreed with Lévi-Strauss that racial prejudice against "negroes in Mexico" was on the rise and attributed this "to Mexico's financial dependence on the United States of America."[67] Expanding on the question of racial discrimination in Latin America, Juan Comas explained that the most recent research showed that it stemmed from two main psychological factors: "a superiority complex in the white people and an inferiority complex in the coloured people." In Mexico, Comas explained, "the superiority complex of the whites no longer existed but—what was more serious—the inferiority complex of the coloured people had survived."[68] As he speculated on the future, Comas suggested that identifying the "root causes" of these complexes was a crucial task and argued that "the inferiority complex of the coloured people" would be harder to "eradicate" because of the "survival of a language difference."[69]

Ironically, as the committee members looked to the future and sought to imagine what race would look like after the demise of racial typologies, they continued to operate within a framework of thought that was settler colonial and imperial in its outlook. Rather than study human genetic variation, future studies of race envisioned by the committee would be composed of comparative studies of differing imperial approaches to racial difference and indigeneity. At stake in these prospectuses were imperial and biopolitical anxieties provoked by the shifting relations between settler, indigenous, and mixed populations in colonial territories.

The imagined studies of the 1949 committee and UNESCO's other social science projects from this period thus share a common thread of inquiry concerned not with human biological variation but rather with race as a category of social analysis, whose reality rests in the attitudes, perceptions, and beliefs held by individuals about themselves and others at a given moment in time. In this research program, conceptions of race were embedded in discussions of social change in various dimensions including the processes of urbanization, modernization, and industrialization. From this frame, international harmony entailed recalibrating the attitudes and psychological dispositions of white elites and "backward" colored peoples to attain equilibrium between feelings of inferiority and superiority. In the speculations of these scientists, the antiracist project of disseminating scientific facts to remove racial prejudice had to be complemented by the project of improving and uplifting the ways of life of indigenous and nonwhite groups through assimilation and acculturation.

CONCLUSION

In the 1950 issue of the *UNESCO Courier* where the 1950 Statement on Race was published, Alfred Métraux, the newly appointed director of UNESCO's "Race Division," wrote an article describing racism as "one of the most disturbing phenomena of the great revolution of the modern world." Echoing the 1949 committee's concern with studies of social change, Métraux described racism as something that prevented "coloured" people at the margins of civilization from being incorporated into its fold. "At the very time when industrial civilization is penetrating to all points of the globe and is uprooting men of every colour from their age-old traditions," Métraux wrote, "a doctrine, treacherously scientific in appearance, is invoked in order to rob these men of their full share in the advantages of the civilization forced upon them."[70]

Like many of the social scientists who drafted the 1950 Statement on Race, Métraux learned his profession during the late colonial period and devoted his career to documenting and improving the cultures of so-called primitive peoples in the Southern Hemisphere. Métraux was a product of the Institut d'Ethnologie founded by Marcel Mauss in Paris and became an avid field researcher who spent considerable portions of his career observing indigenous and Afro-descendant communities in Latin America, Central America, the Caribbean, and the South Pacific.[71] After World War II, he joined the United Nations and then moved to UNESCO where he directed the organization's first technical assistance project in Haiti. Métraux praised the United Nations' technical assistance program for its promise to give "the economically backward countries" the kinds of technical knowledge that would enable them to "raise their standard of living and to have their share in the progress of the highly industrialized countries." Yet he also worried that "progress, in the form in which the United Nations seek to propagate it throughout the world," had the potential to "inevitably destroy the many forms of local culture still surviving on several continents."[72]

In the decades following the 1950 Statement on Race, the projects Métraux identified as having the potential to destroy local cultures flourished into a vast industry known as "international development." Although it blossomed during the Cold War, the intellectual framework of the development industry had already been laid down during the late colonial era by colonial and post-colonial technocrats seeking nonbiological and even antiracist alternatives to "race." The research programs of the committee of experts that drafted the 1950 statement bear the influence of these colonial development schemes. Through their involvement in projects of indigenous acculturation or eugenic projects of mental hygiene, the 1949 committee conceptualized modernization as linked to projects of racial uplift and cultural improvement through notions such as "integration," "assimilation," and "progress." This reformist sensibility can be seen in one of the 1950 statement's key declarations: "genetic differences are not of importance in determining the social and cultural differences between different groups . . . social and cultural changes in different groups have, in the main, been independent of changes in born constitution." In other words, the statement proclaimed, "vast social changes have occurred which were not in any way connected with changes in racial type."[73]

Though it is tempting to read this vision of vast social change as nonracial, the interwar itineraries of the social scientists who drafted the 1950 statement show that they thought of social change in racialized terms, namely as a process of acculturation to white norms. By dismantling racial determinism and celebrating human biological unity, the UNESCO statements thus gave scientists moral license to speak on behalf of "simple peoples" while concealing how they were implicated in racial regimes of truth. With their moral authority bolstered by antiracist declarations and the specter of scientific racism relegated to a lamentable past, social scientists greatly expanded their terrain for intervention and study. As the influence of the social sciences flourished during the Cold War, governments in the North Atlantic and international organizations grew increasingly intrigued by the tools that disciplines such as anthropology, sociology, and economics offered for understanding the fate of nations in the Global South. As new post-colonial states emerged in the Southern Hemisphere, Pentagon officials enlisted social scientists to contain the spread of communism and to predict the paths of revolutions in emerging nations in Asia, Africa, and Latin America.[74] Similarly, the UN system began adopting economic indicators such as annual incomes to diagnose the "Third World" as chronically poor and to propose comprehensive alterations to "Third World" economies that would put them on a path to economic growth and industrial progress.[75] The vast assemblage of agencies invested in development in this period framed the process of becoming modern as a technical process, an anti-politics machine, that could be implemented by "Third World" governments if they followed the right template.[76]

Thus, the Cold War development industry suggests that the retreat of scientific racism did not signify an end but rather an amplification of racial politics.[77] Indeed, during the Cold War era, interest in the futures of "underdeveloped" groups intensified and produced an expansion in the kinds of technical and scientific experts who would intervene in the lives of so-called backward peoples. By critiquing determinism, the UNESCO statements confronted scientific racism yet also legitimated relations of rescue between northern experts, southern elites, and so-called backward peoples. These intensified relations, however, did not obey the impersonal precepts of abstract modernization theories nor did they produce a linear and unimpeded path to Western modernity. Rather, they gave rise to greater and ever more complex encounters between technical experts and those deemed in need of improvement. The proliferation of scientific concepts that sought to replace "race" in the post–World War II period—populations, ethnicities, cultures of poverty—can thus be seen as attempts to address these growing contacts through technical and impersonal means and as attempts to depoliticize what had become politically fraught terrain.

CHAPTER THREE

Vikings of the Sunrise

Alfred Métraux, Te Rangi Hīroa, and Polynesian Racial Resilience

When Alfred Métraux was appointed as director of the "Race Division" of UNESCO's Social Sciences Department (SSD) in 1950, his projected duties were carefully outlined in a letter from Walter H. C. Laves, UNESCO's deputy director-general. Laves's letter explained that Métraux would primarily contribute to the field of "race problems" and thus continue the work initiated by Arthur Ramos. Laves also explained that UNESCO's program was increasingly concerning itself with "the non-mechanized peoples of the world"—particularly in the Near East, the Far East, Latin America, and Africa—and that the SSD therefore needed a "competent anthropologist" to give "aid in the planning and direction work in these areas." In addition to its ongoing work on "race problems," Laves explained that the SSD wished "to make a major contribution to the [UN's] technical assistance program" and was badly in need of a man "with knowledge of the underdeveloped areas and with imagination in respect to the contribution which a social scientist can make in the development of a technical assistance program." With expectations this high, Laves noted that even an anthropologist of Métraux's notable qualifications would face "a challenge of the first order."[1]

If we consider how anthropologists and historians have described Métraux's career, his appointment as director of the SSD's "Race Division" seems odd. Métraux has typically been lauded for his ethnographic skills and rigorous empiricism. For instance, Sidney Mintz praised Métraux as a "fieldworker's fieldworker" uninterested in "grand theory" and whose main talents consisted in "a great gift for languages, the strength and endurance of two ordinary men, an ability . . . to forget all else while pursuing the slightest snippet of data, and a genuine consecration to the field."[2] Similarly, the anthropologist and Brazilianist Charles Wagley, Métraux's close friend and obituary writer, described him as fundamentally a "field ethnologist" and "scholar" whose work combined rigorous knowledge of historical sources from the colonial period with "careful and objective data in the field." According to Wagley, Métraux's fondness of bookwork and fieldwork made him an awkward fit at UNESCO, where he was frequently "irked" by administrative obligations and anxious to retire and pursue "pure" research.[3] More recently, British anthropologist and Amazonian specialist Peter Rivière described Métraux's ethnographic writings as "astonishing" because of their "almost total lack of reference to or concern with any of the theoretical works on [his] subjects other than the occasional foray into cultural history."[4] These portrayals of Métraux and his acumen as a fieldworker help make sense of his prolific anthropological corpus, yet they suggest that his vigorous commitment to ethnographic purity made him ill suited to the bureaucratic demands of his postwar career as an applied anthropologist. The tendency to describe Métraux as a rigorous empiricist with no interest in theory also seems at odds with his UNESCO appointment. After all, as suggested by Laves's appointment letter, Métraux's position with the SSD required him to tackle two of mid-century anthropology's most pressing *theoretical* questions: Do human races exist in any meaningful way? Are all humans equally capable of adopting modern industrial ways of life?

This chapter sheds light on the trajectory that led Métraux to the forefront of UNESCO's engagement with race science and economic development. Métraux's appointment to UNESCO's "Race Division" appears less incongrouous, this chapter argues, when viewed from the frame of his encounters with southern regimes of racial thought— notably his little recognized encounters with anthropometric and demographic studies of Polynesia in the 1930s. Indeed, this chapter argues that Métraux's often overlooked work in Polynesian anthropology gave him a foothold in race science and applied anthropology and prompted

FIGURE 3.1 Alfred Métraux at Candelaria (Argentina), July 1932 (photograph, 1932, Charles Mallison, Wikimedia Commons).

him to move beyond his much-lauded empiricism. Through a chance encounter with the islands of the South Pacific—which took him away from his preferred South American field sites—Métraux entered debates concerning the history of Polynesian settlement, which were saturated in race theory. Métraux ended up in the South Pacific when he was offered a last-minute invitation to serve as the ethnologist on a Franco-Belgian scientific expedition to Rapa Nui (Easter Island) in 1934. After six months of field research in Rapa Nui, he completed a two-year fellowship at the Bernice P. Bishop Museum in Honolulu—the leading center for Pacific studies lavishly funded by U.S. philanthropists—from 1935 to 1937.[5] During his time at the Bishop Museum, Métraux's most important guide through the unfamiliar field of Polynesian anthropology was his officemate—the charismatic Māori anthropologist Te Rangi Hīroa (Peter Buck). In his anthropological work, Te Rangi Hīroa crafted epic narratives about the Māori and about Polynesians as a heroic seafaring Caucasian race—the Vikings of the Pacific—who originally settled the islands of the South Pacific in ancient times. Te Rangi Hīroa's account of Polynesian migrations was grounded in comparative anthropometric studies of South Pacific islanders, collection and analysis of Polynesian folklore, and detailed studies of Polynesian material culture. His account provided Métraux with a useful template for his Rapa Nui research in which he attempted to reconstruct the history of an enigmatic ancient civilization that was seemingly lost to the existing islanders. Indeed, in his studies of Rapa Nui, Métraux made extensive use of comparative anthropometric and blood group data collected by Bishop Museum surveys to argue that Rapa Nuian culture was Polynesian through and through and that the island's native inhabitants were descended from the same racial stock as other Polynesian islanders and were not of Melanesian racial descent. Métraux's ethnological work on Rapa Nui thus served to confirm Te Rangi Hīroa's account of Polynesian settlement, which stressed the dissimilarities between Polynesians and their darker-skinned Melanesian counterparts.

By tracking Métraux's itinerary through Polynesian anthropology, this essay demonstrates the ways in which race science was appropriated by indigenous scholars like Te Rangi Hīroa and used to assert indigeneity before the rise of the global indigenous movement. In recovering the influence of this Polynesian racial agenda on Métraux's research, this chapter also demonstrates how narratives about Polynesian racial and cultural resiliency and plasticity unsettled Métraux's own romantic attachments to an imagined pristine indigene on the verge of disappearance. Through his Polynesian research, Métraux encountered a model of a mobile, adaptive, and economically resilient indigeneity that was easily integrated into the liberal internationalist and antiracist ambitions of UNESCO.

Before his Pacific encounters, Métraux built up a reputation as one of the foremost Americanists of his generation largely through meticulous research focused on archiving the language, historical migrations, myths, and ways of life of indigenous races in the Americas.[6] In this phase of his career, Métraux's research agenda was shaped significantly by colonial institutions—namely the Institut d'Ethnologie and Trocadero Museum in Paris and the Gothenburg Museum in Sweden—that invested heavily in collecting objects from the material culture of Amerindian races before they vanished. Because they assumed that indigenous groups were destined to disappear, European anthropological institutions imbued this work of salvage anthropology with an urgency that Métraux embodied to a point of excess. As Métraux explained to his closest confidantes, his decision to join the Franco-Belgian expedition to Rapa Nui in 1934 was an attempt to find relief from the project that consumed the first five years of his career as a professional anthropologist: the systematic study of the "ancient indigenous races" of the Gran Chaco region in South America before they were transformed through cultural contact and colonization.[7]

The Gran Chaco was Métraux's first and most formative field site. Born to a Russian Jewish mother and a Swiss doctor in Lausanne, Métraux grew up in Argentina in the Andean province of Mendoza where his father practiced as a surgeon from 1907 to 1954.[8] After finishing high school in Mendoza, Métraux moved to Paris where he enrolled at both l'École de Chartes and l'École Pratique des Hautes Etudes, Sections des Sciences Religieuses, and aspired to become a historian and archivist. Yet while training as a historian, he also got his first taste of fieldwork. In 1922 Métraux arranged an eight-month leave of absence from his studies to conduct fieldwork in Argentina to study the Calchaquí (a group that no longer exists). During this leave, he also ventured into the Argentinian Chaco and interviewed a wichí cacique and collected Chorote ceramics.[9] Shortly after this first taste of Chaco field research, he told the surrealist writer Georges Bataille—a lifelong friend from his student days—of his desire to become an ethnographer.[10] He went on to train as an ethnologist at the newly founded Institute d'Ethnologie where he studied with Paul Rivet, Marcel Mauss, and Lucien

Levy-Bruhl and received a doctorate in 1928. His doctoral dissertation—*"La Civilization Matérielle des Tribus Tupi-Guarani"*—was primarily a library-based study that sought to identify the pre-Columbian origins of the Tupi-Guarani tribes found throughout South America and track their migration routes after the Iberian conquest in the fifteenth century. Métraux wrote most of this dissertation during a two-year stay at the Gothenburg Museum in Sweden where he worked closely with the Swedish anthropologist and Gran Chaco specialist Erland von Nordenskiöld and relied heavily on von Nordenskiöld's library collection.[11]

After completing his doctorate, Métraux became the founding director of the University of Tucumán's Institute of Ethnology thanks to a glowing recommendation from Paul Rivet. During his five-year term as director (1928–1933), Métraux frantically endeavored to turn Tucumán into Latin America's leading center for the salvage of Chacoan indigenous culture and purposefully sought to model the institute's proposed museum on Sweden's ethnological Gothenburg Museum directed by Nordenskiöld.[12] Métraux also became the founding editor of Tucumán's Revista del Instituto de Etnologia, whose mission he described as a "work of salvation" oriented to the preservation of the objects, myths, and dictionaries of "primitive races" on the verge of extinction.[13] From the outset of his appointment, Métraux endeavored to use Tucumán's resources to conduct systematic collections and studies of the indigenous groups of the Chaco region such as the Chiriguano, Toba, Pilaga, Mataco, Chorote, and Ashuslay. In letters and funding proposals from this period, Métraux described this envisioned Chacoan research as an urgent project of cultural salvage and sought to impress upon his European colleagues how quickly the window of opportunity for this kind of research was closing due to the incipient and destructive forces of colonization and cultural contact.[14] Métraux's considerable work in this period left an impression on Paul Rivet (who had suggested the creation of the Tucumán institute after visiting Argentina in 1928) and led Rivet to describe it as one of South America's most important centers of "ethnic research." Indeed, Rivet suggested that with more cooperation with French institutions, Tucumán could become the equal of the Bureau of American Ethnology in the United States.[15]

Yet managing a project of salvage anthropology of this scale proved tiring and complex. Although Métraux succeeded in organizing four major expeditions to the Chaco (and numerous smaller trips), his correspondence from this period suggests that he tried and failed to carry out at least ten other expeditions.[16] These failed expeditions all had the common intent of traveling farther and farther away from colonized frontier regions—such as the acculturated Argentinian Chaco—in order to access ostensibly pristine indigenous territories (typically in Paraguay).[17] Métraux's last expedition to the Chaco while director of the institute at Tucumán occurred in 1933 when he traveled to the Pilcomayo—a frontier region between Argentina and Paraguay. On this occasion, Métraux encountered a region devastated by the war that had broken out between the two countries and in which the Toba-Pilagá of the Argentinian Chaco were being decimated by smallpox and left to die by a settler colonial Argentinian military.[18] Until this encounter, Métraux's writings rigorously avoided cultural change as an object of study and instead aimed to salvage the pure and authentic elements of the cultures he visited through oral history and reconstruction through colonial sources. Yet this encounter with war and settler colonial violence shattered Métraux's commitment to ethnographic purity and prompted him to extol the seemingly benign assimilation and evangelization offered by missionaries in the region. Indeed, in his writings from this period, he described the work of English missions as providing the necessary medical support and education that would allow for "race conservation" and population growth.[19] The brutality of both the Argentinian military and Paraguayan military toward the region's indigenous peoples also prompted him to propose to Argentina's minister of the interior the creation of an agricultural colony for the Toba-Pilagá. Métraux also proposed to this minister that he was prepared to leave behind his anthropological work and adopt an official governmental role as "Inspector General de los Indios del Chaco y Formosa."[20] This plan also failed to come to fruition.

By 1934 when his contract with the University of Tucumán ended, Métraux had grown disillusioned with the prospect of further Chacoan studies. In addition to the brutality of the Chaco war, he complained bitterly about the corruption and politics at the University of Tucumán and felt increasingly isolated and distant from his family as well as the friends and intellectual life he had known in Paris.[21] His main mentor, Erland von Nordenskiöld, passed away in 1932 and Métraux became

1. Jeanne Chiriguano, fille du « capitan grande » *Kaynapadyú*. Ses joues sont peintes avec de la poudre de *kalazapa*

2. Indienne Chiriguano de la région de Caruruti, considérée par les indiens comme une beauté. (Photo. A. M.)

FIGURE 3.2 Young Chiriguano, daughter of "capitain grand" Kaynapadyú (photographs, in Alfred Métraux, "Études sur la Civilisation des Indiens Chiriguano," *Revista del instituto de Etnología de la Universidad Nacional de Tucumán* vol. 1, 1929, planche VI, pp. 340–41).

increasingly reliant upon Paul Rivet for guidance and support. As this phase of his career ended, Métraux's attachment to the work of cultural salvage and the pursuit of an authentic and pristine indigenous past had been unsettled. His subsequent publications on the Chaco displayed increasingly sociological themes—such as the study of suicide among the Matako—and his career was thus primed for a regional and epistemic reorientation.[22] It was at this moment that he was called to Polynesia and prompted to engage in anthropometric debates about the racial types of the South Pacific.

RAPA NUI: "THE MOST UNHAPPY OF ALL THE PACIFIC COLONIES"

Recalling his arrival on Rapa Nui's shores in 1934, Alfred Métraux wrote that it "was undoubtedly the most unhappy of all the Pacific colonies."[23] While other islands had "adjusted themselves to modern life" and settled into a status quo between "them and their conquerors," Rapa Nui was largely left to its own devices under Chilean colonial rule and was receiving no aid other than that provided by the Williamson-Balfour Company—a Scottish-owned Chilean company—that managed the island as a sheep farm and provided basic services to its employees on the island. Twenty years later when prompted to revisit the island, Métraux observed that it was showing signs of "some progress" and described an incipient modernization taking root that was spurred by young men who had escaped to Chile as stowaways where they learned technical trades and then brought their expertise back to the island. Yet Métraux showed no concern that this process would erode the islanders' native traditions and instead suggested that such a "transformation is not to be regretted" as it will destroy "nothing that was not already in ruins, for Easter Island civilization died between 1862 and 1870."[24] If the Chaco war unsettled the romantic assumptions that fueled Métraux's cultural salvage work, then Rapa Nui prompted him to consider mechanisms such as acculturation and cultural contact, which hinted at both the creative and destructive potential of cultural change. Yet in Rapa Nui Métraux faced—as he did many times in his career—a considerable ethnographic obstacle. How to reconstruct the past of a civilization that seemed lost to the island's inhabitants? And how to do this when he had no previous experience with the region's languages and history? In confronting these difficult tasks—what some called the "Mystery of Easter Island"—Métraux eventually turned to a new evidentiary source: race science. Yet his adoption of race science in the Pacific should not be seen as an uncritical acceptance of scientific racism. In fact, when he turned to physical anthropology to understand the racial makeup of Pacific islanders, it was with the aim of defending the Polynesianness of Rapa Nuians and thus affirming the history of Polynesian settlement favored by the foremost indigenous anthropologist of this period—Te Rangi Hīroa. Métraux's ethnographic studies of Rapa Nui can therefore be situated within, or close to, the genealogy of academic disciplines like Māori studies and Pacific indigenous history in which Te Rangi Hīroa is a foundational figure.[25]

Studies of the racial composition of Rapa Nui's inhabitants were not part of the proposed mission of the Franco-Belgian expedition that brought Métraux to Rapa Nui. Instead, it was the Parisian intelligentsia's enthusiasm for a new linguistic and archaeological theory that traced the origins of Rapa Nui's civilization to the Indus Valley that prompted anthropologists to raise funds for an exploratory mission to the remote Pacific island. The theory was proposed by the Hungarian linguist Guillaume de Hevesy. De Hevesy claimed that the Rongorongo hieroglyphics found on wooden tablets from Rapa Nui bore a striking similarity to scripts from the Mohenjo-Dara and Harappa civilizations of the Indus Valley (from ca. 2000 BCE) whose remnants had been revealed during much publicized archaeological digs from 1922.[26] Articulated at the height of diffusionist theory in anthropology, de Hevesy's theory speculated that the similarities between these two scripts suggested a common racial origin, which would have thus established a genealogical relation between Rapa Nui's inhabitants and the "pre-Aryan" civilizations of the Indus Valley.[27] Because European savants had long considered the origins of Rapa Nui's inhabitants and especially the giant statues that dotted the island landscape a "mystery," de Hevesy's theory implied that Rapa Nuians "had a right to proclaim themselves the survivors of a glorious past and the scions of a privileged race."[28]

One of the most enthusiastic supporters of de Hevesy's theory also happened to be one of the most influential and well-connected anthropologists in France—Métraux's mentor and erstwhile champion Paul Rivet. Although Rivet was primarily an Americanist like Métraux, much of his research was concerned with issues concerning the historical diffusion of cultural and racial groups to the Americas. Rivet was in fact a prominent proponent of the theory that the indigenous inhabitants of the Americas arrived from Asia across the Bering Strait. In 1926, Rivet also published an article titled "Les Malayo-Polynésiens en Ameriques," which argued that there was linguistic, cultural, and physical anthropological evidence of "Malayo-Polynesian" influence on the indigenous people of the Americas. Rivet also argued for the importance of studying the indigenous groups of the Pacific and argued that they had played a major role in populating the world and in the diffusion of primitive civilizations.[29] When de Hevesy's theories were introduced in Paris in 1932, Rivet became convinced that they held important clues to questions concerning the diffusion of ancient civilizations and promptly organized a public exhibit of the Hungarian linguist's findings and wrote the introductory panel for visitors.[30] Yet because de Hevesy's study was based on an examination of a sole photograph of a Rongorongo tablet, Rivet came to believe that there was immense potential for further discoveries and mobilized his considerable resources to arrange an expedition to look for more tablets on the remote Pacific island.

Rivet's enthusiasm for de Hevesy's theory and for organizing a mission to Rapa Nui was also shared by the Belgian archaeologist Henri Lavachery. In December of 1932, after meeting in the museum of a Belgian Catholic mission that held several of the tablets from Rapa Nui, Rivet and Lavachery began to make arrangements for a joint Franco-Belgian expedition.[31] As planning for the expedition advanced, Rivet decided that the mission would be jointly led by Lavachery as well as the French archaeologist Charles Watelin. But when the Belgian government withdrew its financial support for Lavachery at the last minute, Rivet moved quickly to find a researcher to replace Lavachery while they waited for his situation to resolve. To Rivet's good fortune, Métraux and his family had recently relocated to Paris after his contract with the University of Tucumán had come to an end. Métraux had become disillusioned with his South American experience and was delighted to reconnect with his friends from the Trocadero Museum and pined for a post that would allow him to remain in the fold of Paris's anthropological scene. With the promise that he would help Métraux find a job in Paris, Rivet convinced Métraux to act as Levachery's replacement and to lead the mission along with Watelin.[32] Although Métraux had no previous experience with Oceania or any familiarity with the Rapa Nui language or script, Rivet trusted in his skills as an ethnographer and linguist and argued that his familiarity with South America and ability to speak Spanish would be useful for dealing with Chilean authorities. Rivet's decision to include Métraux proved perceptive. As the expedition made its way to Rapa Nui aboard the French Navy vessel *Rigault-de-Genouilly*, the French archaeologist Charles Watelin succumbed to pneumonia during a hunting party in Tierra del Fuego and died.[33] In addition to Métraux and Lavachery (who later caught up with the mission), the expedition included the Chilean physician Israel Drapkin, who was sent by the Chilean government to study the problem of leprosy on the island and also made studies of the demography and blood groups of the island inhabitants.[34]

As conceptualized by Rivet, Lavachery, and Watelin the main purpose of the Franco-Belgian mission to Rapa Nui was to search for more Rongorongo tablets with the hope of shedding light on the puzzles surrounding the island's history and also the broader history of human diffusion. Yet at this stage in his career, Métraux had lost interest in such diffusionist speculation and instead took much greater interest in the present inhabitants of the island and their ways of life. Indeed, Métraux mocked the ideas of his colleague Charles Watelin, whom he described as "indifferent to the modern Easter Islanders and the traditions that might still survive" and consumed by the illusion that he would "see the walls of ancient cities—similar to Mohenjo-daro—emerge beneath his pick." Métraux described himself, by contrast, as "attracted by these few hundred Polynesians who have survived so many disasters and continue to speak their ancient language and hand down the legends and stories of their distant ancestors." In his numerous writings on Rapa Nui, Métraux described a history of civilizational collapse and loss of ancient tradition epitomized by the colossal and enigmatic stone statues for which the island was best known. Métraux himself traced this narrative of collapse back to the eighteenth century when the island was visited by European colonial explorers such as Cook, La Pérouse, and others. For these first explorers, Métraux wrote, "the contrast between the monuments, indicative of a flourishing and skillful population, and the desolation they found about them was a peculiar enigma." To make sense of this enigma they speculated about "cataclysms, or volcanic eruptions, that might have changed the course of the island's history" and precipitated the collapse of its civilization. Métraux also argued that

Rapa Nuians' connection to this "lost civilization" was further eroded in the nineteenth century when the islanders were subject to further disaster. In 1859, Peruvian slave-raiders attacked the island and "kidnapped the king, a large number of the nobles and priests, and many hundred commoners, all of whom were carried off to the guano islands of Peru to work as slaves." Though some of the kidnapped were repatriated by a French ship, they brought smallpox and tuberculosis contracted in Peru and "within a few years most of the native population and with them the vital links with the past were wantonly destroyed."[35] When the first Christian missionaries arrived on the island in 1864, they found that the islanders had little knowledge of the origin of the statues or how they were transported. According to Métraux, this "ignorance, combined with the state of primitive poverty into which they had fallen, again emphasized their enigmatic relation to the lost civilization, of which the statues and great stone mausoleums as well as other finely wrought remains of the past were mute evidence."[36]

Although he viewed the history of the island since European contact as profoundly marked by civilizational collapse and depopulation, Métraux argued that the island's more recent history showed signs of a rapid demographic resurgence. Yet in contrast to his studies of Amerindian groups where he combined extensive bibliographic research, recent statistical records, and local oral histories, Métraux struggled to find sources for tracking the island's history. In lieu of his habitual sources, Métraux turned to accounts from the first European explorers who made population estimates on the basis of the number of inhabitants they saw during the time they spent anchored on the island (a few minutes in some cases). Although he questioned the accuracy of these figures, he noted that estimates of the island's population ranged from 600 to 3,000 in the period between 1770 and 1871. While figures from this period were imprecise, Métraux argued that the numbers tracked by missionaries after 1864 offered a more reliable portrait of the population decline propelled by the Peruvian slave raids and introduction of foreign diseases such as syphilis, smallpox, and tuberculosis. The steady nineteenth-century decline in population culminated in 1871 when most of the native islanders left for Tahiti and Mangareva. Yet, after this nadir, Rapa Nui's population enjoyed a steady growth leading up to the arrival of the Franco-Belgian expedition in 1934. During his stay on the island,

Métraux identified a population of 456 islanders and observed that they were confined to a 2,000-hectare area encompassing Hanga-Roa, the island's only village. Although the islanders cultivated the fields around Hanga-Roa, Métraux argued that this was not particularly fertile land and that the "natives could produce far beyond their needs if they wanted to." Indeed, on the basis of his observations concerning the present relation between land-use and population density, Métraux speculated that "eight or nine times [the current population] could have made a comfortable living on that entire island." As such, Métraux concluded that the population of the island must have been about "3000 before the coming of the white men," which gives a density of "13.7 persons to one square kilometer, about the same number as that for Tahiti."[37]

Métraux's study of Rapa Nuian demographics shows how the island's enigmatic history prompted him to complement his well-honed ethnographic empiricism with the conjectural possibilities afforded by population forecasting. Yet given the questionable accuracy of his demographic data, Métraux also turned to the phenotypic markers of the islanders for clues to the island's history. Like many of the European explorers who preceded him, Métraux described a lurid fascination with the islanders' appearance upon first arrival. When he first saw the faces of Rapa Nui's inhabitants, Métraux "could not decide whether these men were a heterogeneous crowd of European beachcombers or real Polynesians, the sons of the sea rovers who had colonized the island." Though he was certain that "European blood flowed in their veins," he also detected "something exotic in all of them and traces of old Polynesian descent could be seen in their black, wavy hair, in the strange, vivacious dark eyes, in the high foreheads."[38] Eventually, through typically fastidious analysis of Rapa Nui's language, material culture, and folklore, Métraux arrived at the conclusion that the culture of the island's inhabitants was Polynesian through and through. Yet to strengthen his argument, Métraux turned to racial data collected by the U.S. physical anthropologist Harry Shapiro. In fact, in *Ethnology of Easter Island*— Métraux's major scientific publication on the island published by Honolulu's Bishop Museum—Métraux included a chapter by Shapiro on the "Physical Relationships of the Easter Islanders" and insisted that Shapiro's anthropometric data proved that the inhabitants of Rapa Nui were of Polynesian and not Melanesian descent.

Shapiro's chapter made a strong case for the Polynesian origins of Rapa Nui's inhabitants and explained how previous researchers had been led astray. Shapiro described how racial distinctions between Polynesians and Melanesians were first introduced by eighteenth-century European explorers influenced by enlightenment thought.[39] But whereas the first European explorers arrived at the "logical conclusion" that Rapa Nuians belonged to the Polynesian race, by the late nineteenth century anthropologists were fooled by "inadequate data and by complex but erroneous statistical manipulations" into entertaining the "hypothesis that the Easter Islanders form a racial complex formed by successive migrations of Melanesians and Polynesians."[40] According to Shapiro, it was the French physical anthropologists Armand de Quatrefages and Ernest Hamy who introduced this Melanesian hypothesis after studying skulls brought back to France from Rapa Nui and finding similarities to Papuan skulls. Quaterfages and Hamy's thesis was then amplified by the German physical anthropologist Wilhelm Volz, who published an analysis of crania from Rapa Nui in 1895 and concluded that the islanders represented a racial amalgam of "several successive migrations." Although the skulls suggested extensive amalgamation, Volz claimed to have isolated the various racial elements of Rapa Nui's population through a statistical procedure that showed the "chief strain was Melanesian, both an east and a west variety." In 1923, Roland B. Dixon—a U.S. anthropologist and student of Franz Boas—expanded Volz's analysis through a survey of the existing data on the "length-breadth, length-height, and nasal indices" and concluded that Rapa Nui was settled first by a "proto-negroid people identified with 'Melanesia.'"[41] According to Dixon, this "strain" contained not only "protonegroid" but also a "proto-Australoid element and a 'dash of Caspian.'" Shapiro also noted that anthropologists had offered important critiques of the calculations and conclusions of the studies supporting the Melanesian hypothesis. Yet he also cautioned that although these critiques refuted a possible genealogical relation between Rapa Nuians and "non-Polynesian groups," they also did not confirm the islanders' "Polynesian origin." The existing "craniological investigations of Easter Island racial affinities" thus failed "to offer the solution, which occurred long ago to the untutored explorers, that the Easter Islanders are Polynesians."[42]

In contrast to these studies conducted by scientists working with skulls brought back to museums in metropolitan centers, Shapiro emphasized that his racial analysis of Rapa Nuians was based on *in situ* measurements from "the living population." Shapiro's contact with the "living population" of the island occurred just a few days after the Franco-Belgian expedition left the island, when he arrived on the island as part of the Templeton Crocker expedition aboard the *Zaca*.[43] During his visit to the island, Shapiro gathered a "series of anthropometric measurements on 22 Easter Island adult males who were rigidly selected out of a total population of 456 for their purity of descent."[44] As such, Shapiro stressed that "everyone suspected of mixture with foreigners was excluded from the series." On the basis of this study of "pure" Rapa Nuians, Shapiro concluded "the Easter Islanders are definitely Polynesians of a somewhat specialized and exaggerated type isolated by migration and intensified by inbreeding." To buttress this thesis, Shapiro included a table comparing the islanders' anthropometric measurements with those of other Polynesian groups and suggested that the data showed that Rapa Nuians fell within the "Polynesian pattern," although they occupied an extreme position when it came to several traits, notably the "cephalic proportions." Shapiro also argued that the association between Rapa Nui and "Melanesian and Australian stocks" made little sense from the vantage of genetics of skin color and hair form. "There is little doubt that the dark pigmentation of Melanesians and Australians and the frizzly [*sic*] hair of the former," Shapiro wrote, "are traits which are often dominant in racial crossing and are never completely recessive."[45] Yet the "Easter Islanders" did not bear any trace of these distinctively "Melanesian or Australian" features. Although this conclusion was plain from casual observation, Shapiro included comparative tables of skin color measurements (based on the Von Luschan scale) and of hair form to show that the Easter Islanders conformed to a Polynesian pattern on both counts.

In his previous work, Métraux had paid little attention to anthropometric evidence. Yet for his Rapa Nui research, Métraux turned to Shapiro's racial arguments with great enthusiasm. In fact, before including Shapiro's chapter in *Ethnology of Easter Island*, Métraux published an article in *Mankind* (the journal of the anthropological society of New South Wales) that leaned heavily on Shapiro's then unpublished data to attack the Melanesian hypothesis.[46] According to Métraux, the theory of Melanesian origins was no more than a piece of "dogma" that could not "rest on weaker grounds." In his searing rebuttal of this theory, Métraux took aim at the British archaeologist and curator of the Pitt Rivers Museum, Henry Balfour, whose arguments for the cultural similarities between Rapa Nui and Melanesia were influenced by the "mistaken statements of physical anthropologists" and rested on a series of supposed cultural parallels that in fact "appear to be superficial resemblances of a few traits selected at random in an area conspicuous for the divergence of its local cultures."[47] Like Shapiro, Métraux argued that the cranial statistics used by late-nineteenth-century physical anthropologists in Europe (i.e., Quatrefages, Hamy, and Volz) was at odds with what could be observed *in situ*. Moreover, Métraux insisted that anyone who visited the island would see the ridiculousness of anthropometric arguments linking Rapa Nui with Melanesia. "But how is it that the Easter Island population does not show any markedly Melanesian feature such as kinky hair or dark skin?" Métraux asked and noted that "modern Easter Islanders have always been described by travelers as true Polynesians."[48] In this piece, Métraux also argued that the publication of Shapiro's "extensive anthropological research" would provide a definitive answer to the racial status of Rapa Nuians. He also revealed that Shapiro had authorized him to paraphrase his conclusions, namely that "Easter Island skulls show a specialized type . . . [that was] produced by long in-breeding, does not exhibit negroid characteristics, [and] can be easily related to other Polynesian groups." Métraux also drew parallels between Shapiro's portrait of a distinct racial type molded by endogamous forces and the linguistic and cultural traits of Rapa Nuians. In his conclusion to *Ethnology of Easter Island*, Métraux wrote that one of the main lessons conveyed by his book was "the fact that cultures are not static, that they do not need external stimulus and influence to develop and reach some degree of perfection." He also closed his book by inviting readers to marvel at the fact that "on the most solitary inhabited island in the world, Easter Islanders were able to develop and perfect the culture which they received from their Polynesian ancestors to the West." And the fact that the resilient islanders had perfected their culture, Métraux argued, shed light on one of the island's greatest mysteries—the giant statues, which, Métraux believed to be an amplified version of existing Polynesian folk art (in the same way that Shapiro described Rapa Nuians as an exaggerated Polynesian type).[49]

In the southern expanse of the Pacific ocean, Métraux thus became a race scientist. He also revealed a dimension to his work that has not typically been recognized by anthropologists from the Americanist tradition. Describing the method Métraux used in his research on the material culture of the Tupi-Guarani, the Brazilian anthropologist Eduardo Viveiros de Castro described him as a rigorous diffusionist. "Métraux's method," wrote Viveiros de Castro, "consists of comparing the material traces of a culture, in terms of its diffusion, and of establishing logical inferences about its transmission routes."[50] Through this comparative method, Viveiros de Castro argued, Métraux believed he could establish the essential traits of Tupi-Guarani material culture and thereby discern which "populations are 'authentic' [and which are] acculturated Tupi-Guarani." For tracking the history of the Tupi-Guarani, one of the most widespread families of indigenous languages in South America, Métraux's method proved highly successful. Yet in the context of Rapa Nui—where multiple waves of colonialism had left the island's inhabitants in a precarious state—this project of identifying the core features of a culture through its material history was one that vexed researchers and remained shrouded in mystery. Like the settler colonial violence he encountered in the Gran Chaco before he left, the seeming collapse of Rapa Nui's "civilization" unsettled his romantic quest for a pure and uncontaminated indigene. Yet in Rapa Nui, he encountered not only a paucity of sources for the island's history but also a population that appeared to be highly admixed with origins that were difficult to discern. And it was this series of challenges that prompted Métraux out of the methodological comfort zone he had developed in South America. Instead of a strict reliance on colonial records coupled with ethnographic observation, in Rapa Nui he enthusiastically embraced evidence afforded through anthropometric measurement, demography, and even blood group data. Yet he continued to insist on the primacy of data gathered *in situ* and embraced data on racial types insofar as it aligned with his findings from material culture, linguistics, and folklore. As such, the race science that Métraux embraced represents a rejection of European accounts from the nineteenth century and an attempt to highlight the endurance of Polynesian peoples.

TE RANGI HĪROA: RACIAL ORIGINS AND CULTURAL CHANGE

By forcefully defending Rapa Nui's Polynesian heritage, Métraux confirmed the expansive narrative of Polynesian settlement held by one of the era's most prominent authorities—Te Rangi Hīroa, the Māori anthropologist who served as director of the Bernice P. Bishop Museum in Honolulu from 1934 until his death. Métraux's embrace of Te Rangi Hīroa's work is no coincidence. After his six months of fieldwork in Rapa Nui, Métraux secured a research fellowship at the Bishop Museum, where he spent two years (1936–1937) writing up his field research as a series of articles and eventually a book. At the Bishop Museum, Métraux encountered an ideal location to retool his ethnographic skills and to complement the bookish diffusionism he honed in South America (via Paris and Gothenburg) with the four-field approach then in vogue with Pacific anthropologists.

When Métraux arrived in Honolulu in 1936, the Bishop Museum had become one of the leading centers of Pacific scientific research and was graced with ample financial support from Yale University and U.S. philanthropists. The museum was created in 1889 by the U.S. businessman and philanthropist Charles Reed Bishop as a memorial to his wife, Princess Pauahi, last of the Kamehameha family of the chiefs of Hawai'i. Bishop and Pauahi created the institution for the purpose of studying "Polynesian and kindred antiquities, ethnology, and natural history."[51] During this first stage of its existence, the museum amassed an impressive collection of books and artifacts that established its reputation as an important research center. In 1920, thanks to a formal agreement with Yale, the museum expanded its role to serve as the central base for a series of anthropological expeditions described as "the first comprehensive attack on a large scale of the problem of Polynesian origins."[52] Financial support for the expeditions came from the New York stockbroker Bayard Dominick, who made a generous donation to Yale (his alma mater) that was transferred to the Bishop Museum. The agreement that the Bishop Museum secured with Yale also included the appointment of Herbert E. Gregory—a professor of geology at Yale—as director of the museum. Promptly after his appointment, Gregory convened the first Pan-Pacific Science Congress, which took place in Honolulu in 1920 and brought together dozens of scientific researchers from throughout the Pacific to survey the state of Pacific research and propose new directions. Conference participants placed special emphasis on conducting comprehensive archaeological and anthropological surveys of the islands of the Pacific with the hope of solving the "problem of Polynesian origins."[53] In the years following the Honolulu conference, Gregory made use of the generous funds at his disposal to organize a series of wide-ranging surveys known as the Bayard Dominick Expeditions.

The Bayard Dominick Expeditions not only turned the Bishop Museum into an active center of operations for Polynesian research but also offered a new research model emphasizing international and interdisciplinary collaboration. In designing the plans for fieldwork, Gregory consulted some of the leading luminaries in Boasian anthropology—notably Clark Wissler, Roland B. Dixon, and Alfred L. Kroeber—as well as prominent Pacific scholars in New Zealand including the founder of the Polynesian Society, Stephenson Percy Smith, and Chancellor of the University of New Zealand, John Macmillan Brown. With this input from scholars across the Anglophone Pacific, Gregory selected four Polynesian areas for study—the Marquesas, Tonga, Australs, and Hawai'i—and assembled four field teams made up of energetic graduate students from elite U.S. universities. The fieldworkers in each team traveled to their site in pairs with the idea that one would conduct archaeological research and the other would focus on studying the social organization and religion of the islanders.[54] The research teams also took calipers and standardized anthropometric cards made by the physical anthropologist Louis R. Sullivan for recording data about the "physical characteristics of as large a number as possible of the native inhabitants."[55] The teams collected "somatological" measurements and shared them with Sullivan, who analyzed the data and published a series of articles describing "race types in Polynesia" and how difficult it proved to identify a uniform "Polynesian type."[56] In the two decades after the Bayard Dominick Expeditions, the Bishop Museum continued to serve as a central hub for dozens of similar expeditions that sought to comprehensively map the archaeological history of Polynesian settlement, the social organization of the multiple island communities, and the biological process of race formation through miscegenation.

By the time Métraux arrived in Honolulu, Bishop Museum researchers and staff had become convinced that they were close to resolving the mystery of Polynesian racial origins.

For Métraux, the museum's formidable and interdisciplinary research agenda proved an alluring counterpart to the diffusionist work in vogue in Paris. In correspondence with his friend Yvonne Oddon, the U.S.-trained librarian at the Trocadero Museum in Paris, Métraux bemoaned the state of French anthropology and spoke glowingly about his colleagues at the Bishop Museum. Whereas in Paris he felt constantly depressed by his fruitless search for employment, in Honolulu he described himself as brimming with optimism and delighted to be in a well-functioning museum where he worked in a large office with big windows and could take daily swims at the beach. In one letter Métraux wrote that "with Rivet all is lost and ruined" and complained bitterly that he was constantly being overlooked in favor of mediocre researchers and that money was being wasted on "stupid and vain enterprises like the Dakar-Djibouti mission—an archaic ethnographic endeavor of the sort that was practiced around 1850."[57] In contrast to the dismal state of anthropology in France, Métraux marveled at the rigor and efficiency of his Bishop colleagues Kenneth Emory, Ernest Beaglehole, and Edwin Burrows. He also confessed to Oddon that he felt "much less prepared than the worst [US] graduate [student]," and that he was doing his "best to learn and complete my rather poorly done education."[58] In particular, Métraux lamented his lack of training in physical anthropology yet aspired "to fill these gaps in a few years and be as well prepared as the guys here."[59]

Although he felt years behind his Bishop colleagues, Métraux wrote to Oddon that he felt redeemed by his charming officemate Te Rangi Hīroa, whom Métraux described as a "half Maori who also claims to be entirely of his race."[60] Shortly after Métraux arrived in Honolulu, he became delighted to learn that Te Rangi Hīroa was confirmed as Gregory's successor and appointed director of the museum. Thankfully, this newfound responsibility did not prevent Te Rangi Hīroa from having daily conversations with Métraux, who was smitten by his "marvelous nature" and "extraordinary charm."[61] Métraux also confessed that he could not stop listening to his Māori officemate and explained that "with him I learn more about Polynesian ethnography than I would by devouring that enormous literature on Oceania."[62] Striking a more romantic note, Métraux described Te Rangi Hīroa as graced by the "profound intuition of an indigenous soul" and as the Polynesian version of the Inuit-Danish anthropologist and explorer Knud Rasmussen.[63]

In Te Rangi Hīroa, Métraux happened upon one of the most prominent and respected figures in the field of Polynesian anthropology. More important, through Te Rangi Hīroa Métraux encountered a regime of racial thought that asserted the vigor and bright future of the Māori people and refused European prognostications of their inevitable extinction.[64] Indeed, when seen from the perspective of Te Rangi Hīroa's triumphalist accounts of Polynesian settlement and Māori revival, Métraux's emphatic insistence on the Polynesian origins of Rapa Nui and celebration of its demographic revival demonstrate the impact Te Rangi Hīroa the Māori expert had on his scholarship. Yet Te Rangi Hīroa's Polynesian-centric vision of the Pacific also shared some of the pitfalls the race science that it relied upon. Indeed, celebration of the strength and resilience of the Māori and Polynesian races invoked their proximity to the white race and relied on comparative judgments about the supposed inferiority of their darker-skinned neighbors, the Melanesians.

Te Rangi Hīroa was born in a working-class village in New Zealand in 1877 to a Māori girl who died soon after his birth and an Irish father who was a laboring contractor.[65] He worked with his father on a sheep station before attending Te Aute College, an Anglican boarding school that aimed to uplift the Māori race by preparing young boys for professional careers.[66] Te Aute College opened the door to a political and scientific career. While studying at the college, he formed a close relationship with Apirana Ngata (who would go on to become one of the most influential Māori politicians) and became active in the Association for the Amelioration of the Condition of the Māori, which later became the Young Māori Party. After Te Aute, Te Rangi Hīroa studied medicine at the University of Otago and eventually graduated with a medical doctorate in 1910 after defending a dissertation on Māori medicine in ancient and modern times. In 1905, he was appointed a Māori officer of health, a position he held for several years under the guidance of the Māori physician and politician Maui Pomare. After the sudden death of a Māori member of parliament in 1909, Te Rangi Hīroa stepped in as his replacement and served in the New Zealand Parliament, where he was a member of the Native Affairs Committee, until 1914.[67] When World War I broke out, Te Rangi Hīroa decided not to seek reelection and instead accepted a position as the medical officer for a volunteer Māori battalion stationed in the Middle East. When he returned to New Zealand after the war, Te Rangi Hīroa was appointed director of Māori Hygiene in New Zealand's Department of Health. In this role as director of Māori Hygiene, he worked to improve the sanitation infrastructure in Māori communities and to persuade Māori leaders to work closely with nurses and medical officers in ongoing efforts to prevent the spread of infectious diseases, especially after the influenza pandemic of 1918, which had a very high Māori death rate.[68]

Although this early phase of his career was marked by his medical and political service, Te Rangi Hīroa also developed a fascination with the material culture and folklore of indigenous communities and began to transform himself into what he called a "home-made anthropologist."[69] In spite of having no formal training in anthropology, through his medical service Te Rangi Hīroa developed close relationships with Māori communities throughout New Zealand and with indigenous communities in the Cook Islands and other neighboring islands. In correspondence with Apirana Ngata, his friend from his Te Aute days, Te Rangi Hīroa also described how his Māori ancestry gave him privileged access to aspects of Polynesian cultures not available to his Pakeha colleagues and how this allowed him to quickly gain anthropological expertise.[70] During this time, Te Rangi Hīroa also began to develop relationships with leading scholars and institutions focused on the study of Polynesian history and culture. For instance, in 1907 he joined the Polynesian Society and became a keen student of Percy Smith, the society's founding director, longtime editor of its journal, and an authority on Māori and Polynesian origins and migrations.[71] When he returned from the war, he increasingly devoted his attention to documenting Māori material culture and went on several field trips to record the culture and music of Māori communities. In the years that he served as Māori officer of health, Te Rangi Hīroa published his first anthropological essays in the *Journal of the Polynesian Society*, and by the mid-1920s Te Rangi Hīroa published his first monographs, namely *The Evolution of Maori Clothing* (1926), which was published as a Polynesian society memoir, and *The Material Culture of the Cook Islands (Aitutaki)* (1927). In his

FIGURE 3.3 Te Rangi Hīroa/Peter Henry Buck (photograph, ref. 1/2-078259-F, Alexander Turnbull Library, Wellington, New Zealand, records/23202767).

quest to become a homemade anthropologist, Te Rangi Hīroa developed a distinctive ethnographic style oriented toward a careful study of the construction techniques and function of the everyday technologies used by Māori and Polynesian communities. His studies tended to be meticulous and were accompanied by his skillful line drawings. Around the time that he published his first monographs, Te Rangi Hīroa also began to attract attention on the lecture circuit, particularly through

FIGURE 3.4 Eric Ramsden, Peter Buck (Te Rangi Hīroa) studying Paratene Ngata making an eel basket (photograph, ref. 1/2-037930-F, Alexander Turnbull Library, Wellington, New Zealand, records/22495223).

his lecture on Māori migrations titled "On the Coming of the Māori." As his popularity rose as a lecturer and social scientist, Te Rangi Hīroa also found the opportunity to become a full-time anthropologist. In 1923 he met Herbert E. Gregory, the director of the Bishop Museum, at the Pacific Science Congress in Melbourne. Gregory funded Te Rangi Hīroa field research to the Cook Islands in 1926 and then offered him a five-year fellowship at the Bishop Museum, which allowed him to devote himself entirely to anthropology. By the time Métraux arrived in Honolulu and witnessed Te Rangi Hīroa succeed Gregory as director, he had already worked as a full-time anthropologist at the museum for seven years and had also spent a brief period as visiting professor of anthropology at Yale University.

In addition to his meticulous studies of the material culture of Polynesian communities, Te Rangi Hīroa also developed a keen interest in the anatomical specificity of Polynesian types and how they were shaped by historical and environmental forces. The beginnings of this interest in physical anthropology stretch back to his days as a medical student in Otago where the anatomist J. H. Scott had published a landmark paper on Māori and Moriori crania and established a tradition of physical anthropology and osteology.[72] Although he initially expressed revulsion at this tradition's reliance on Māori skeletal remains, Te Rangi Hīroa eventually found a more palatable way of doing this research by conducting anthropometric studies on living subjects.[73] In 1919, as World War I came to an end, Te Rangi Hīroa took advantage of his return trip from Britain to New Zealand aboard the H.M.T. *Westmoreland* to conduct a systematic anthropometric survey of the Māori battalion under his medical care. He went on to publish the results of this study in a series of articles titled "Māori Somatology" in several issues of the *Journal of the Polynesian Society* in 1922–1923. His anthropometric studies of the Māori race were based on measurements taken of 814 adults "with Māori blood in their veins" of which Te Rangi Hīroa deemed 424 to be "full blooded Māoris." Through detailed measurements of skin color, eye and hair color, weight and height, head, and the precise dimensions of other body parts, Te Rangi Hīroa aimed to delineate the basic features of the Māori type in order to understand its relationship to other branches of Polynesians and to understand what was lost and preserved through racial mixing.[74] As Te Rangi Hīroa explained in his acknowledgments, his study also relied on the support of prominent British scientists—notably the comparative anatomist Arthur Keith and the biostatistician Karl Pearson—who counseled him on appropriate methods (such as the cephalic index and use of the Von Luschan scale for skin color measurements) and lent him instruments including the "Karl Pearson head-spanner."[75]

Te Rangi Hīroa's study of Māori somatology also paid careful attention to Māori beauty standards and the practices through which Māori families supposedly preserved their distinctive physical attributes. In one article, he devoted multiple pages to a discussion of "nose measurements" and Māori nose aesthetics. Whereas the New Zealand scholar J. Macmillan Brown had described the flat nose as the Polynesian ideal of beauty, Te Rangi Hīroa explained that for Māoris the term flat nose (*ihu parehe*) was a "term of opprobrium."[76] Instead of flat noses, Māoris admired the "well formed, well bridged nose" to such an extent that mothers habitually engaged in the practice of massaging and molding the noses of their babies shortly after birth. The nostrils, Te Rangi Hīroa wrote, "were compressed with the fingers to correct excessive width, which was regarded with disgust." Te Rangi Hīroa suggested that this disgust with flat noses stemmed back to the "Māori colonists from Eastern Polynesia" and the people they first encountered when settling in New Zealand, which "according to tradition" were "woolly-haired, dark-skinned, thin-legged, with shifty eyes and *very flat noses.*"[77] As a result of this first encounter, for Māoris "savageness" became associated with flat noses (and dark skin) to such a degree that in some districts, people took to identifying "future fighting warriors" from the flatness of infant's noses. In such cases, the nose would be massaged to increase the flatness of the nose and thus increase the perceived fierceness of the warrior. Te Rangi Hīroa concluded that these practices and oral traditions offered compelling evidence of "Melanesian or Negroid intermixture in the past" as well as evidence of how ostensibly Melanesian traits became either objects to be removed or exaggerated depending on the desired effect.

In the subsequent years, Te Rangi Hīroa folded his study of "Māori somatology" into a broader research program focused on deciphering the origins and migration history of the Māori and Polynesian race. In the same way that Te Rangi Hīroa sought to refine and update the less savory aspects of the anthropometric research done by his Otago mentors, his work on Māori migrations borrowed from and modified the work of his predecessors, namely the British diffusionists that predominated in the Antipodes during the 1920s and 1930s before the rise of functionalist anthropology.[78] In New Zealand, the prominent turn-of-the-century diffusionists S. Percy Smith and Edward Tregear (who cofounded the Polynesian Society) were determined to trace Māori origins to a Middle Eastern or Indian homeland and described the Māori as an Aryan race.[79] Although Te Rangi Hīroa eventually came to believe that Smith and Tregear's theories relied on dubious oral records, he clung to their conviction that Polynesians had a "Caucasian" origin and had traveled to the Pacific from Southeast Asia.[80] Yet as Te Rangi Hīroa refined his theories concerning Polynesian migration during the 1920s and 1930s, he found that many of his white colleagues viewed Polynesians as their inferiors. Much to his chagrin, scholars based their arguments for the inferiority of Polynesians on data gathered during the anthropometric surveys of the Pacific sponsored by the Bishop Museum. For instance, on the basis of Bishop Museum surveys, anthropologists (such as Sullivan and Dixon) concluded that Polynesians migrated to the Pacific through Melanesia and were an inferior and poorly mixed race that incorporated both "Negroid" and "Mongoloid" elements.[81] Such dim views of Polynesians and associations with darker-skinned and supposedly inferior races not only clashed with the discourse of Māori improvement that Te Rangi Hīroa inherited from his Te Aute days but also affected his legal standing within the United States. In fact, after becoming director of the Bishop Museum in 1934, Te Rangi Hīroa applied for U.S. citizenship but was denied because he was not deemed to be sufficiently white. At this time, U.S. citizenship laws restricted citizenship to those deemed to be more than 50 percent "Caucasian." Although Te Rangi Hīroa was deemed to be 50 percent Caucasian because of his Irish ancestry, U.S. officials classified Māoris and Polynesians as "Oriental," which precluded Te Rangi Hīroa from passing the 50 percent threshold.[82] So it is important to recognize that Te Rangi Hīroa articulated his theories of Polynesian migration in a context in which all of his professional and scientific acumen mattered little to the U.S. state, which did not consider him to be white enough for citizenship.

In his own theory of Polynesian migrations, Te Rangi Hīroa traced a route that scrupulously avoided contact with Melanesia and thus removed any hint of Melanesian influence on Polynesian society and culture. Indeed, Te Rangi Hīroa theorized that Polynesians descended from an ancestral "Caucasian" stock that left Central Asia because of climatic changes. For instance, in his 1928 article "Races of the Pacific," Te Rangi Hīroa offered a sweeping history of the settlement of the Pacific Rim by different races that positioned Polynesians as a tall and courageous Caucasian race blessed with innate navigational expertise. In this account, Te Rangi Hīroa surmised that "man originated in Central Asia" yet dispersed in several migrant waves in response to dramatic climatic changes. Implying a Spencerian struggle for survival, Te Rangi Hīroa speculated that "the weakest and most primitive were the first to go."[83] He described this first primitive wave of migrants as "pedestrian" by nature and associated them with "one of the most primitive races of man, the Tasmanians," who walked their way from Asia to

Tasmania and were later followed by other "primitive races" including the "Australian Aboriginal," the Negritos, and the "Oceanic Negroes" (Papuans and Melanesians). Although they traveled considerable distances, Te Rangi Hīroa pointed out that this first wave of "pedestrian" races was unable to conquer "the true Pacific" by which he meant the "many remote isles set in the vast expanse of ocean known as Polynesia." Left without human inhabitants after this first wave of migration, the "[Polynesian] islands awaited the coming of a race that could not only invent or adopt a form of ocean transport, but that had the courage to venture out on unknown depths and across vast expanses of speckless sea" he mused.[84] The race with the "courage and initiative" for this undertaking was bred from a "Caucasian wave" that did not follow the rest of their stock to Europe and instead traveled west eventually coming into contact with "people of Mongoloid stock" somewhere near Indonesia. "The resulting Polynesian," Te Rangi Hīroa concluded, "inherits physical characteristics in various degree from his Caucasian and Mongoloid ancestors."[85]

During his time as director of the Bishop Museum, Te Rangi Hīroa polished this account into an expansive book, *The Vikings of the Sunrise* (1939), written for a wide public audience. The book was lavishly illustrated, included Māori and Polynesian proverbs for every chapter, and had several maps, notably one titled "The Polynesian triangle, with the northern Micronesian route and the rejected southern Melanesian theory." The chapter titles—which included titles such as "The Great Ocean," "Whence Came They?," "On the Trail of the Rising Sun"—also added to the book's framing as a travel epic. Although written in a popular style, Te Rangi Hīroa account in *Vikings of the Sunrise* relied heavily on the scientific surveys of Polynesia produced by the Bishop Museum in the preceding decade. For instance, in the book's second chapter—"The Manner of Men"—Te Rangi Hīroa summarized the results of the Bishop Museum's extensive anthropometric surveys of Polynesia and how they supported the hypothesis that Polynesians are of "Europoid" ancestry. Thus, Te Rangi Hīroa argued that the Bishop Museum was "the first scientific institution to study the problem of measuring living Polynesians on a comprehensive scale."[86] Beginning with the Bayard Dominick Expeditions in 1920, Te Rangi Hīroa noted that Bishop Museum researchers had produced numerous studies of the racial characteristics of Polynesians that were "based on the measurements of 2500 living people from representative parts of Polynesia."[87] According to Te Rangi Hīroa what was evident from these extensive Polynesian surveys "is that the master mariners of the Pacific must be Europoid for they are not characterized by the woolly hair, black skins, and thin lower legs of the Negroids nor by the flat face, short stature, and drooping inner eyefold of the Mongoloids."[88] For Te Rangi Hīroa such measurable physical differences between the "Europoid" Polynesians and "Negroid" Melanesians served as compelling evidence that the great migration of Polynesians into the Pacific passed through Micronesia and not Melanesia.

POLYNESIA AND THE ROAD TO MODERN PROGRESS

After his encounter with Rapa Nui and Te Rangi Hīroa's research program at the Bishop Museum, Alfred Métraux not only changed his conception of anthropology but also transformed his views on the indigenous peoples of South America. Exhausted from the relentless pace of his Gran Chaco research, Métraux found Te Rangi Hīroa's charisma and confidence a welcome respite. He also developed a surprising fondness for the crafty and resilient inhabitants of Rapa Nui. In his writing, Métraux argued against the misconception that Rapa Nuians represent a degenerate population with "nothing in common with the people who carved the statues and inscribed the tablets." To the contrary, Métraux argued that they were "in many ways highly gifted" and he admired "their ingenuity and remarkable talent for assimilation." Métraux not only praised the islanders' intelligence; he also argued that their ability to adapt to changing circumstances heralded their place with the modern world. "No European village has given me the impression of more intelligent adaptation to a changing world," Métraux effused and described how Easter Islanders "are constantly on the lookout for new ideas, new fashions—and also new vices." In comparison to the Rapa Nuians, Métraux described the indigenous groups he had previously studied in South America as shackled by their traditions and suspicion toward others. Thus, in his first communiqué to the Société des Americanistes after his return from Rapa Nui, Métraux engaged in a trait-for-trait comparison between the Pacific islanders and the inhabitants of the

Gran Chaco. "One can hardly imagine two more different mentalities," explained Métraux—"among the [Gran Chaco] Indians there is a rather slow intelligence, an incredible tenacity, a furious traditionalism, in the others [Rapa Nuians] a lively and awakened intelligence, an extraordinary mobility of spirit and a perverse taste for change."[89] In other words, while the inhabitants of Rapa Nui displayed a mentality that would allow them to thrive as Western civilization spread to its shores, the indigenous groups of South America appeared destined to continue resisting its ever encroaching presence.

Métraux's emphasis on the adaptive resilience of Rapa Nui's inhabitants also resembles Te Rangi Hīroa's account of Polynesian—and especially Māori—resurgence. In fact, much of Te Rangi Hīroa work was a response to late-nineteenth-century prognosticators who predicted that the Māori race was on the verge of extinction. Like his work on the migrations and material culture of the Polynesian "race," Te Rangi Hīroa's arguments for the demographic resilience and bright prospects of the Māori emerged from his medical service. The clearest and most direct statement of Māori resurgence is found in his 1924 article "The Passing of the Māori," which took stock of the "present condition of the Maori race" given that the time when many predicted the collapse of the Māori race had already passed. In this article, Te Rangi Hīroa historicized arguments for Māori extinction by situating them in the context of European conquest. Using Captain Cook's "very rough" estimate of the Māori population as 100,000 as his baseline, Te Rangi Hīroa demonstrated that the Māori population rapidly declined with European conquest and ebbed to its lowest point in 1871. Yet instead of taking this demographic decline as evidence that the Māori were being supplanted in some kind of natural race struggle, Te Rangi Hīroa offered a more precise etiology. The introduction of firearms by "civilized traders," Te Rangi Hīroa argued, "altered the whole aspect of Maori warfare";[90] what was once just "manly physical exercise" degenerated into "killing expeditions to avenge old defeats and to acquire new territory."[91] In addition to European guns, Te Rangi Hīroa cited the "introduction of epidemic and venereal diseases, and the abuse and misuse of European alcohol, foods, and clothing," all of which "played their part in the decimation of the race."[92] It was at the moment when these destructive forces were at their peak that European scholars began to anticipate the demise of the Māori race. Yet when the colonial government began to conduct "proper" censuses every five years beginning in 1906, the Māori population displayed a steady increase between 1906 and 1921 despite hundreds of Māori troops being absent for war service in 1916 and the impact of the 1918 influenza pandemic, which claimed "over 1000" Māori victims. With these figures, Te Rangi Hīroa rebuked Archdeacon Walsh, who claimed in 1907 that the Māori race had reached a point of finality. Writing sixteen years after Walsh's fatalistic claim, Te Rangi Hīroa demonstrated that there had been "an actual increase of 5020" and that Walsh's hypothesized "vanishing-point" had been hopefully "deferred for ever as extinction is concerned."[93]

Instead of attributing the previous population decline to "an implied law that all dark-skinned races die out after contact with civilization," Te Rangi Hīroa theorized it as stemming from the "confusion that followed the clash of two cultures [Maori and Pakeha]."[94] The decline began with the Māori of the early nineteenth century, who "was unable to distinguish the good from the evil in the two systems" and "voluntarily commenced the disintegration of his own system of culture" by "adopting European weapons, food, and clothing, and becoming Christianized." Yet in the twentieth century, Te Rangi Hīroa explained, the fortunes of the Māori race reversed "due to the gradual elimination of the factors that caused decay."[95] By eliminating the causes of decay, Te Rangi Hīroa hinted that the twentieth-century Māori began a process of modernization. Indeed, he painted a portrait where intertribal warfare with European weapons had ceased and epidemic diseases no longer went unchecked. Although he saw the causes of Māori decay as European in origin, he also argued that they were in some cases best treated with European solutions. In the same way that European populations were benefiting from public health measures geared toward the prevention of disease, Te Rangi Hīroa argued that the establishment of "Maori Health Councils, the appointment of Native Health Nurses and Sanitary Inspectors, and the setting up of a Division of Maori Hygiene in the Department of Health" as well as major improvements in sanitation all had a positive effect for the Māori race by "lowering the heavy mortality due to epidemic diseases."[96] Although population increase came with the loss of traditional Māori practices—such as "the old Maori haka (war dances) and poi dances"—Te Rangi Hīroa cautioned against becoming attached to these "amusements of people living together and spending their evenings in a communal meeting-house."[97] Instead, Te Rangi Hīroa urged a pragmatic acceptance of change. "The Maori is adapting himself to changed circumstances, to a changed environment," he mused, and explained that "the dirge of the lament and the rhythm of the dance will disappear with the communism that brought them into life." While such loss of tradition might be seen as "a pity from the point of view of sentiment," Te Rangi Hīroa insisted that "sentiment alone will not provide for man's material welfare."[98] With little trace of sentimentality, Te Rangi Hīroa explained that "to see old Maori men of the present day changing into pyjamas ere ensconcing themselves between clean sheets" was a symbol of the long and arduous "road to modern progress" that the Māori had traversed.[99]

In keeping with his interest in physical anthropology, Te Rangi Hīroa's celebration of a dawning Polynesian modernity also took stock of the impact of such changes on the Māori racial type. In fact, as he interrogated the Māori's incipient modernization, Te Rangi Hīroa speculated about its relation to the evolution of the Polynesian race. Citing the work of the British evolutionary anthropologist Robert Marrett, Te Rangi Hīroa argued that the evolution of life is a process marked by a "partial rigidity" together with a "partial plasticity." In the back and forth of this evolutionary process, "race" stands for "the stiffening in the evolutionary process."[100] And yet, Te Rangi Hīroa noted, if the fate of the Māori is viewed exclusively through the prism of "race" as a form of evolutionary "stiffening," then the Māori are inevitably conceptualized as doomed to extinction. Instead of viewing Māori evolution solely through the "rigidity" side of the evolutionary equation, Te Rangi Hīroa argued for greater attention to "the element of partial plasticity in the evolutionary process" and how it served "as an avenue of escape for the Maori."[101] This "superadded measure of plasticity," he argued, existed apart from "the racial factor" and reflects the side of human existence that responds "to the effects of environment and culture." The Māori, in fact, offer a striking example of how the interaction between the evolutionary forces of continuity and change—rigidity and plasticity—play out over time. "As the environment has been changing," explained Te Rangi Hīroa, "so the Maori, whilst maintaining his race, has been changing with it." Indeed, the five centuries that the Māori spent in the "temperate climate" of New Zealand "toughened his constitution, sharpened his mentality, and altered his material culture." By contrast, Polynesian islanders "in the tropics" with their "open houses, scanty tapa clothing, and food without labour, were left far behind the Maori" who built warmer houses, invented weaving, and developed sophisticated agricultural and forestry practices. By stimulating these advances in their material civilization, the "more vigorous climate" of temperate New Zealand caused the Māori "to shed the indolence of the tropics."

According to Te Rangi Hīroa, the Māori were not only saved from indolence and extinction by their vigorous climate but also redeemed from extinction by their "happy disposition." In making this psychological argument, he again established a stark contrast with Melanesians, whom he described as hapless and prone to extinction. Citing W. H. R. Rivers's infamous investigations of Melanesian depopulation, Te Rangi Hīroa explained that the greatest factor identified by the British anthropologist "in the nearing extinction of some of these people" is a "psychological cause in the lack of the incentive to live." Similar lackluster dispositions could be found among Polynesians and even Māoris of the past; yet for more recent generations of Māori, argued Te Rangi Hīroa, the "healing hand of time" had "effaced such destructive pessimism." Indeed, it was the Māori's sense of humor that "saved him from undue depression," and Māori life was now marked by "his amusements and manly games, his hopes and aspirations, and every desire to prolong life."[102]

The proud optimism Te Rangi Hīroa developed in the 1920s amid debates about the depopulation of indigenous peoples in the Pacific was one he carried forward to the period after World War II. As Jane Carey has compellingly shown, Te Rangi Hīroa's forceful arguments against settler theories predicting Māori extinction entailed reinterpreting "racial fusion" between Māori and Pakeha peoples as a mechanism for Māori resurgence rather than passive absorption. Te Rangi Hīroa first articulated this celebration of racial fusion in the "Passing of the Maori" and argued that it would only serve to uplift the Māori race if it was accompanied by a revitalization of Māori culture and the restoration of Māori racial health and pride. Te Rangi Hīroa reiterated this argument in the postwar period, notably in the 1949 edition to his book *The Coming of the Maori*, in which he concluded by wishing for the continuation of the "happy blending" of Māori and Pakeha blood that was producing a future—and culturally and physically superior—New Zealander. Yet Te Rangi Hīroa did not view the emergence of a new and racially blended New Zealander as entailing the disappearance of the Māori but rather as a pragmatic means for perpetuating Māori people's best features. In fact, Te Rangi Hīroa forcefully argued against the work of those like the Pakeha scholar Ernest Beaglehole who suggested that Māoris were trapped by a clash of cultures and that their future was best served by fully assimilating to the "rugged individualism" of modern Pakeha culture. In response to Beaglehole, who called for an end to segregated schooling so as to encourage Māori integration, Te Rangi Hīroa stressed the sense of "racial pride" that he acquired through his Te Aute education. "Though co-education seems sound enough in theory," Te Rangi Hīroa explained, "I can never forget that Te Aute gave me a start in life which I would have never obtained from any pakeha secondary school."[103] Te Aute taught Te Rangi Hīroa "loyalty to race" and also brought out "the kindness, courtesy, and good manners" that he saw as inherent in the Māori people but "sadly lacking in the aggressive pakeha world of today." Te Rangi Hīroa also applauded the fact that the Young Maori Party movement originated at Te Aute and asserted that Māori secondary schools, in general, give adolescent students "self-respect, courage, new values, and motivations for study that could not be obtained elsewhere."[104] In a final rebuke to Beaglehole's clash of cultures thesis and policy proposals, Te Rangi Hīroa concluded that he did not believe that "a *complete* change over to the pakeha culture-structure is desirable as a racial policy." Echoing the contestations of European civilizational hegemony that Michael Adas has identified as a prominent current in anticolonial thought during the interwar period, Te Rangi Hīroa observed that "we know only too well that in some parts of Europe, pakeha culture sank to lower depths than that of any race living in a state of primitive savagery."[105] "The standard of lower-class or middle-class pakeha culture," Te Rangi Hīroa continued, "is not attractive enough to induce us to give up all that we cherish." Indeed, he urged his Māori readers to consider that "maybe we have some qualities that might improve pakeha culture, which is so obsessed with the urge to hoard up money to buy social status."[106]

CONCLUSION

In 1952, the Wenner-Gren Foundation held an "epic" and multidisci-
plinary International Symposium on Anthropology attended by more
than eighty participants from different regions of the world. The ambitious event was organized with the aim of taking stock of the global
accomplishments of "anthropological science" and imagining its future
course. As part of this initiative, Alfred L. Kroeber (the symposium's
president) invited Alfred Métraux to present an article that was eventually published in *Anthropology Today*—the major publication and
"encyclopedic inventory" that was the end product of the symposium.
Although Métraux had recently been a major contributor to another
encyclopedic project—Julian Steward's *Handbook of South American
Indians*—Métraux's article did not focus on the indigenous peoples of the
Americas. Instead, Métraux discussed a topic—"Applied Anthropology
in Government: United Nations"—befitting his new role as an applied
anthropologist. In this article, Métraux offered a characteristically circumspect assessment of the role played by anthropology in the United
Nations' economic development schemes. Métraux lamented that, thus
far, anthropology had played a small role in the work "of the United
Nations and the specialized agencies." Yet he also offered some cause for
optimism and cited the United Nations' "expanded program of technical assistance for the economic development of underdeveloped countries" as a domain where anthropology could play an important role.
Métraux's cautious optimism toward technical assistance—which he
defined as a UN scheme "to bring to the economically backward countries the technical knowledge and methods that will enable them to
have their share in the progress of the highly industrialized countries"—
reveals his debt to Te Rangi Hīroa and the inhabitants of Rapa Nui
that he became so fond of. As this chapter has shown, the seeds for
Métraux's shift to applied anthropology and his eventual appointment
to UNESCO were sown when his ambitious project of salvage ethnography in the Gran Chaco collapsed amidst the outbreak of war. More
important, it was Métraux's encounter with Rapa Nui and Te Rangi
Hīroa's work on Māori resurgence that prompted him to pragmatically
accept modernization as an inevitable—and even desirable—force in
the lives of indigenous communities. Métraux brought this pragmatic and

cautious acceptance of economic development to his job as a UNESCO social scientist and his reflections in *Anthropology Today*. "The leisurely well-ordered rhythm of country life," Métraux wrote, "has all too often been replaced by joyless, soul-deadening toil."[107] And this joyless existence, Métraux cautioned, was at risk of spreading to decolonizing nations like the Philippines where zealous elites decried a "romantic attachment . . . for indigenous culture" and instead embraced industrial development "at a dangerous rate" and left anthropologists and UN experts with the difficult task of restraining the "impulse toward premature innovation." With little regard to his privileged location, Métraux lamented that it was often the most "cultivated representatives of the colored races who protest most vehemently when white men advise them to maintain their traditional customs." They even regard anthropologists with "suspicion" and accuse them of being agents of an "insidious imperialism, which, under the cloak of respect and affection is striving to perpetuate its supremacy and to debar the colored races from all access to power and happiness."[108] Such ill-timed and premature pursuits of modernization came at a devastating cost, insisted Métraux. Evoking the ruination he observed in Rapa Nui, Métraux warned that "the impact of mechanization is appalling in its leveling-down effect." Imagining the darkest outcomes, he further cautioned that when social change occurs at a "vast scale, the original culture may be shaken to its foundations or even destroyed."

Yet Métraux also offered hope that such barren futures could be avoided and pinned his hopes for the future of "backward" peoples to the dissemination of anthropological knowledge. With the "experience and acquired instinct of the anthropologist," Métraux wrote, UN economic development schemes can anticipate the repercussions "any slight change may have on a society as a whole." And "when [the] concept [of culture] has penetrated the public as a whole," Métraux augured, "it will no doubt breed a spirit of tolerance between the various nations and cultures which is far from existing at present." With the culture concept ingrained in public consciousness, Métraux continued, "the idea of biological predestination will give way to the idea that change is unending and that man is master of his fate."[109] Though the culture concept was by this point ubiquitous in anthropology, the portrait of relentless change and plasticity that Métraux evokes is of a kind with the narratives of Māori resilience that Te Rangi Hīroa honed amidst his travels in the South Pacific. Yet this model of indigenous resilience was not without its pitfalls. As this chapter has shown, Te Rangi Hīroa narrative of Māori demographic resilience opposed scientific racism insofar as it prophesied Māori extinction. Yet his work was also crafted through and often retained the hierarchies assumed by race science and especially its valuation of whiteness and corresponding devaluation of blackness. Like Te Rangi Hiroa and his predecessor at the SSD—Arthur Ramos—Métraux was not insulated from the hierarchical conceptions of human difference that structured scientific knowledge in this period. As we will see in the following chapters, even with all his cautiousness and commitment to ethnographic rigor, Métraux resigned himself to the view that a transition to the trappings of Western civilization represented the best possible outcome for the world's "backward" peoples.

PART TWO

Race in the Tropics and Highlands and the Quest for Economic Development, 1945–1962

CHAPTER FOUR

A Tropical Laboratory

Race, Evolution, and the Demise of UNESCO's
Hylean Amazon Project

In its early days, when it was brimming with imperial optimism, no project seemed too big for UNESCO. During UNESCO's formative years, the project that best exemplified this imperial grandeur was the International Institute of the Hylean Amazon (IIHA). In 1948 the project was poised to move from abstract idea enthusiastically debated by UNESCO experts to functioning institution with a working staff and budget. The project quickly moved to the top of UNESCO's priority list. As described in UNESCO Courier reporting from that year, no other project had "attracted more interest than this proposed creation of the IIHA."[1] From the frame of the late colonial and post-colonial ideologies this book has tracked, it makes perfect sense that the IIHA generated such intense interest within UNESCO. If it were to succeed, the IIHA promised to do something no European empire or Latin American nation had been able to do: to populate and control the Amazon River basin and exploit its abundant human and natural resources.

According to the *UNESCO Courier*, the IIHA's promise lay in a "broad programme for the study of a huge, but very little known area of the world."[2] And the *Courier* article made sure to emphasize the stunning geographic scale of this "little known" area. "Extending from the Andes to the Atlantic and from the River Orinoco to the Mountains of Bolivia," the *Courier* article proclaimed, "the Hylean Amazon—the vast wooded region of the Amazon river basin—is some 7,000,000 square kilometers in area."[3] In describing the potential for this immense region, the *Courier* article also evoked the classic imperial doctrine of *terra nullius* (land belonging to no one) and brazenly described the region as sparsely inhabited and thus ripe for occupation. "Except for a few settlements mainly along river banks," the article explained "the only inhabitants of this region are about three hundred thousand Indians, whose conditions of life in many cases are extremely primitive." To further emphasize the Hylean Amazon's sparseness, the article asserted that "the density of population of the region is one of the lowest in the world."[4] And it was this seemingly remote and barely inhabited tropical region that UNESCO conceptualized as a fertile testing ground for its cosmopolitan ambitions.

The *UNESCO Courier* was not alone in placing such lofty expectations on the IIHA. As it was debated, UNESCO experts from various disciplines and regions wagered that, if successful, the IIHA would serve as a model for how to use collaboration between international scientists to spark economic development in regions once thought doomed to tropical stagnation. The idea for the IIHA was first proposed to UNESCO in Paris by the Brazilian biochemist Paulo Carneiro during the first session of UNESCO's general conference in 1946. After it was proposed, the IIHA was also championed by the British biologist Julian Huxley, who was UNESCO's first director-general, and the British biochemist Joseph Needham, who was the first director of UNESCO's Natural Sciences Department. The basic aim of the IIHA project was to create an international scientific laboratory in the heart of the Amazon River valley that would serve as a research center focused on the natural and social sciences. UNESCO planners also envisioned the laboratory acting as a "fundamental education" center that would uplift the regions inhabitants through tutelage in modern hygienic practices. By choosing an immense and complex site like Amazonia, UNESCO officials indicated the lofty scale of their ambitions as well as the central role they saw their organization playing in building a lasting global peace.

The IIHA illustrates how UNESCO widened its focus during its first few years of existence. As World War II came to a close, the Conference of Allied Ministers—a group composed of politicians from allied countries who met semiregularly during the war to plan how they would rebuild their education systems—first imagined UNESCO as an entity for reconstructing the parts of Europe and Asia that "suffered cultural distortion or destruction at the hands of Axis powers."[5] Once World War II came to an end, UNESCO's program expanded to include the natural and social sciences; and with the launch of the United Nations' technical assistance program in 1949, UNESCO's mission further expanded from cultural reconstruction in the Northern Hemisphere to economic development in the Southern Hemisphere. During these formative years, UNESCO championed international scientific cooperation as a mechanism for restoring peace between nations and laid the groundwork for the creation of international scientific societies.[6] In contrast to war-torn Europe where their task was to restore the technical and cultural order that was lost, UNESCO officials conceptualized the IIHA as an opportunity to design an entirely new society in a dense tropical region that, in their minds, offered a blank civilizational slate.

Yet, as Brazilian scholars have thoroughly documented, the IIHA's lofty cosmopolitan ambitions proved oblivious to national politics. As the project unfolded, UNESCO's redemptive vision of taming the Amazon through international scientific cooperation collided with the developmentalist designs of nationalist politicians and experts from Brazil who denounced the project as an imperialist exercise threatening Brazil's territorial and economic sovereignty. Existing histories of the IIHA have examined the precise causes for the project's failure, the controversies generated by the IIHA, and the impact of the failed project on Brazil's development policies in the Amazon. For instance, in a meticulous and illuminating study, Brazilian historians of science Marcos Chor Maio and Magali Somero Sá argue that the project failed because the actors involved had conflicting views on the social function of science and because UNESCO failed to establish strong links with relevant scientific experts in Brazil.[7] In another study, Chor Maio and Rodrigo Cesar da Silva Magalhães argue that the anti-internationalist arguments used by Brazilian politicians to undermine the IIHA have reinforced a possessive and nationalistic approach to the development of the Amazon that has persisted into the present. Chor Maio and da Silva Magalhães thus argue that the project "consolidated a development model for the Amazon that continues to guide the initiatives in this region today."[8]

As we saw in previous chapters, the retreat of racial typologies signified by the UNESCO Statements on Race did not lead to a retreat
of racial thinking in the realm of development and modernization
discourse. While the previous chapters demonstrated how the socioeconomic development of supposedly backward groups was often a
prominent theme in the work of social scientists who challenged rigid
typological conceptions of race, the next two chapters demonstrate how
conceptions of race stubbornly persisted in some of the most ambitious
development projects to emerge within the UN system after World War II.[9]
Questions concerning the politics of development are crucial to understanding the IIHA's short history. Yet this chapter demonstrates how,
in the Amazonian context, discourses of development did not concern
economics and territorial sovereignty alone and were often intertwined
with long-standing concerns about the relationships between race, physiology, and extreme environments. Although the economic exploitation
of the Amazon served as an important horizon for the various actors
who supported and opposed the IIHA, this chapter highlights how such
anticipated economic development rested on a prior project of reconceptualizing the relationship between race and tropical geographies.
Since the Iberian conquest of the Americas in the sixteenth century
European natural philosophers, scientists, and artists had often conceptualized the tropics as a geographic area diametrically opposed to the
temperate environment of Europe.[10] As a result, European savants had
long raised doubts as to whether "civilization" could flourish outside of
temperate regions and questioned whether Europeans could preserve
anatomical and cultural distinctiveness in tropical environments.[11]

Historical geographers and postcolonial science studies scholars
have introduced the term "tropicality" to describe this potent colonial
discourse that constructs the tropical world as the West's environmental
other. According to the geographers Daniel Clayton and Gavin Bowd,
tropicality can be conceived as "a discourse—or complex of Western ideas,
attitudes, knowledges and experiences—that, since the fifteenth century,
has both created and been shaped by distinctions between temperate and
tropical lands, with the temperate world routinely exalted over its tropical counterpart, and tropicality becoming central to the definition of the
West as a temperate (moderate and hard-working rather than extreme
and indolent) human as well as physical environment."[12] As we saw in

chapter 1, confronting and reconceptualizing the racial implications of "tropicality" was central to the post-colonial nation-building strategies of scientists and physicians like the Tropicalistas, Nina Rodrigues, and Arthur Ramos. The postwar debates concerning the IIHA represent a similar confrontation with tropicality. Much like the Tropicalistas, who called for a research program focused on the medical distinctiveness of the tropics—and the context-specific solutions this entails—IIHA enthusiasts saw the project as an opportunity to mobilize collaborative scientific research for the purpose of discerning the cultural, biological, and geographic specificity of the Amazon and thereby unlocking its economic potential. Proponents of the IIHA, as well as the Brazilian politicians who opposed it, thus reconceptualized "the tropics" as "under" or "yet-to-be" developed geographies and recoded them as regions of great anticipatory import.

By tracking the genesis, development, and demise of the IIHA, this chapter demonstrates how conceptions of race from multiple scientific disciplines permeated the earliest articulations of technical assistance and modernization in UNESCO's early agenda. In contrast to later economic development projects that were primarily framed through social science rubrics, the ambitious scope of the IIHA project entailed a high degree of interdisciplinary collaboration that enlisted experts from across the natural and social sciences. This chapter highlights the contrasting conceptions of race and development articulated by experts working in different disciplinary traditions. Experts from UNESCO's Natural Sciences Department who were steeped in positivist theories of progress and well versed with colonial development projects in British Africa (namely the Brazilian chemist Paulo Carneiro, the British biologist Julian Huxley, and the British chemist Joseph Needham) envisioned the proposed Amazonian institute as a mechanism for spreading Western science and technology to peripheral regions in order to accelerate the modernization of so-called backwards societies that had predominantly "primitive" races. Although they rejected the idea that "races" represent fixed biological types, UNESCO's natural scientists conceptualized social change in evolutionary terms and continued to position non-European races as culturally and possibly biologically inferior and in need of improvement through technical assistance from Europeans. They also took inspiration from development schemes in sub-Saharan Africa, thus demonstrating how UNESCO's early technical assistance projects operated within a trans-imperial framework in which "race" persisted through notions of civilization and social evolution.

In contrast to the natural science experts, the anthropologists involved with the IIHA—namely the Latin American specialists Alfred Métraux and Charles Wagley—envisioned a path to modernity for the Amazon region through a process of physiological, cultural, and technological adaptation to Western ways of life. Accordingly, Métraux and Wagley proposed anthropological studies of racial adaptation and acculturation to the tropics as central pillars of the IIHA project. Their proposed research drew heavily from the area studies methods that were become increasingly fashionable among social scientists in the United States as well as the Lusotropical ideologies of Brazilian elites that celebrated racial mixing as a key pillar of postcolonial nation building. Through their anthropological research on the Amazon, Wagley and Métraux confronted the notion that tropical regions were unsuited for civilized races and instead proposed that the Amazon could be modernized through a cultural and physiological fusion between indigenous and European bodies and cultures.

EVOLUTIONARY INTERNATIONALISM AND SCIENCE

For prominent figures in UNESCO's Natural Sciences Department, the project of creating an international laboratory in the Amazon epitomized the faith they placed in science as an inherently democratic social practice. In conceptualizing science this way, Natural Sciences Department officials assumed that distinctions such as race, gender, class, and nationality were irrelevant to the pursuit of scientific truth. Accordingly, they adopted a politics of color blindness and assumed that science was necessarily antiracist becuase it was inherently democratic in its methods.[13] Yet this democratic conception of science paradoxically rested on a raced geography that modeled the world through tidy developmental categories such as central-peripheral, advanced-backward, and bright-dark. Within this diffusionist framework, Natural Sciences Department officials adopted the language of the British biochemist Joseph Needham, who was the department's first director and sought to "integrate" peripheral scientists from "dark zones" into global networks and thereby combat scientific "isolation." Concretely, this involved creating scientific field offices in "peripheral" regions of the world with the hope that they would serve as places where scientists from central regions would transmit their knowledge on to those who were comparatively isolated. Not only would such collaborations uplift scientific institutions in peripheral regions, Natural Sciences Department officials wagered, they would also trickle down to nearby communities and redeem local populations from backwardness by sparking economic development. Within the context of these raced schemas of progress, the Amazon—long conceptualized as the epitome of tropical regions where white civilization struggled to exist—stood as the ultimate test of UNESCO's philosophical principles. If science, technology, and modernity can flourish in the Amazon, then why not anywhere else?

Yet the Hylean Amazon did not simply represent an imposition of scientific ideas from the Global North. In fact, it was the Brazilian Paulo Carneiro who put forward the first proposal for the IIHA, which was met with much enthusiasm during the first session of UNESCO's General Conference held in Paris in 1946. During this meeting, conference delegates stressed the importance of prioritizing original projects with ambitious yet realistic geographic scales during UNESCO's first year and cited the Brazilian proposal for the Hylean Amazon institute as a promising example. For instance, René Cassin, the French jurist and coauthor of the Universal Declaration of Human Rights, argued that UNESCO's role should not be limited to that of a "co-ordinating" and "stock-taking" agency and that it must therefore have "other aims, creative aims, original aims, modest as they may be. This is neither too vast nor too ambitious a project."[14] As examples of such creative projects, Cassin cited "the great project of the Tennessee Valley" and praised the fact that "our programme contains a notable and well-defined project for the Amazon Basin."[15] Admiring the way that, in the past, "astronomers co-operated to draw a map of the sky," Cassin insisted that it was now time for "Unesco to become a laboratory of new enterprise, capable of developing initiatives which are entirely independent of what any nation, or group of nations has ever conceived."

Brazilian delegates to the General Conference skillfully presented the Hylean Amazon project as precisely the sort of concrete initiative that exemplified UNESCO's philosophical commitments to universality and international cooperation. For instance, during a session focused on the program for the Natural Sciences Department, the Brazilian physiologist Miguel Osorio de Almeida described the Brazilian delegation's proposal for the Hylean Amazon institute and argued that "Unesco could do far more in the scientific field than in any other" and that "in order to take effective action in favor of peace, projects of a concrete nature must have priority over others." According to Osorio, the Hylean Amazon project was exactly the kind of concrete endeavor that embodied UNESCO's ambitions and insisted that it "was not of regional, but of international importance." Indeed, Osorio argued, Brazil and other South American countries were not the only ones who stood to gain something from increased research on the Amazon basin: all countries with "geographical conditions similar to those of America" stood to profit from such research. During his pitch, Osorio de Almeida also made the case that the Hylean Amazon project could also address the question of "the limited scientific development of the so-called 'dark' zones" and argued at length that UNESCO could play a role in "ascertaining the reasons for this defective development and for trying to find a remedy."[16]

During the heady days of UNESCO's infancy, the IIHA and its focus on human welfare through applied science perfectly embodied the late colonial and post-colonial philosophies of progress that shaped the careers of so many of its key figures. For instance, for Paulo Carneiro, the project's main enthusiast, the IIHA grew out of his lifelong commitment to the Positivist Church of Brazil (Igreja Positivista do Brazil). Carneiro was born in 1901 to a prominent middle-class Brazilian family in Rio de Janeiro who were adherents of the Positivist Church—a church that practiced the secular "religion for humanity" developed by French philosopher Auguste Comte.[17] After his birth, Carneiro was baptized in the Templo de Humanidade (the seat of the church, where his parents also married), and his parents chose Cândido Rondon—a prominent figure in the positivist movement who was a military officer and served as first director of Brazil's Indian Protection Service—to be his godfather.[18]

As a member of the Postivist Church, Carneiro studied Comte's writings closely during his childhood and chose a career in science. In the early 1920s, Carneiro trained as an industrial chemist at the Escola Politécnica do Rio de Janeiro then completed a doctorate in biochemistry at the Pasteur Institute in Paris in 1931.

The research Carneiro conducted during his time at the Pasteur Institute and in the following years fit within a broader program geared toward cultivating Brazil's economic independence by growing its industrial sector and making greater use of its natural resources. For his doctoral work, Carneiro studied the chemistry and medical uses of Guaraná—a plant with stimulant properties that was traditionally consumed by indigenous communities in the Amazon.[19] After receiving his doctorate from the Pasteur Institute, Carneiro moved to Pernambuco, Brazil, where he worked as a research scientist under the Getulio Vargas administration, first in the Ministry of Agriculture and then the Ministry of Technology.[20] When Vargas first came to power following the 1930 revolution, Brazil's economy was suffering a downturn precipitated by the global economic recession. Because of the recession, Brazil's oligarchy in São Paulo lost much of its wealth and influence, which allowed Vargas to consolidate power in Rio and govern with bold promises of restoring social order and rapidly transforming Brazil into a modern industrial state.[21] For Vargas, the Ministries of Agriculture and Technology where Carneiro worked served as crucial cogs in the developmentalist machinery his administration was keen to build. During his time at the Ministry of Technology, Carneiro spearheaded a project to create a new research section dedicated to discovering novel or better industrial applications for the vegetables, plants, and animals within Brazil's borders.[22] In November 1937, Getulio Vargas announced a supposed communist plot to take over the government and pushed through a new constitution that concentrated power in the office of the president and thus ushered in a period of authoritarian rule. Fearing that his ideas concerning the need for agricultural cooperatives could be construed as communist, Carneiro secured a research fellowship at the Pasteur Institute and fled to Paris where he continued his research on the physiological effects of curare. During World War II, Carneiro was held captive twice by the Nazi forces that occupied France and returned to Brazil when he was freed in 1944.[23] Upon his return to Brazil, Carneiro joined the Brazilian delegation that participated in the various meetings and conference leading up to the creation of UNESCO in 1945. And during the first session of the General Conference in 1946, UNESCO delegates elected Carneiro to the organization's executive committee.[24]

After the war, Carneiro's faith in science as a tool for social progress coupled with his difficult experiences under the Vargas and Vichy regimes deepened his commitment to international collaboration. With a worldview profoundly shaped by positivism, Carneiro theorized UNESCO and the IIHA as projects poised to trigger a process of social evolution and international harmony at the planetary scale.[25] Indeed, in a short booklet describing the purpose and aims of the IIHA, Carneiro boldly declared that "it is the law of general biology that organisms, as they improve, increase their subordination to their surroundings."[26] Yet Carneiro did not restrict this process to living organisms. "The evolution of civilizations is subject to the same principle," Carneiro maintained. As civilizations become more advanced, they become more dependent "on the whole of humanity," and the measure of progress is thus "the planetary character that gradually covers all human relations." With both organisms and societies in their full development, Carneiro argued, what can be observed is a state where "everything conspires, everything converges, everything competes." And it was precisely this "organic consensus" that Carneiro defined as "the quintessential attribute of social evolution."[27] Carneiro was also at pains to present international collaborations as safeguards of national sovereignty. And he adopted a racialized language of evolution to describe this process. Efforts to impede nations' natural tendency toward collaboration could only result in a disastrous involution of the basic structures of the planetary civilization that was emerging and lead to "hunger, war, revolution."[28] Instead of threatening national sovereignty, Carneiro argued that the "constant reaction between the whole and its parts" is the best guarantee of the "independence and liberty of each country."[29] By contrast, Carneiro described nations that resisted this "organic consensus" as primitive and pathological. Looking back to the preceding decades, Carneiro pronounced "Italian fascism, German Nazism, and Russian Bolshevism" as manifestations of a "social pathology conditioned by the breaking of international ties."[30] By severing their international relations, argued Carneiro, these pathological nations produced both psychological and political "aberrations" such as "the hypertrophy of pride and a demolishing instinct, an imperialist aggressiveness, the compression of freedoms, and a contempt for the human person."[31] "Only primitive peoples can live in isolation" Carneiro wrote.[32]

In contrast to the depths to which nations fell during the war, Carneiro heralded the formation of the United Nations in 1945 as marking the birth of a "new world order."[33] More important, with the creation of the United Nations and its allied agencies—WHO, FAO, and UNESCO—Carneiro saw an opportunity for nations to join forces and tackle the most urgent problem threatening human existence: an impending geopolitical disaster due to global population growth outpacing the rate of food production. Yet although Carneiro framed this problem in Malthusian terms, he did not see measures to control population growth as a viable solution. Instead, Carneiro argued that, on the whole, humanity was using land inefficiently and could stand to increase its rate of food production. According to Carneiro's calculations, humanity was using only 15 percent of available arable land to sustain food production for a global population that was growing at a rate of 20 million people per year. The rates of malnutrition throughout the world suggested that the percentage of land used was insufficient, and, to make matters worse, much of the arable land used was being damaged. In fact, Carneiro warned, "each year brings about the destruction of 80,000 km^2 of arable land, transformed by ignorance and negligence into unusable laterites [clay]."[34] Ruling out birth control for moral and political reasons, Carneiro argued that nations must therefore join forces to "increase [food] production on our planet."[35]

It was from this southern Malthusian perspective that Carneiro invested immense hope in the UN system and the future it promised to bring about. Like his Brazilian colleague Josué de Castro—a nutritionist who became director of the FAO—Carneiro viewed the United Nations' specialized agencies as institutions that could offer technical solutions to humanity's most pressing existential problems without infringing upon the reproductive freedoms of individuals.[36] At the same time, Carneiro cautioned that increasing the global percentage of arable land could be disastrous without careful scientific planning. If done improperly (i.e., without concern for conserving forests and other natural, mineral, and animal resources), humans ran the risk of transforming large swaths of arable lands into arid lands. Worse still, mismanaged land could disrupt the delicate "biotic equilibrium" of the planet and put the "survival of our species, and of the plant and animal communities that are inseparable from us, at risk."[37] It was in the context of these looming ecological crises that Carneiro regarded the IIHA as a project that could help solve many of humanity's problems. Given the sheer size of the Amazon basin, the region represented a vast reservoir of potential arable land as well as a compelling example of organic interdependence. For Carneiro, the Amazon also served as a paradigmatic example for all tropical regions. "The ecological conditions of Amazonia reproduce in their broad lines the characteristics proper to all humid equatorial plains" Carneiro wrote.[38] As such, Carnerio proposed that the kinds of studies conducted through the IIHA should be comparative and analogical in scope with an eye to exporting findings to other tropical regions or drawing insights from studies of tropical life elsewhere. Indeed, Carneiro argued that a "comparative method" would be "the most fruitful and efficient" and the most likely to shed light on "the problems" of tropical regions.

Carneiro was certainly not alone in conceptualizing the UN system in such cosmic and racially tinged terms. Indeed, his views on the international system are not unlike those of UNESCO's first director-general, the British biologist Julian Huxley. But whereas Paulo Carneiro approached the international from the standpoint of Brazilian positivism, Huxley's outlook was indebted to British imperial thought. As Glenda Sluga has shown, the particular brand of liberal idealism that Huxley espoused and which guided him in his role as UNESCO's first director-general was shaped by political views that stemmed from British imperialism. Since Huxley's first field mission in 1929 when the British Colonial Office sent him to British East Africa to study the possibilities of improving colonial education in biology, Huxley embraced the view that liberal imperialism was a positive force insofar as it allowed white men to tutor "backward peoples" and thus bring them civilization and education, which would help modernize their societies. During the World War II, as Britain was faced with strong resistance from anticolonial uprisings in its colonies, Huxley argued that the best way for Britain to regain control of its colonies would be through modernization programs designed to uplift colonial peoples out of their state of primitiveness and raise them to a level of material equality. Huxley's interest in projects of racial uplift continued after World War II, and during his years as director-general of UNESCO he championed the view that UNESCO's "fundamental education" program should seek to level up the educational, scientific, and cultural facilities in the "dark areas" of the world populated by "darker races" and less privileged classes.[39] In order to implement "fundamental education" projects at UNESCO, which were modeled on the "mass education" projects of the British Empire that Huxley had helped to design during World War II, Huxley drew upon his experiences in colonial filmmaking and recruited experts from the British Colonial Film Unit.[40]

Another aspect of Huxley's political thought that was shaped by his entanglements with the British Empire was his advocacy of scientifically planned economies as an antidote to the excesses and inefficiencies of laissez-faire capitalism. Huxley's belief in the importance of scientific planning and governance were influenced by the popularity and growth of ecological thought and by the rise of population genetics in the context of the British Empire of the interwar period.[41] In response to the economic depression of the 1930s, Huxley argued for the need to halt laissez-faire individualism in favor of an ecologically engineered economy that would be in balance with nature. In order to accomplish this, Huxley believed that the mathematical models of genetics and natural selection crafted by the British statistician and eugenics supporter R. A. Fisher served as tools for the rational planning of society and "improvement of the race" and advocated for the importance of planned economies in colonial settings.[42]

During his time as UNESCO's director-general from 1946 to 1948, Huxley further elaborated his political philosophy in a booklet called *UNESCO: Its Purpose and Its Philosophy* published in 1946. In this monograph, Huxley described UNESCO's purpose through the lens of an evolutionary narrative positing increasing complexity in all domains of life as the hallmark of progress, and where increased scientific and technological complexity was seen as an impartial yardstick for evaluating progress and evolution. For Huxley, the directional movement toward greater complexity could be seen in the distinction between primitive and modern: "the elaboration of a modern state, or of a machine-tool factory in it, is almost infinitely greater than that of a primitive tribe or the wooden and stone implements available to its inhabitants."[43]

The IIHA's goals of international scientific cooperation and the comprehensive study of the Amazonian ecology fit well with Huxley's ideas. For Huxley, the comprehensive scientific study of the Amazon region and the cooperation between scientists and nations, which the proposed center promised, would solve many of the region's problems. In a 1947 article published in the Rotarian magazine, Huxley suggested that the institute was to serve as a "clearing house of information" where international research groups could study "the countless scientific and social problems of the Amazon."[44] One of the more obvious problems, according to Huxley, was the primitive state of the region's inhabitants: "a measure of the problem can be glimpsed when it is realized that the 300,000 population of the Amazon, scattered in tribal villages along the 2,000-mile reaches of the river, are among the last surviving examples of Stone Age society."[45]

The concern with scientific internationalism and the language of centers and peripheries at play in the IIHA also bore the imprint of the first director of UNESCO's Natural Sciences Department—the British biochemist and historian of science and technology in China, Joseph Needham. During World War II, Needham had spent five years in China as head of the Sino-British Science Cooperation Office. While in war-torn China, Needham observed that his Chinese colleagues were often excluded from scientific exchanges with thriving scientific communities in Europe and North America. He thus began to conceptualize non-Western scientists as languishing in an epistemic "Dark Zone" characterized by geographic isolation and intellectual neglect from their colleagues in the "Bright Zone," namely the West.[46] These disparities led Needham, who was also a committed socialist, to play an active role in promoting scientific internationalism along with Julian Huxley through participation in conferences on scientific international exchanges organized by the British Association for the Advancement of Science (BAAS) and the Association for Scientific Workers (AScW). Like many of their British and French colleagues, Needham and Huxley mobilized against the rise of fascism and the scientific racism of Nazi Germany during the 1930s and became concerned with issues involving the application of science for peace and well-being and the development of international scientific cooperation.[47] After the war, Needham acted as one of the main advocates for the inclusion of "science" into UNESCO's program, which led to a brief appointment as director of the Natural Sciences Department from 1946 to 1948.[48]

Needham developed his approach to scientific internationalism during World War II and then honed it within UNESCO's Natural Sciences Department. Like Arthur Ramos, during the last years of the war Needham articulated a vision for postwar science through a series of lectures and memoranda, which he wrote while still in China. In articulating this vision for postwar international science, Needham spoke from the experience of seeing his colleagues marginalized in China and argued that though scientists generally welcomed cooperation with colleagues around the world, it could not be assumed that international cooperation would emerge through a "laissez faire organization of science."[49] For Needham, true scientific international cooperation would have to be strategically planned through the creation of durable links between scientists in the "Bright Zone" and the "Dark Zone"—a concept that he came to call "the periphery principle." As director of UNESCO's Natural Sciences Department, Needham attempted to put the periphery principle into action. Although international associations and conferences might stimulate international cooperation, Needham wagered that the creation of what came to be called Field Science Cooperation Offices (FSCOs) would be a more effective means of creating durable collaborations. During his time at UNESCO, Needham thus made the creation of these FSCOs one of his top priorities and modeled them on the British Empire's Scientific Liaison Offices (SLOs), which had proven to be an effective means of stimulating the exchange of staff and scientific information between Allied nations during World War II. For the postwar era, Needham thus saw UNESCO as playing an analogous role to the British Empire by expanding international scientific networks beyond the realm of Allied nations and encompassing the "Dark Zones," which in his mind included both the "Far East" and the Southern Hemisphere. During his brief time as director of the Natural Sciences Department, Needham witnessed the creation of several FSCOs in "peripheral" zones, namely in Cairo, Egypt; Rio de Janeiro, Brazil; Nanjing, China; and New Delhi, India.[50]

Carneiro's, Huxley's, and Needham's differing conceptions of scientific internationalism shared common themes that were shaped by British imperialism on the one hand and Latin American positivism on the other. All three of these natural scientists viewed science and technology as tools that could be harnessed for the sake of social progress and international cooperation. And an important aspect of this vision was a conception of science as a democratic and collaborative enterprise that thrives through international cooperation and does not discriminate on the basis of race. This "ecumenical" conception of science, which in Needham's case was informed by socialist principles, ultimately proved to be out of sync with the ideological polarization that emerged in the subsequent years as a result of the Cold War. In this sense, Carneiro's, Huxley's, and Needham's internationalist postwar visions were late colonial and post-colonial products that owed much more to the imperial and internationalist projects of the interwar period than they did to the liberal internationalism that came to shape UNESCO during the Cold War. It is thus not surprising that their lofty internationalist visions were often tinged with the ecological, evolutionary, and racial language of European imperialism.

PORTABLE IMPERIAL EXPERTISE

Once the Brazilian proposal for the IIHA project was accepted, UNESCO officials faced the enormous task of putting its internationalist principles into action. From the comfort of Parisian boardrooms, international scientific cooperation seemed a perfectly laudable goal for UNESCO officials. Yet from the vantage of the Amazon River basin's complex geography, evolutionary internationalism often became an abstract and intangible ideal. As they laid the groundwork for the IIHA, UNESCO officials faced daunting logistical and political challenges that tested their ideals, stamina, and organizational ability. For Latin American scientists, Needham's periphery principle, which presumed that scientific expertise would travel seamlessly across regions, raised difficult questions about the imperial nature of UNESCO's activities. Moreover, in Amazonia—an immense region with comparatively little by way of transportation and communication infrastructure—IIHA organizers found themselves traveling long distances through difficult and unfamiliar terrain and negotiating with a wide range of scientific and political actors from the various nation-states encompassed by the Amazon basin. It is not surprising then that the experts who conducted fieldwork for the project—like the tropical botanist Edred John Henry (E. J. H.) Corner and Alfred Métraux—were experienced field scientists initially trained to serve the colonial administrations of England and France during the interwar period. In its quest to accelerate modernization of the Amazon through scientific internationalism, UNESCO reached into its imperial toolkit and mobilized forms of expertise honed through colonial conquest of tropical regions.

The first step in UNESCO's plan for the IIHA was to create a regional field office. After the 1946 conference in Mexico City, UNESCO officials created the Latin American Scientific Cooperation Office (Lacso) in Rio de Janeiro and made planning for the establishment of IIHA the office's top priority. The Rio office was one of several scientific cooperation offices UNESCO created. The offices were set up in so-called *dark zones*, and their main function was to put Needham's periphery principle into action. Needham also had a close hand in appointing the directors of the field offices. In the case of Lacso, Needham turned to his contacts at Cambridge University and in the British Colonial Office to recruit E. J. H Corner as Lacso's first director. Although he had no previous experience with Latin America, Corner did have a reputation as an accomplished tropical botanist. Corner trained in botany in the 1920s at Cambridge University. After receiving his degree, Corner applied for a position in the Colonial Service and left Cambridge. From 1929 to 1946, Corner worked as assistant director of the Singapore Botanical Gardens—which was under British colonial control—and produced important botanical works documenting the Malayan forest.[51] In 1947, when he was invited by Needham to serve as the principal field scientific officer for UNESCO in Latin America, he surely looked forward to the change of pace.

Whereas Needham and Huxley likely looked at Corner's storied botanical career in Asia as evidence of relevant experience for the Lacso post, their Latin American colleagues were less impressed. Shortly after Corner was appointed as director of Lacso, Paulo Carneiro wrote to Huxley and cautioned that, without help from established experts from Latin America, the British botanist would be out of his depth. Carneiro also stressed that Corner should be in close contact with representatives from Amazonian countries and should endeavor to have them accompany him during his field trips whenever possible. Indeed, Carneiro warned Huxley that diplomacy of this sort was necessary because South American countries don't enjoy being treated like "colonies" and studied by foreign scientific missions.[52] Huxley's solution to Carneiro's concerns was to appoint a "young scientist" to assist Corner. Yet Carneiro viewed this solution with suspicion and believed that Huxley's preference for a young and impressionable scientist tipped the scales in favor of British imperialism.[53] As such, Carneiro insisted that instead of appointing a young and inexperienced scientist to assist Corner, the Natural Sciences Department needed to place an experienced Latin American scientist at the forefront of the IIHA and volunteered himself to act as a "special consultant" to Corner with this in mind.[54] In the following year, as preparations for the IIHA turned to organizing a preparatory conference in Belém, Carneiro also recruited Alfred Métraux as an anthropological consultant to the organizing team.

During the northern spring of 1947, Corner embarked on the first of many long journeys in preparation for the establishment of the IIHA. Corner's main destination was Rio de Janeiro, yet his itinerary included several stops along the Eastern Seaboard (Boston, Chicago, New York, Washington, Miami) and in the Caribbean (Puerto Rico and Trinidad) to meet with scientists and politicians in an effort to raise support for the creation of the IIHA.[55] Once in Rio, Corner and Paulo Carneiro organized a meeting at the Itamaraty Palace (Brazil's Ministry of Foreign Affairs) where they described their plans for the IIHA to a group of scientists, politicians, and intellectuals.[56]

A month after the Itamaraty meeting, Corner traveled to Belém— the capital city of Brazil's northern state of Para, which sits at the confluence of the Atlantic Ocean and the Para River, which is part of the Amazon system—for the purpose of meeting Alfred Métraux and scouting out a potential headquarters for the IIHA. By this point the Brazilian government had proposed the Goeldi Museum in Belém as a potential headquarters for the IIHA.[57] The UNESCO secretariat was also intrigued by the Goeldi Museum and was convinced by the recommendation letters it had received that it would be approved by other countries in the region as a suitable site for the IIHA headquarters.

As established field scientists in tropical botany and anthropology, Corner and Métraux were thus charged with taking stock of the Goeldi Museum's facilities and assessing its suitability as a headquarters for the IIHA.

Once in Belém, Corner and Métraux spent two weeks inspecting the Goeldi Museum's collections, meeting with local politicians and scientists, and visiting local sites including the city's botanical gardens, orchid gardens, medical school, national consulates of Peru and France, and the Indian Protection Services (Serviço de Protecçao aos Indios).[58] After their visit, Métraux wrote a report about the Goeldi Museum that painted an unflattering portrait of the institution and underscored the role UNESCO could play in transforming it into an important research center. Although he marveled at the museum's valuable archaeological collections from the lower Amazon, Métraux lamented that they were exhibited in a small and crowded room with "antiquated glass cases" and that the "technique of presentation is antiquated and provincial and does not do justice to the value, both artistic and scientific, of the specimens."[59] Métraux offered a similar assessment of the museum's "ethnographic collections." As such, Métraux suggested that the museum needed more space and better storage facilities as well as an infusion of "new collections" to fill up the existing geographic gaps; the museum, he noted, had "practically no objects from the Amazonian areas outside of the Brazilian territory." One of the main tasks of the IIHA would thus be to preserve the museum's Amazonian collections as a "monument to the native populations of this area," especially "since they are the best proof of the technical ingenuity and artistic taste of the Indians."[60]

In the months that followed, Corner and Métraux collaborated with Carneiro to organize a meeting of the IIHA's International Commission—a committee composed of delegates from Amazonian countries interested in creating the international scientific center. The International Commission met in Belém from August 12 to 18, 1947, and included delegates from Brazil, Colombia, Ecuador, France, Peru, Venezuela, the United Kingdom, the United States, UNESCO, WHO, FAO, the Pan-American Union, and the Pan-American Sanitary Bureau.[61] Corner served as executive secretary for the meeting, and the U.S. epidemiologist Fred Soper, who was also the director of the Pan-American Sanitary Bureau, served as the meeting's "president." In addition to presiding over the conference, Soper was given the task of submitting a report to UNESCO's director-general with recommendations for the purpose and organization of the IIHA and suggestions for its research program in the years to come.

Soper's report illustrates how the IIHA's envisioned research program built on colonial conceptions of the Amazon and Needham and Huxley's imperial internationalism. "Since the discovery of Amazonia," Soper wrote, "it has been continuously explored by scientific missions of many nationalities with the aim of drawing up its botanical and zoological inventory, of becoming acquainted with the state of social development and organization of its native tribes, of determining the essential characteristics of its climate and soil . . . and finally of opening up the economic wealth and and exploring the demographic possibilities of its vast area."[62] Yet, echoing Needham, Soper described how such pressing scientific tasks were often "fruitless" because of lack of a "permanent centre or body to co-ordinate them, follow them up and pass them on to succeeding generations." Without such a center of calculation, Soper explained, knowledge accumulated slowly and in a piecemeal manner, and the materials collected "were often lost" while important reports and documents were "largely scattered and forgotten." Such failure to pass knowledge on to succeeding generations prompted Soper to argue that "only an international body, jointly maintained by the countries of the Amazon forest and those . . . specially interested in the problems of the natural and social sciences peculiar to it, is capable of ensuring lasting results to such an undertaking."[63]

The research program that Soper and his colleagues drafted during the Belém meetings was also framed in the language of imperial internationalism. In his report, Soper explained that the Belém commission decided to focus on ecological problems "relating to the natural environment and conditions of life of man in the equatorial forest" and proposed a series of stocktaking surveys oriented toward the subjects of "physiography, pedology, botany, zoology, agriculture, social sciences and education." By comprehensively taking stock of the Amazon basin's ecology, the Belém commission sought to lay the groundwork for the creation of an educational and public health infrastructure for inculcating modern hygiene practices in the region's inhabitants.[64]

With a basic research agenda in place, UNESCO representatives then faced the task of bringing the Hylean Amazon institute into being.

The results of the Belém Commission were submitted to the Second Session of the General Conference of UNESCO in Mexico City in 1947, at which the conference participants passed a resolution instructing the director-general to take steps to "bring the Institute into being in 1948."[65] The first step was to assist the Brazilian and Peruvian governments in organizing a conference for the establishment of the institute, which was to be held in Iquitos, Peru. Shortly after the Mexico meeting, Corner was appointed secretary of the conference for the establishment of the IIHA, and the Peruvian writer and ex-justice and labor minister Luiz Alaysa y Paz Soldan was appointed as "special advisor" to UNESCO for the conference. A team composed of UNESCO-appointed experts and delegates sent by the Brazilian and Peruvian governments was also assigned to Latin America to assist with preparations.

The Iquitos conference met from April 29 to May 11. During this time, conference delegates worked closely with a legal advisor from the United Nations and drafted a convention that would serve as a guiding legal document for the operation of the IIHA. The delegates also approved a budget for the IIHA's first year of operation and established an "Interim Commission" to serve as the IIHA's leadership team. During conference breaks, Peruvian delegates and politicians treated the conference participants to evening parties and other "diversions" such as a river trip in a gunboat (the *Amazonas*) provided by the Peruvian Naval Forces; the conference delegates also participated in a symbolic ceremony at the Plaza de Armas where they planted a "Tree of Amazon Confraternity" and laid the first stone of a monument to "commemorate the Conference."[66] The French delegation headed by Paul Rivet was the only one to fully ratify the conference convention and budget. All the other delegations signed the convention "ad referendum," that is, pending approval from their governments. According to Corner, thanks to the "energetic leadership" of Rivet—a key architect of some of France's biggest colonial ethnographic missions—the conference delegates also approved a proposal to immediately begin scientific surveys conducted by research teams working in "limited areas."[67]

Even though the Iquitos convention was not approved—it required ratification by five Hylean Amazon states to come into force—the type of surveys proposed by Rivet began immediately after. The first of these surveys was a "community study survey" led by the U.S. anthropologist

Charles Wagley and a research team composed of his Brazilian wife, Cecilia Wagley, and the Brazilian anthropologist Eduardo Galvão from Rio's National Museum and his wife, Clara Galvão. The purpose of this "pilot" study was to conduct an ethnographic survey of a "typical" Amazonian town and to report back on the possibilities of conducting similar studies on other communities throughout the Hylean region. After visiting Wagley and his research team, Corner reported back on the "sound work which had been accomplished" and argued that the IIHA should sponsor similar studies "on the same lines but on a much more extended scale, until the whole region of the Hylean Amazon had been fairly sampled." The second survey conducted after the Iquitos conference was a "bibliographic survey" conducted by the Colombian botanist Enrique Perez-Arbelaez. In a period of five months, Perez-Arbelaez used libraries in Washington and Bogotá to generate a large card-index with a preliminary catalogue "of some 3000 entries" that would serve as the basis for a comprehensive bibliography to be kept in Manaus.

The third survey was a scientific survey of the Huallaga River valley—a tropical valley in Peru—which was recommended by members of the Interim Commission who envisioned it as a "pioneer project in team-work."[68] The research team was charged with "obtaining a general picture of the state of the Rio Huallaga Valley" and with "advising, in view of the probable economic and social development of the valley on the action which should be taken by the IIHA to record its original state, to preserve its natural features and to assist in its development." To get to the Huallaga valley, the expedition traveled by car across the Andes from Lima to Tingo Maria where they established a base with the director of colonization. They then descended the river on balsa rafts in two groups led by the anthropologists and geographers while Coronel Gerardo Dianderas conducted an aerial survey in an airplane. Despite the difficult travel conditions, Candidó Bolivar, a Mexican zoologist who led the mission, enthusiastically described the Huallaga valley as one "wherein health conditions are so favorable for human life." Appraising the valley's future economic potential, Bolivar also evoked the doctrine of *terra nullius* and described "wide ares of land suitable for cultivation" that were "largely untouched by man" and reasoned that "the valley itself constitutes a major reserve which could easily be opened up to civilization and might become a source of immeasurable wealth for Peru, once an adequate system of roads has been built through the valley."[69]

Although experts entrusted with the IIHA's future dutifully wrote their reports with the required enthusiasm, they also expressed doubt about whether the Amazon was suited for the sort of scientific cooperation envisioned by Needham and Huxley. These doubts were most forcefully expressed in the reports of an increasingly fatigued E. J. H Corner. Indeed, although Corner described the interest the project generated among people in the Amazon valley, he also questioned the grandiose scale of the project. "It was not merely the immense geographical extent of the Project," Corner wrote, "but, paradoxically, its intangibility which was baffling." How an institute staffed by a handful of scientists would render such an immense geography legible seemed a genuine source of uncertainty for Corner. "Who could advise and comprehend the country and its people from the Andes to the Atlantic, from the arrow-head to the paper-mill, or from the pavements of Rio to the sanctuary of Quito?" Corner asked. As such, Corner stressed that "there was in no country a centre of information about the Hylean Amazon, or a person, or group of persons" who was "able to advise on all the many scientific and political aspects which rose up on every side whenever it was attempted to plan the Project in operation."[70]

In the absence of a central base, Corner accumulated thousands of travel miles and grew increasingly fatigued. Describing his travels, Corner wrote that "it was necessary to travel far and wide in order to visit the numerous officials, scientists and historians . . .; in order to gain a first-hand information about the Amazon plain, rivers, valleys and towns; and in order to discuss the development of the Project with state officials." Although he was ostensibly hired by UNESCO to serve as the person responsible for the Science Cooperation Office in Rio, Corner complained that the excessive travel demands of the IIHA project prevented the development of said office. In fact, soon after his arrival to the office in Rio, UNESCO changed his title from field scientific officer (FSO, LA) to principal field scientific officer (PFSO, LA), thus freeing him up to focus solely on the IIHA while a newly appointed FSO took charge of the Rio office. In a bid to convey in a manner "better than words the difficulty of concentrating effort on the Hylean Amazon," Corner drew up a comprehensive list and summary of the journeys he had taken and explained that out of the total of 545 days since he left Paris in 1947, he had spent "in the aggregate 186 days in Rio de Janeiro, 83 days in Manáus and 276 days (or slightly more than half the total) in travel."[71]

Corner's frustrations illustrate one among many tensions that led to the IIHA's downfall. As a field scientist tasked with both developing one of UNESCO's field offices and organizing the creation of the IIHA, Corner quickly found himself overwhelmed by such lofty internationalist expectations. As we will see in the rest of this chapter, other field scientists who became involved with the IIHA—notably the anthropologists Charles Wagley and Alfred Métraux—shared some of Corner's concerns yet ultimately articulated a different approach to internationalism and to modernizing the Amazon—one that eschewed the lofty evolutionist framing of Carneiro, Huxley, and Needham in favor of a more circumspect approach that emphasized the region as a unit of anlysis and processes of acculturation.

AMAZONIAN PHYSIOLOGIES

In an article about the future of the Amazon, the U.S. science-fiction writer and solar-energy advocate Peter Van Dresser enthusiastically described the IIHA as a "hemispheric TVA [Tennessee Valley Authority]."[72] While previously "wave after wave of [Western] expeditions of exploitation" had slaughtered the Amazonian rain forest and wasted dollars and lives, the IIHA, according to Van Dresser, heralded a brighter future. Instead of reckless attempts to "gouge raw material and food out of the untamed forest," UNESCO's project signified a prudent endeavor to "apply the skill and weapons of modern science to making the jungle habitable and fruitful."[73] Yet in working toward this harmonious ecological vision, UNESCO faced formidable obstacles. The most pressing problem, Van Dresser argued, was "establishing and maintaining a vigorous and stable population in the region."[74] "Without such a population," Van Dresser cautioned, "'development' in any sense—mercantile, economic, or cultural—is impossible."[75] While the "Westernized colonist" had typically approached settlement by recklessly "clearing the jungle," Van Dresser insisted that future Amazonian societies would have to be "forest-based" and envisioned "a pattern of numerous, well-distributed communities, each largely self-sufficient in the production of the bulk necessities of life, each harvesting the rich and varied crops of the surrounding forest." Van Dresser did not imagine this distinctively Amazonian society as made up exclusively of colonists. Rather, he insisted there was no "better parent stock for the breeding of such a forest adapted culture . . . than the 300,000 to 400,000 native Indians in the region[.]"[76] Van Dresser thus imagined a hybrid Amazonian society and claimed that the IIHA's social science program, which aimed to "develop a technique whereby the science and skill of Western civilization may be grafted intimately and understandingly into these indigenous folkways," was poised to lead the way in bringing this vision to fruition.[77]

Van Dresser's vision was closely aligned with the other group of experts who played an important role in articulating the IIHA's research program: anthropologists. From the outset, IIHA project organizers decided that anthropological knowledge would be crucial to modernizing the Amazon. During the 1947 Belém meeting, participants formed a Committee for Social Sciences and Education (CSSE) that drafted a list of general principles for the IIHA and gave anthropological expertise a prominent role. This emphasis on anthropology is not surprising given the prominent role anthropologists played during the Belém meeting. In addition to Alfred Métraux, who acted as one of the meeting organizers along with E. J. H Corner, the Brazilian anthropologist Heloísa Alberto Torres, who was then director of Brazil's National Museum in Rio, played a prominent role in the meeting. She served as rapporteur for the CSSE and was later elected as president of the Interim Commission charged with establishing the IIHA. During the late 1930s and 1940s, Alberto Torres transformed the National Museum into an important transnational research hub that hosted French, German, and U.S. anthropologists and prepared them for fieldwork with indigenous groups in the Amazon and other remote regions. During their fieldwork, foreign researchers worked closely with young Brazilian scholars in interdisciplinary teams including zoologists, botanists, geographers, and anthropologists, which also helped to foster a new generation of trained Brazilian researchers.[78]

The principles articulated by the CSSE resemble the kinds of work done by the interdisciplinary research teams that Alberto Torres sponsored as well as the intellectual itineraries charted by transnational anthropologists like Alfred Métraux. In fact, the CSSE principles reflect the conflicting projects of "salvaging" and "modernizing" indigenous peoples that informed much of the work done by Alfred Métraux and the social scientists who drafted the 1950 statement on Race. As the first principle that the IIHA should follow, the CSSE recommended obtaining "the technical advice of anthropologists in all aspects of its activities which may reflect directly or indirectly on the life of the natives."[79] In addition to making anthropologists essential to the functioning of the IIHA, the CSSE called for a series of measures designed to foster and protect the ways of life of the Amazon's native inhabitants. For instance, the CSSE called for the IIHA "to study the cause of the depopulation of native centres in the Amazon and the necessary resources to stop its continuation; to foster by all possible means the integrity of native cultures and to limit any interference except for the establishment of hygienic conditions and technical conditions favorable to their development; to promote studies in . . . ethnozoology and native medicine and to stimulate among the Indians the resurgence of native arts and crafts; to study folklore and linguistics and to try to reassert in the mind of the natives the value of their own original culture."[80]

Yet in addition to studying the Amazon basin's native populations and fostering their cultural integrity, the CSSE also drafted principles concerned with modernizing the region. These principles suggested the need for educational and applied social science projects geared toward lessening intercultural tension and for cultivating modern professionals and technicians. For instance, the CSSE argued that it would be important for the IIHA "to contribute to the lessening of tensions in the relationship between groups of different origin and culture and to suggest better means to harmonize such relationships." Although the CSSE declaration did not explicitly state the factors behind these tensions, the subsequent principle hinted at a perceived imbalance in the educational levels between indigenous groups and their caboclo (mixed indigenous and European ancestry) counterparts. Thus, the CSSE called on the IIHA "to recommend to the governments of the Amazonian countries that they provide scholarships to acculturated indians so that they might continue their technical and professional educations." This last statement drafted by the CSSE also pointed to one of the guiding concepts for much of the economic development work done by UNESCO and other UN organizations: acculturation. In fact, the document that outlined the CSSE principles concluded by pointing out that these should be thought of as "general aims," and that the "main purpose" of the Anthropological Department of the IIHA will "be the study of the native populations of the Hylea and the investigation of some of the problems raised by their acculturation."[81]

Other IIHA documents from Métraux's archive also confirm the centrality of "acculturation" as an object of study within the proposed institution and how it functioned as an alternative narrative to the evolutionary internationalism envisioned by Carneiro, Huxley, and Needham. Although these documents do not have an author attributed to them, the fact that they are housed in Métraux's papers and correspond to common themes within his research career suggest that he was the author. These documents also point toward an expansive imperial narrative—similar to the one articulated by Arthur Ramos during his short tenure with UNESCO—that described the history of the modern world as one marked by the spread of European culture throughout the globe. This narrative was one that also described the disappearance of indigenous cultures as the inevitable consequence of the continual expansion of "Western" civilization. And it was with this backdrop of European colonial dominance in mind that anthropologists like Métraux conceptualized "acculturation" as involving the processual study of cultural contacts between Europeans and non-Europeans. For instance, in a document describing the "main objective" of the IIHA's proposed anthropological section, Métraux argued that the immense territories of the Amazon region represent a vast "terra incognita" that poses a large "array of problems and few elements for their solution."[82] The most difficult challenge that this vast region posed to the anthropologist, according to Métraux, was the painstaking empirical task of "recording" the existing cultures before they disappear. Indeed, Métraux stressed that "specialists of South American ethnology have become sadly aware of the rapidity with which Indian cultures disintegrate and collapse" and explained that it was very difficult to "find groups sufficiently large to present a front against the impact of white civilization." It was precisely because of this dynamic, Métraux explained, that "for more than fifty years anthropologists have raised their voices to stress the urgency of field work in South America."[83]

For Métraux, the geographic vastness of the Amazon region coupled with the fragility of indigenous groups in the face of European civilization meant that the region offered "unequaled opportunities for studies in acculturation." In the Amazon, anthropologists could expect to find a wide spectrum of indigenous groups marked by differing stages of adaptation to European culture. Métraux thus described how researchers might encounter tribes "practically unaffected by European civilizations" that offer "a picture of Indian life in its aboriginal setting"; yet Amazonian researchers could also observe other groups who were in contact with either "Indian tribes of a different culture," "Negroes," "assimilated or acculturated Indians," or "are in contact only with missionaries, Indian posts, soldiers, settlers or traders." For Métraux, this typology of cultural contacts on display in the Amazon region offered vast opportunities for comparative studies that could yield "a wealth of data which will be useful to those who will deal with Indians or who will have to take measures on their behalf."[84] Métraux's report thus evoked common imagery of Brazil and South America as a kind of anthropological laboratory. But whereas these tropes typically described Brazil as a laboratory of race mixing, in Métraux's report we see the question of "cultural contacts" and "acculturation" given a central role.

Yet the question of race mixing was not far from Métraux's mind. In addition to the acculturation studies, Alfred Métraux also drafted a research proposal for the IIHA focused on physical anthropology and physiological adaptations to tropical environments. The physical anthropological studies outlined in the report evoked Métraux's experience in the South Pacific and the Bishop Museum during the 1930s. But in contrast to his study of Rapa Nui where he used demographic and anthropometric data for the purpose of reconstructing the island's enigmatic past, Métraux's proposed research program in physical anthropology for the IIHA concerned the question of how to populate the Amazon in the present and future. Like the Bahian Tropicalistas of the nineteenth century who championed rigorous studies of tropical diseases as a way of confronting damaging perceptions of Brazil's future, Métraux's research proposal also sought to confront long-standing perceptions that the tropics were unsuited to white settlement. His proposal thus emphasized that the Amazon region offers an ideal site for studying questions of human adaptation to tropical conditions and stressed the importance of scientific knowledge of the region for future colonization. For instance, Métraux argued that in the Amazon region, one could find, "within a short radius," groups belonging to "numerous racial and cultural types"; these conditions were thus ideal, continued Métraux, for conducting "comparative studies" of "general interest economically" and within a "relatively short period of time."[85] "[I]f this region of South America is to be opened to colonization," Métraux argued, physiological studies would allow "science [to] help in the expeditions an[d] healthful settlement of the colonists." By so doing, Métraux argued, "many otherwise unavoidable errors could be foreseen and overcome."[86]

Métraux also insisted that his proposed physical anthropology program would be closely tied to sociological studies of Amazonian towns. Métraux described how the IIHA's research program for 1948 included a proposed study of a typical Amazonian town and explained that studies of "existing communities would be of the greatest importance" insofar as they offer a "full understanding of the social life and health conditions of representative human groups" and allow researchers to create a portrait "of [the] present conditions and future needs in the area." Echoing the team surveys of Polynesia he encountered at the Bishop Museum, Métraux argued that in addition to these community studies, the IIHA should promote physical anthropological studies of Amazonian populations conducted by teams made up of "one or two cultural anthropologists, one or two social psychologists, a rural sociologist and one or two physical anthropologists." In a passage that evoked the kind of work done by Te Rangi Hīroa, Métraux also suggested that, ideally, one of the physical anthropologists on these teams should be medically trained and that one of the members of the team should also be "familiar with the broader problems of food habits and nutrition."[87] Through its emphasis on interdisciplinary team research and the importance of medical perspectives, Métraux's template for physical anthropological research in the Amazon not only harkened back to his training in France but also to his transformative period in the South Pacific.

In this research proposal, Métraux was also careful to characterize the discipline of physical anthropology as having gone through an antiracist turn. "Today physical anthropology is no longer concerned exclusively with anthropometric measurements and the description of human groups," Métraux explained. Instead of its prior obsession with cranial measurements and comparative anatomy, Métraux argued that physical anthropology was in fact increasingly drawing upon methods from the fields of physiology and medicine. Physical anthropology was now focused, Métraux argued, on "physiological functions and race" and on objects such as "metabolic rates, energy output, susceptibility to disease, hormone balance and the activity of ductless glands, nervous excitability, perceptions, etc." In particular, Métraux flagged "the establishment of precise energy ratings of the various groups within the Amazon basin and the determination of the influence of climate and diet on the energy quotient" and "the influence of climate on metabolism" as pressing research problems in need of scientific attention.[88]

The interdisciplinary research program that Métraux sketched out for the IIHA also promised to rectify many of the pitfalls of previous studies concerning race and heredity as well as divisions between the social and life sciences. Because it encouraged close cooperation between social anthropologists, physical anthropologists, and those with medical training, the proposed research augured benefits to both social and life scientists. While social scientists stood to gain greater awareness of the health problems faced by indigenous groups, physical anthropologists would benefit by gaining access to "cultural data which are too often neglected because of a lack of time and training." Through cooperation between these two branches of the human sciences, Métraux argued, "many misconceptions currently existing about race and heredity could be dispelled and many of the possible factors could be taken account of in a realistic manner."[89] From the expansive southern standpoint that he acquired through his wanderings during the interwar period, Métraux thus signaled the possibility of a reformed antiracist science of race.

Other documents from Métraux's archive reveal how the IIHA organizers' joint interest in tropical physiology and the social organization of Amazonian towns was tied to a broader economic project of populating the region and harnessing its abundant natural resources. In fact, other documents show how IIHA organizers conceptualized "populating" the Amazon not simply as a project of increasing the region's human population but also as a project of establishing the cultural and material hallmarks of Western civilization. For example, one document suggested that the Amazon forest is "suited for increased production and population" given that jungle products such as rubber are abundant and that there is a large amount of land suitable for cultivation. Although the report describes the Amazon as brimming with natural resources and economic potential, it also suggests that the project of populating the Amazon is complicated by racial factors. A "pure white culture or race [is] unsuited to [this] environment" the report explains. Indeed, the report notes that "white colonization on permanent self-supporting basis" has never been "successful in this or analogous regions."[90] Yet, although it ruled out the possibility of a "pure white" settlement of the Amazon, the report also ruled out the region's indigenous inhabitants as candidates for populating the region. At 300,000–500,000 people, the report observes, the "Indian Population" was too small and "culturally too different from Western Civilization to expect complete eventual conversion." According to the logic of the report, the "Indian" population's incommensurability with Western civilization rendered it unsuitable for "populating" the region.

The report's implied meaning of "populating" the Amazon was thus creating a population base that would perpetuate Western civilization in a tropical environment. Yet like many of the southern racial projects described so far in this book, the report did not see pure whiteness, in the phenotypic sense, as crucial to this project. Rather, the report conceptualized populating the Amazon as an open question and asked "what physical types can form [the] basis of [the Amazon's] population, labor force, consuming market, etc.?" And the report's response to this question shows how southern experts favored the production of racial hybridity as a solution to the quandaries of tropical settlement. In fact, the report hypothesized that "a mixed population: Indian-white, negro-white; Indian-negro-white" was potentially best suited for populating the region and that a mixed population of this kind "would combine Indian or negro physical adaptation to tropical environment with nervous energy of whites." Yet the report was careful to present this as a hypothesis and emphasized that "studies [are] needed to settle this point."[91]

The report's favorable portrayal of race mixing should not be confused with a project that recognized the culture of each of these purported races as having equal value. In the same way that the report argued that no single race was suited to populating the Amazon, it also posited that the region's development depended on a successful blending of European and non-European cultures. Envisioning a future where the Amazon was thoroughly populated by a mixed-race population that fruitfully combined the "nervous energy of whites" with the tropical adaptation of "Indians" and "Negroes," the report asked "what kind of culture can these people have which combines cultural adaptation to environment together with participation on a self-supporting basis with white civilization?"[92] Like the previous question about the physical types suited to Amazonian settlements, this question reveals the Eurocentric logic behind the IIHA's purported internationalism. The non-European elements of the envisioned syncretic culture are coded here as the raw materials that provide adaptations to the environment. Yet the more important elements, the question implies, are the European elements that will allow for participation with "white civilization." The hypothesis offered in response to this question is even more revealing. Indeed, the report suggests that the best culture that the envisioned mixed-race inhabitants could have would similarly be "a mixed culture," specifically one that combines "Indian subsistence techniques with European economic and political values and procedures," although the report concedes studies are "needed to settle this point." Once studies settle these questions about what racial and cultural types are best suited to the Amazon region, the reported claimed, IIHA organizers would then be able to "go ahead with intelligent planning for its development economically, politically, intellectually, culturally, or what you will."[93]

ACCULTURATION IN AN AMAZON TOWN

Because of the IIHA's eventual demise, the ambitious anthropological research program that Métraux sketched out did not materialize. Yet, as we saw in the previous section, two major anthropological studies were completed in Brazil and Peru as part of the preparatory work for the IIHA. Although the IIHA collapsed, the anthropological studies that survived the failed institution, especially the Brazilian community study conducted by a research team led by the young U.S. anthropologist Charles Wagley, illustrate how antiracist projects of applied social science imagined a future for the Amazon region where its indigenous groups were fully absorbed into the economic life of their nation and region. For Wagley, the anthropological survey he conducted under the IIHA's auspices formed the basis for his monograph *Amazon Town: A Study of Man in the Tropics* (1953), which became a landmark Amazonian ethnography that served to cement his status as a Brazilianist. And as we will see later in this book, Wagley's IIHA fieldwork also paved the way for his later involvement with UNESCO's landmark studies on Brazilian race relations. Wagley's research trajectory before and after World War II thus offers another example of the ways that mid-century antiracism was shaped by the intellectual and political issues of the Southern Hemisphere.

In Wagley's work, as in the anthropological research proposals sketched out by Métraux, "acculturation" was the concept most frequently invoked to describe the social transformations taking place in the Amazon region. Yet in IIHA studies, "acculturation" also began to acquire a new meaning aligned with visions of modernizing the tropics. In chapter 1, we saw how the concept of "acculturation" was central to Arthur Ramos's ambition to reformulate Nina Rodrigues's research on Afro-Brazilians along antiracist lines. In the work of Ramos and other researchers concerned with African survival in the Americas, "acculturation" indexed attempts to map the elements of African culture that survived the transatlantic passage and the cultural changes precipitated by enforced slavery and contact with other groups. By contrast, in Wagley's work and in IIHA research proposals, "acculturation" acquired meanings associated with scientific efforts to alter (and efface) indigenous cultures. To understand how "acculturation" figured in Wagley's IIHA work, we have to go back to late 1930s and observe Wagley's transformation into a Brazilianist.

Raised in a poor family in Texas, Wagley completed his doctoral training in anthropology at Columbia University in 1941 and worked with Franz Boas, Ruth Benedict, and Ralph Linton.[94] For his doctoral dissertation, Wagley conducted an ethnography of a Mayan town in Guatemala named Santiago Chimaltengo and studied the town's economy, religious orientation, social organization, and interethnic relations. After completing his doctoral fieldwork in 1939 at the tender age of twenty-five, Wagley decided to pursue further research in South

America (instead of looking for work in what was then a poor job market) even though his ethnographic knowledge of the region was "scanty." As he was looking for a new research direction, his mentor Ralph Linton along with other notable Latin Americanists—notably Robert Redfield and Melville Herskovits—were promoting "acculturation" as a new anthropological research agenda focused on studying the cultural changes that arise when individuals from different cultures come into continuous firsthand contact. In 1936, Linton and his collaborators published a memorandum, based on research funded by the Social Sciences Research Council in the United States, that proposed a basic framework for this research agenda, and in the following years Linton organized a series of workshops and publications that positioned acculturation as a central research topic in U.S. anthropology.[95] In theory, acculturation encompassed contacts between all cultures. Yet it was often defined as the study of the "impact of continuous contact of western civilization on primitive peoples."[96] And it was this framing of acculturation that Wagley found most appealing.

At this point in his career, Wagley had never been to Brazil and did not speak Portuguese. Brazil did not seem to be a likely research site. It was only through Ralph Linton and Alfred Métraux's guidance that Brazil became Wagley's chosen main research area. Yet when he first traveled to Brazil, he did not travel on the pretense of becoming a Brazilianist. Rather, Wagley traveled to Brazil with the narrower goal of conducting an "early-contact acculturation study." In theory, such a study entailed observing a group that had recently come into contact with "western civilization." Linton endorsed this proposed research and quickly found funds for the project. Wagley then had to find himself an "auspicious locale."[97] While searching for this locale, Wagley met and befriended Alfred Métraux at a research seminar at Yale where Métraux was about to join the anthropology department for the year as "Bishop Museum Professor." Through this fortuitous meeting, Wagley found someone with extensive networks in the field of Americanist anthropology.

Métraux's regional knowledge proved pivotal as Wagley searched for a site to conduct his first acculturation study. In fact, it was Métraux who convinced Wagley to study the Tapirapé "Indians" of central Brazil because his mentor—Erland von Nordenskiöld—believed them to be "the remnants of the now extinct Tupinambá."[98] The Tapirapé were an isolated and little-studied group that had only recently been "contacted" by Brazilian settlers. As such, they seemed an ideal group for the "early-contact acculturation" study that Wagley discussed with Linton.[99] Wagley also learned from William Lipkind—a Columbia PhD student who was in Brazil doing fieldwork with the Carajá—that the Tapirapé had recently been seen by missionaries and that they lived in several villages. Métraux's grounding in Tupinology thus provided Wagley with the clues to his first Brazilian research site.

Intrigued by the prospect of studying the remnants of an "extinct" group, Wagley set sail for Brazil in January 1939 aboard the SS *Argentina*

FIGURE 4.1 Alfred Métraux aboard the SS *Argentina* where he prepared Wagley for fieldwork in Brazil (photograph, Charles Wagley Papers, University Archives, Special and Area Studies Collections, George A. Smathers Libraries, University of Florida, Gainesville, FL).

where he spent twelve days accompanied by Métraux before reaching Rio de Janeiro. During this trip, Métraux regaled Wagley with stories from his fieldwork and briefed him "on conditions in the field and on people whom I would meet and upon whom I would depend."[100] Métraux also gave Wagley "letters of introduction to several people" and introduced the young Wagley to his colleagues at the National Museum in Rio, which became his "research base for many years to come."[101]

The National Museum offered Wagley training in Brazilian anthropology that was not available at Columbia. When he arrived, the museum had an informal agreement with Columbia to "co-sponsor ethnological studies in Brazil." The agreement was secured by Heloísa Alberto Torres—the National Museum's "director and patroness"—who corresponded with Franz Boas and requested that young scholars like Wagley be sent to Brazil. As soon as he arrived in Rio, Alberto Torres extended a warm welcome to Wagley and became his mentor. According to Wagley, Alberto Torres "felt a personal obligation" toward visiting scholars and used her "great prestige and wide network of friends" to guide visitors through "the intricate bureaucracy, which called for registration of aliens, permission to carry out a scientific expedition, and various official papers."[102] In addition to mundane administrative support, Alberto Torres used the museum's collections to teach "Brazilian ethnology" to visiting scholars like Wagley and pointed them to books that were not available in New York. Most important, though, "Dona Heloísa" taught visiting anthropologists "proper Brazilian manners" and the "wonders of her country." Before traveling to the Tapirapé, Wagley spent a month at the museum soaking up everything "Dona Heloísa" could teach him and reading, listening, and studying Portuguese.

After his crash course in Brazilian anthropology, Wagley traveled to central Brazil and spent fifteen months conducting an acculturation study with the Tapirapé. The Tapirapé did not prove to be ideal subjects for studying the process of acculturation. Acculturation studies were supposed to reveal how the value systems and cultures of groups change through continuous firsthand contact. Instead, Wagley's study revealed that even fleeting and indirect contact with the Western world had devastating effects on the Tapirapé.[103] In one of the few articles he published after this first taste of Brazilian fieldwork, Wagley observed that the Tapirapé contracted diseases—such as smallpox, yellow fever,

FIGURE 4.2 Charles Wagley at the National Museum in Rio. From left to right: Claude Lévi-Strauss; Ruth Landes of Columbia University; Charles Walter Wagley of Columbia University; Heloísa Alberto Torres of the National Museum; Luís de Castro Faria of the National Museum; Raimundo Lopes da Cunha of the National Museum; Edison Carneiro (Wikimedia Commons).

and a strain of influenza from the worldwide outbreak in 1920—after their first brief encounters with European missionaries and Brazilian hunters during the first half of the twentieth century. The introduction of these diseases proved devastating and led to the demise of four of the five villages that existed. As a result of this "wholesale depopulation from disease," the Tapirapé experienced major changes to their social organization and cultural practices. For instance, once confined to one village, all the customs attached to "inter-village relations" disappeared, and ancient inter-village antagonism was reduced to occasional taunts. Instead of offering insights on acculturation, Wagley's Tapirapé research showed how even an ostensibly remote and isolated group was seriously affected by global forces at play within a region.

During World War II, Wagley continued doing "acculturation" studies with Brazil's indigenous groups as well as public health and "applied medical anthropolog[ical]" work in the Amazon through the Serviço

FIGURE 4.3 Charles Wagley at work in his hammock during his first trip to the Tapirapé (photograph, Charles Wagley Papers, University Archives, Special and Area Studies Collections, George A. Smathers Libraries, University of Florida, Gainesville, FL).

Especial de Saúde Publica (SESP)—a joint Brazil-U.S. program created during the war.[104] By the end of World War II, when he began teaching at Columbia University, Wagley had acquired considerable regional expertise. During the 1940s, Wagley had incrementally increased the geographic scale of his research, beginning first with a small Tapirapé community and then expanding to cover the Amazon region. His studies considered how the interactions between diverse social groups yielded social change. Wagley's experience with multiple field sites and his transnational collaborations, with Alfred Métraux and with Brazilian

FIGURE 4.4 Charles Wagley getting face-painted during Tapirapé fieldwork (photograph, Charles Wagley Papers, University Archives, Special and Area Studies Collections, George A. Smathers Libraries, University of Florida, Gainesville, FL).

scholars like Dona Heloísa, gave him an ideal standpoint from which to claim regional knowledge. Although at the onset of World War II he was an inexperienced, poorly connected, and vulnerable novice, by its end he had become an accomplished and respected Tupinologist and Amazonian specialist.[105]

Although it failed, the IIHA provided Wagley with an opportunity to cement his standing as an expert on processes of cultural adaptation. In 1948, UNESCO recruited Wagley to conduct one of the IIHA's proposed "social surveys," which ended up providing him with the source material for *Amazon Town*. For this survey, Wagley spent four months living in Itá, an Amazonian town he called Gurupá in order to protect its identity. While there, Wagley rented a local house and formed a research team with Cecilia Wagley, his wife, and Eduardo Galvão, his mentee, and his mentee's wife, Clara Galvão. During their stay, the two couples immersed themselves in the community and made detailed observations of daily life.

Participant observation was one of several methods used by Wagley's research team. Reporting to UNESCO, Wagley described that his team relied on "direct observation, participation in local affairs, and long interviews with numerous members of the community." They also collected "considerable statistical data" on "the cost of living, on property, on exports and imports, on agricultural holdings, on nutrition, and on many other aspects of Gurupá social [and economic] life."[106] Wagley also stressed that direct observation coupled with "detailed explanations of a local expert" were crucial to capturing the "normal daily pursuits" of the community, including "techniques of agriculture, of food preparation, of rubber collecting, of house building and of fishing."[107]

Wagley's UNESCO report also described interdisciplinary cooperation as vital for future studies of "the present cultural adaptations of man to the Amazon region." In fact, Wagley's report argued that cultural adaptation could serve as a "central orienting problem" for the IIHA, which would facilitate cooperation between experts from the natural and social sciences.[108] With a nod to an SSRC conference on area studies from 1947, Wagley added that "many of the same problems involved in the planning and execution of the so-called area studies" could guide the IIHA, supported by citations to the U.S. geographer and Japanologist Robert B. Hall's pamphlet, *Area Studies: With Special Reference to Their Implications for Research in the Social Sciences* (1947), as well as his own report.[109]

After completing his ethnographic survey for UNESCO, Wagley adapted it into his book *Amazon Town*. Wagley's *Amazon Town* is a striking example of how southern conceptions of race were adapted to the modernizing imperatives of southern elites. *Amazon Town* represents a blend of the Lusotropical philosophy promulgated by Gilberto Freyre and the anti-typological agenda of Columbian anthropology. Yet it must be understood in the context of the "march to the west"—the ambitious program for modernizing the Amazon pushed forward by the Vargas regime.

In the 1940s when Wagley conducted his formative Brazilian research, political and intellectual elites in the country's industrial states (São Paulo, Minas Gerais, and Rio Grande do Sul) began reassessing their attitude toward the Amazon. The Amazon had long been viewed by Brazil's coastal elite as a national burden that conjured images of untamed wilderness and uncontacted "savages." Yet during the Estado Novo period (1937–1945) of Getulio Vargas's authoritarian regime, Brazil's industrial elites began viewing the region through a nationalist lens and conceptualized its vast and untapped resources as having the potential to unlock a prosperous future for the nation.[110] When Vargas centralized political control and formed the Estado Novo in 1937, he also unveiled an ambitious state-led development project to colonize and develop the Amazon known as the March to the West. As part of this project, Vargas's government created a series of agencies whose purpose was to populate and economically develop the Amazon by distributing plots of land to farmers who were supported with technical assistance from government experts.[111] Scientists from these agencies conducted epidemiological surveys of the region as well as research on Amazonian flora and soils and on methods for cultivating fungus-resistant rubber trees that produced high yields.[112] When Wagley conducted his first Brazlian anthropological studies in the Amazon, the Vargas regime was thus in the midst of an ambitious campaign to develop and unlock the Amazon region's economic potential.

It was in this developmentalist context that Wagley produced *Amazon Town*. In both his UNESCO report and *Amazon Town*, Wagley evoked the Lusotropicalist ideology of Brazilian elites. During the 1940s, the writings of the sociologist Gilberto Freyre became an important source for Brazilian social scientists and intellectuals, who argued that Brazil's system of race relations could be labeled a "racial democracy" because of its comparative lack of racial prejudice. Freyre's writings—especially his major study *Casa Grande e Senzala* (1933)—served to justify this idea of a Brazilian "racial democracy" by arguing that, compared to other European colonizers, the Portuguese were remarkably benign and tolerant toward other races. Indeed, Freyre argued that the Portuguese were more tolerant toward darker-skinned peoples because of previous contact with the Moors in the Iberian Peninsula. He also described the Portuguese approach to colonization as "luso-tropical" and argued that it involved a deliberate attempt to create hybrid and tropically adapted societies by mixing freely with natives and creating racially mixed populations. Wagley framed much of his analysis in *Amazon Town* in similarly "luso-tropical" terms. For instance, Wagley argued that the local community he studied served as a microcosm of a broader pattern of cultural adaptation to the tropical conditions of the Amazon region. Wagley also described his study as a study of a region— namely the Brazilian Amazon—where "a distinctive tropical way of life has been formed by the fusion of American Indian and Portuguese cultures during the last three centuries."[113]

Wagley's book also described the Amazon region as one that was difficult to develop. As such, Wagley described his work as an important case study of "the adaptation of man to a tropical environment" and analysis of a "backward" and underdeveloped area.[114] To illustrate the region's backwardness, Wagley offered data that purportedly demonstrated the region's "wretched health conditions," "deplorably low standards of living," and the inhabitants' "lack of industry."[115] For instance, Wagley described how 60 percent of the Brazilian Amazon's inhabitants were illiterate, how infant mortality rates in the main cities were as a high as 108 per 1,000 or even 303 per 1,000, and how death rates from infectious diseases such as typhoid, tuberculosis, and malaria were incredibly high. Wagley also observed that most families failed to meet the recommended caloric intake of 2,400 calories per day, and that "dietary deficiency and even semi-starvation" were common throughout the valley. As evidence of the region's "primitive and stagnant" economic life, Wagley described the persistence of "slash-and-burn" agricultural techniques "inherited from the native Indians," a network of transportation and commerce that took place along the river by "slow river boats," and a paltry or nonexistent infrastructure for things such as "waste disposal, electric lights, and water supply."[116]

Yet staying within the frame of his Boasian training and Lusotropical ideology, Wagley insisted that the racial background of the Amazon's inhabitants had nothing to do with the region's economic stagnation. In a survey of scientific literature on the Amazon, Wagley identified two distinct camps of authors: those who considered the Amazon a "green hell" where civilization was doomed to decay because of the forbidding climate; and those who celebrated the region as a "green opium" whose alluring potential gives rise to grandiose dreams of "great cities, rich agricultural lands, and thriving industries."[117] As exemplars of the "green hell" camp, Wagley cited "tropical racists" like the Yale geographer Ellsworth Huntington, who argued in his 1915 book *Civilization and Climate* that tropical environments lead to laziness, drunkenness, irascible temper, and sexual indulgence and that tropical climates could only be suitably inhabited by dark-skinned races. According to "tropical racists," Wagley explained, the Amazon and other tropical regions were inevitably doomed to "a lower level of cultural development."[118]

As a rebuttal to tropical racism, Wagley offered an argument that evoked Freyre's Lusotropicalism and the UNESCO Statements on Race. Wagley thus noted that "all research to date in anthropology and related sciences has shown that all racial groups and all mixtures of racial groups have the same capacity for cultural achievement" and cautioned that those "who warn of the dangers of race mixture and racial barriers to cultural development . . . are simply making dangerous social propaganda."[119] In contrast to the geographic determinism of tropical racists, Wagley argued that humans are the most adaptable of all animals to varying temperatures because of their "warm-bloodedness" and use of "cultural devices," such as clothing and fire, and the modification of their practices and habits. Although evidence existed that tropical climates did produce biological effects such as lower stature stemming from slower physiological growth and development, Wagley was quick to point out that "neither height nor body weight has any correlation with mental capacity and cultural achievement."[120] Instead of looking to the racial makeup of the population or the tropical environment for clues as to the Amazon's backwardness, Wagley argued that reasons had to be sought in "Amazon culture and society, and in the relationship of this region with the centers of economic and political power and with the sources of cultural diffusion."

Ultimately, *Amazon Town* was a celebration of the plasticity and adaptability of human beings through cultural and technological innovation and of the Amazon region's future economic and societal potential. Yet in celebrating the Amazon's vast potential, Wagley's analysis also emphasized Brazil's own technocratic ambitions for the region. To this end, Wagley quoted at length from Getulio Vargas's famous 1940 "Amazon Address" in order to demonstrate how Brazilians imagined the internal development of their own country through the "full extension of modern technological skills and knowledge to an under-developed area."[121] In this address, Vargas boldly declared that "nothing will prevent us from accomplishing in this spurt of effort which is the twentieth century the highest task of civilized man: the conquest and the domination of great valleys of equatorial torrents, transforming their blind force and extraordinary fertility into disciplined energy. The Amazon, with the fecund impulse of our will, of our effort, and of our work, will not remain simply a chapter in the history of the earth but, on the same basis as other great river systems, will become a chapter in the history of civilization."[122]

Though he could not match Vargas's bombast, Wagley outlined a similar vision for the Amazon. Wagley described the Amazon as an "immense area" with "enormous potentiality" that remains "not only an important Brazilian frontier but also a New World frontier." Given its relative isolation and few inhabitants, Wagley also speculated that "the development of such a region with modern technology is one answer to the horrid specter offered by neo-Malthusians who warn of the exhaustion of the world's food supply in the face of a growing population." Yet in order for this "conquest" of the Amazon valley to succeed, Wagley argued that it would require the adaptation of existing technologies to local Amazonian realities. Indeed, Wagley argued, many existing modern technologies were designed with the aim of "controlling and exploiting the Temperate Zone."[123] But in order to exploit the full potential of the Amazon, Wagley suggested that "an adaptation of our Temperate Zone way of life" was necessary and urged that "a whole new field of applied science must be developed for tropical conditions."[124] Though he did not elaborate on what such applied science would look like, the fact that he stressed the geographic specificity of the Amazon and the need to adapt to its regional demands speaks not only to his maturation as an area expert but also to his indebtedness to the Lusotropical and developmentalist discourses of Brazil's elites.

The IIHA's emphasis on international cooperation led to its demise. Though in 1948 Brazil's president Gaspar Dutra recommend that Brazil's National Congress approve the Iquitos convention, the proposal ended up triggering fierce opposition from a diverse range of actors led by Brazil's former president and staunch nationalist Arthur Bernardes. Though they didn't necessarily oppose the basic goals of the IIHA, Bernardes and his allies (which included Brazilian military officers, members of Brazil's Communist Party, economists, and poets) vociferously argued that an *international* laboratory posed a major threat to Brazil's territorial and economic sovereignty.[125]

Although UNESCO officials imagined the IIHA as a cosmopolitan initiative designed to foster international cooperation, Brazilian politicians saw the project as an imperialist exercise and ultimately stopped it from being approved by Brazil's National Congress. The internationalist ideals that buoyed Carneiro, Huxley, and Needham's ambitions for the IIHA proved out of sync with the nationalist interests of a rapidly industrializing Brazil, whose politicians sought to expand territorial and economic control into the Amazonian region so as to keep up with the demands of a rapidly growing urban population.[126] The project also proved to be out of sync with emerging Cold War politics and in particular with the strategic interests of a U.S. state that increasingly saw Latin America as a strategic region in its rivalry with the USSR. The failure of the IIHA highlights the limitations of UNESCO projects in this period, which were narrowly focused on cultural transformation as a vehicle of societal change and proved to be inadequate for contending with the conflicting economic and political imperatives of this period.

Indeed, the IIHA met with significant resistance from Brazilian politicians when Carneiro originally proposed the project in 1945 to the Brazilian Ministry of Agriculture. In his original proposal, Carneiro envisioned an international foundation under the care of the French and Brazilian governments, with two research centers in the Amazonian towns of Belém and Cayanne and a strong emphasis on the systematic study of the indigenous populations of the region. However, the project was rejected by the Ministry of Agriculture, whose director, Felisberto

Camargo, likened it to the kinds of projects carried out by scientists of the colonial era such as Alexander von Humboldt.[127]

Even after Carneiro took his proposal to UNESCO and had it accepted during the 1946 General Conference, the project was met with resistance particularly from the United States. During UNESCO's second General Conference in Mexico in 1947, American diplomats cautioned against moving the project forward too quickly. Later that year, the U.S. national commission declared it would withdraw financial support for the project. Although the reasons behind this decision are unclear, Joseph Needham suggested that it was because the United States wanted to protect its regional interests in Latin America and because it had decided to reduce the budget of projects not directly related to meeting its Cold War objectives.[128]

The project, and UNESCO's handling of it, also drew sustained criticism from E. J. H. Corner, the British botanist who was appointed to act as UNESCO's Latin American field officer and given significant responsibility for organizing the project on the ground. Before a conference was held in Iquitos, Peru, in 1948 for the purpose of drafting a convention for the IIHA, Corner sent a letter to UNESCO officials expressing his concerns with the direction of the project. Corner worried that an international institution within the boundaries of sovereign states had the potential to be seen as a foreign intrusion and a form of scientific imperialism, and hence he urged UNESCO to make sure that the founding documents for the institute clearly stated how it will collaborate with local institutions and how it will incorporate local scientists into its program.

Despite Corner's concerns about the language of the Iquitos conference, the delegates of the conference ratified the proposed convention for the institute. Subsequently, Paulo Carneiro took the Iquitos convention to the Brazilian president and National Congress and asked for its approval. Carneiro's congressional proposal was strongly opposed by Brazilian politicians and members of the Brazilian elite who raised concerns about the imperial dimensions of UNESCO and the IIHA project. The most vocal critic was the former president of Brazil, Arthur Bernardes, who was then a deputy in Brazil's Chamber of Deputies. Bernardes argued that the IIHA opened the door to imperial domination of the Amazon hidden under the pretext of scientific and economic objectives. In order to shore up nationalist opposition to IIHA, Bernardes sought the advice of the Instituto Brasileiro de Geopolitica, which published a report condemning the proposed institute and described it as a threat to national sovereignty and as an attempt to colonize Brazil. Further, the report suggested that the task of "colonizing" the Amazon should be left to Brazilians and stressed that Brazil was sufficiently equipped to do so on its own.[129] In the end, Bernardes's concerns proved decisive, and the Brazilian government withdrew support for UNESCO's initiative. Without the support of the Brazilian state, UNESCO was unable to bring the project to fruition.

Although the IIHA failed, it also served to crystallize a model for international cooperation that was later used for other regions in South America. Indeed, as the next chapter shows, the UN system adopted a similar international approach a few years later for an economic development project that sought to transform the Andes region.

CONCLUSION

For natural scientists like Carneiro and Huxley, the question of how to populate—and develop—the Amazon was linked to a broader Malthusian issue concerning the survival of the human species and the interdependence between nation-states. Yet in contrast to the IIHA's natural scientists who envisioned modernization in evolutionary and Malthusian terms, the social scientists involved with the IIHA conceptualized Amazonian modernization as a process of biological and cultural adaptation, or, rather, acculturation to the tropics. The IIHA's proposed agenda of social science research reframed the long-standing issue of how to populate the Amazon region through an antiracist knowledge regime that emphasized processes of "acculturation" and "physiological adaptation" that were heavily studied by anthropologists during the 1930s and 1940s. Through a research agenda that included physical anthropological studies and ethnographic surveys, IIHA social scientists approached the modernization of the Amazon as a process of racial adaptation and technological acculturation to a distinctive tropical environment. By emphasizing that both race and culture are malleable, IIHA social scientists thus confronted the geographic determinism and "tropicality" shaping many of the scientific discourses about the Amazon and its people that they inherited. Through lofty assessments of the Amazon region's economic and cultural potential, the IIHA's social scientists joined Brazilian and U.S. political elites in their belief that a prosperous future could be unlocked for Brazil and South America by modernizing an area that many thought of as a "green hell." Although Brazilian, U.S., and international interest in the economic modernization of the Amazon intensified during World War II, the research questions that the IIHA's social scientific program mobilized harkened back to many of the basic problems that prominent Brazilian researchers like Arthur Ramos inherited from turn-of-the-century debates about Brazil's racial heritage. Like Nina Rodrigues and the Tropicalistas from the late nineteenth century, one of the central issues that the IIHA's anthropological research program tackled was whether all races were equally suited to tropical life and the degree to which modernity could flourish in the unique environmental conditions of the tropics. Yet while Ramos articulated a research program centered on African survival and the fusion between European and African races, the social scientists involved with the IIHA—notably Alfred Métraux and his protégé Charles Wagley—were situated within a tradition of research concerned with tracking the history of the indigenous inhabitants of Brazil and South America. Indeed, both Métraux and Wagley approached the study of indigenous "acculturation" as one that required detailed knowledge of the history of contact between indigenous groups and Europeans since the Iberian conquest. And as we will see in the next chapter, despite the struggles of the IIHA, the frameworks of scientific internationalism and acculturation as mechanisms of development would find a new life in the Andes region during the 1950s and continue to be haunted by the remnants of race science.

CHAPTER FIVE

"Peasants Without Land"

Race and Indigeneity in the ILO's Puno-Tambopata Project

"If the plan now under consideration does bear fruit," wrote Alfred Métraux in 1955 for the *UNESCO Courier*, "its first result will be to bring to the surface the latent energy, sobriety and courage of the Andean Indian."[1] The plan in question was an initiative, sponsored by Peru's military-backed government and the International Labor Organization (ILO), to colonize the Tambopata Valley with "Aymara Indians from the overcrowded banks of Lake Titicaca where population density may reach 150 persons per square mile." In Métraux's estimation, this scheme could ease the "Land Hunger" afflicting the Titicaca region and accelerate the economic development of Aymara Indians by furnishing them with fertile agricultural terrain. In Bolivia, the revolutionary government of Victor Paz Estenssoro, which had adopted an ambitious program of agrarian reform, implemented a similar scheme with ILO assistance. Yet in Peru, Métraux lamented, the colonization of the Tambopata Valley—a semitropical region near the Amazon basin—provoked controversy because it had become "an accepted belief—almost a dogma—that life in the warm lowlands is fatal to Andean Indians."[2]

According to Métraux, the main scientific proponent of these dogmatic beliefs was Dr. Carlos Monge Medrano—the "famous Peruvian physiologist" who had devoted a "lifetime to proving that the lungs and other organs of Indians living [at high altitude] have been modified by evolution and adapted to rarified air."[3] While Métraux wagered that the resettlement project would unleash the latent economic potential of Lake Titicaca's indigenous inhabitants, Monge warned that the Tambopata Valley's tropical climate could unsettle Andean highlanders' delicate state of "racial acclimatization" and disrupt ancient biosocial customs that dated back to the Inca Empire. If transplanted to the tropics, Monge cautioned, Andean Indians were prone to suffer from "tropical fatigue" and were likely to contract "particularly acute forms of pulmonary tuberculosis."[4] Andean Indians' economic development would be better served, Monge insisted, by keeping them in the highlands.

By invoking the physiological specificity of Andean Indians and their right to remain in the highlands, Monge echoed the work of interwar *indigenistas*, such as Luis Valcarcel and Jose Mariategui, who championed Andean Indians and the Inca Empire in opposition to intellectual elites from Lima who viewed Andean highlanders as racially degenerate. Yet from the vantage of UNESCO's race campaign that Metraux had directed since 1950, Monge's belief in a distinct Andean racial type struck a discordant note. Instead of denouncing racial typologies as the 1950 and 1951 UNESCO Statements on Race had done, Monge's arguments seemed to reintroduce new physiological ones and thus revive the very race science that UNESCO experts were attempting to bury.

This chapter examines this contrast between Métraux and Monge to understand how Cold War development discourse interacted with racial and eugenic logic after World War II. On the face of it, Métraux and Monge's differences suggest divergent strands of racial thought—one invested in the malleability of indigenous bodies and cultures and the other in their fixity and conservation. Yet despite this dissimilarity, Métraux and Monge shared the conviction that the cultural and economic integration of indigenous highlanders into Peruvian society was necessary for modernizing the Andes region. In fact, for all his concern with the ILO's plan, Monge heartily endorsed modernizing indigenous Peruvians and in fact served as co-director of the ambitious development project known as the Cornell-Peru Project (CPP). Lasting from 1952 to 1966 and framed as a bold experiment in "participant intervention," the

CPP was jointly run by U.S. and Peruvian anthropologists who purchased an hacienda named Vicos on behalf of Cornell University and became the estate's de facto bosses who assumed control over the indigenous peasants that were contractually bound to the land.[5] Though they disagreed on the existence of racial types, Métraux and Monge shared paternalistic attitudes toward indigenous groups in the Andean highlands and framed them as populations devastated by colonial conquest and suffering from chronic poverty, poor health, and illiteracy. As such, they both believed that concepts in the human sciences such as "acculturation" offered tools for rehabilitating indigenous peoples and negotiating their inclusion into industrial society and modern agricultural economies. Their shared views thus illustrate how Cold War development discourse accommodated racial and eugenic logics from Latin America and the Global South for the purpose of indigenous improvement.

In the past twenty years, scholars have transformed modernization theory and economic development into objects of historical inquiry as evidenced by an expansive literature from numerous fields.[6] A key concern of these scholars is historicizing development theories and practices as products of Cold War ideologies.[7] Influential studies have focused on how U.S.-based social scientists crafted modernization theories that became ideological instruments of U.S. foreign policy. This literature has also generated important debates about the degree to which the content of the social sciences during the Cold War, particularly in the United States, was dictated by geopolitical interests.[8] Yet despite this interest in mid-twentieth-century developments, eugenics and race science are topics the field has not typically engaged.[9] Implicit in this silence is the once canonical North Atlantic narrative of race science, which is structured by two assumptions: (1) that race science is a relic of the nineteenth century concerned with fixed biological differences, and (2) that rigid—"typological"—conceptions of race were purged from the social sciences and human biology after World War II. Given that Cold War development discourse was primarily concerned with how to engineer social change, the static conceptions of race described in the retreat narrative appear to be an incommensurable epistemological formation.

Yet the retreat of race science and eugenics appears quite different from the perspective of Latin America and the Global South.[10] In her pathbreaking book *"The Hour of Eugenics": Race, Gender, and Nation in Latin America*, Nancy Stepan argued that, unlike in North America, eugenics suffered "no sudden rupture in Latin America" and remained "true to type" after World War II. In North America, eugenic practitioners radically reconstructed the discipline in response to Nazism's rise in the late 1930s and 1940s. Yet in Latin America, no such project was necessary because the softer, Lamarckian approach did not share the hard hereditarian assumptions of Nazi eugenics. As evidence of the persistence of eugenics in Latin America, Stepan identified "traces of a lingering Lamarckism" in the work of prominent geneticists in Brazil, Argentina, and Mexico in the 1940s.[11] Stepan did not look to Cold War development discourse as another site where we might see lingering traces of eugenics, but given that Latin eugenics was so often articulated as a project of race improvement and social modernization, post-1945 development discourse would seem to be another apt location to observe continuity.[12]

EUGENICS AND *INDIGENISMO* IN INTERWAR PERU, 1920–1940

Métraux and Monge's disagreement mirrors the regional perspectives that crystallized in Peruvian racial thought during the first half of the twentieth century. In a country with stark geographic and socioeconomic differences, the project of creating a national racial identity proved difficult. These regional divides were most pronounced during the interwar period. According to Marisol de la Cadena, in this period intellectuals from the supposedly "backward" Andean highlands began confronting the racial theories and authority of Lima-based academics as part of a broader hemispheric movement known as *indigenismo*, which aimed to redeem the indigenous peoples of the Americas.[13] The debates that later emerged in the context of the Puno-Tambopata project reflect unresolved tensions in Peruvian racial discourse from this earlier period.

During the first half of the twentieth century, professional experts in Lima, like many Latin American intellectual elites, promoted a neo-Lamarckian discourse of racial improvement that endorsed whitening through race mixture (mestizaje) and cultural education. This discourse emerged in the late 1860s when Peru's coastal cities began to flourish economically and prompted Lima-based scholars, like the historian Manuel Atanasio Fuentes, to celebrate race mixing as an important cause for Lima's flourishing. In crafting this narrative, Fuentes confronted European race theorists like Herbert Spencer and Paul Broca who feared that racial hybridity led to degeneration. Instead, Fuentes constructed an opposing image of a vibrant and cosmopolitan Lima whose multicolored upper classes rivaled European citizenry in erudition and sophistication. Yet he did not extend these cosmopolitan virtues beyond the charmed circle of Limeño elites. In fact, Fuentes argued that Lima was besmirched by a vagrant underclass of "blacks, Indios, zambos, and cholos" who showed few signs of progress precisely because they remained pure and unmixed.[14]

In the decades after Peru's defeat by Chile in the War of the Pacific (1879–1884), Limeño elites continued to emphasize mestizaje as a means of civilizing the national population. Yet instead of conceptualizing mestizaje in biological terms, Limeño elites began to envision a process of "acculturation" that entailed educating non-mestizos in Western values to convert them into mestizos who, along with whites, were believed to be compatible with modernity. As Andres Rios-Molina has argued, the mestizaje discourse adopted by Limeño elites at the turn of the century did not amount to a systematic plan for interracial coupling but rather "a state project to eliminate cultural and linguistic diversity, with the fundamental objective that indigenous people would assume a Western lifestyle, values and cultural practices, which would favor their insertion into the nation."[15] The turn to "cultural mestizaje" also coincided with a second economic boom that further concentrated power and wealth among the Lima-based oligarchy.[16] At the outset of the twentieth century, Peru had thus developed into a country with stark regional divides—a comparatively wealthy and urbanized coastal region characterized by racial mixture where economic and political power was concentrated and a rural highlands with a large indigenous population that worked primarily as servants on latifundia estates.

Such a divided society offered fertile terrain for the emergence of homegrown eugenic discourse. During the interwar period, Limeño medical experts replaced the previous generation's emphasis on education as a vehicle for mestizaje with one emphasizing hygiene and sanitation as a means of racial improvement. And it was in this context that medical experts based in Lima began to adopt the labels of eugenics and "social medicine" to describe an ambitious public health program to modernize Peru's "backward" regions and populations. Perhaps the most prominent Peruvian eugenicist was Carlos Enrique Paz Soldán (1885–1972), who served as a professor of hygiene and director of the Institute of Social Medicine (ISM) at the National University of San Marcos in Lima. Like his Limeño peers, Paz Soldán celebrated Peru's cosmopolitanism but also worried that the prevalence of infectious diseases like malaria, tuberculosis, and smallpox had caused racial degeneration among Peru's "infinite variety" of "races, castes, and ethnic mixtures" and rendered the white race "impotent" in its ability to assert its "ethnic superiority" and "aptitude" for "nation-building."[17] Paz Soldán therefore promoted a holistic conception of "social medicine" that took "settler ethnic groups" and their relation to "geo-cosmic-social environments" as its object of study. By understanding the relation between Peru's different populations and their environments, Paz Soldán believed that social medicine could decipher the general laws that presided over the "adaptation, persistence, and perfection of humanity."[18] For him, the basic logic that underpinned social medicine was one that shared with eugenics a concern for the biological enhancement of the human race.

From this expansive conception of social medicine, Paz Soldán led South American efforts to confront U.S. eugenic projects and their racist logic. As Nancy Stepan noted, Paz Soldán played a prominent role in the Pan American sanitary conferences beginning in the late 1920s where Latin American medical experts rallied together to resist what they viewed as U.S. scientists' narrow and hereditarian conception of eugenics.[19] During two Pan American eugenic conferences in Havana (1927) and Buenos Aires (1934), Paz Soldán argued that the rigid Mendelian conceptions of race and eugenics that formed the basis of U.S. immigration restriction policies were incompatible with Latin American reality. Although he believed that the term "eugenics" should be used when discussing issues concerning heredity, he strongly opposed U.S. eugenicists' racism and sterilization policies. He also worried that the arrival of "Anglo-Saxon" style eugenics in Latin America would clash with his version of social medicine. In contrast to U.S. eugenicists' emphasis on controlling heredity, Paz Soldán argued that Latin American countries should reorient their health programs to focus on prevention and include social medicine as a specialty within the medical curriculum.[20] For Paz Soldán, in other words, the eugenic concern with heredity figured as a minor component within a broader project of social medicine that conceptualized individuals as part of an organic whole and worked toward racial improvement and societal development through disease prevention and sanitation.

This vision of social progress through national "integration" was not shared by all interwar intellectuals. Scholars from Andean provinces sought to de-center Lima by celebrating the highlands and the Andean "Indian"—which they considered a distinct racial type—as occupying the nation's symbolic core. In paying tribute to Peru's indigenous groups, highland intellectuals joined a hemispheric movement known as *indigenismo* that began gaining traction in the 1920s, especially in nations with significant indigenous presence such as Mexico and Guatemala. In Peru, one of the key ambitions of *indigenismo* was to restore the greatness of the Inca Empire and thus regenerate the "indigenous race."[21] In conceptualizing indigenous groups in this way, *indigenistas* rejected rigid taxonomies that implied fixed differences between racial groups. Instead, they embraced Marxist and neo-Lamarckian conceptions of social progress, which supported their ambition to regenerate the former glory of the Inca race.

José Carlos Mariátegui, a Marxist intellectual from the highlands and founder of Peru's Socialist Party, was perhaps the most influential and outspoken *indigenista* scholar during the interwar period. In his diagnosis of what was commonly referred to as "the Indian problem," Mariátegui insisted that the problems relating to indigenous people were primarily socioeconomic and that nothing short of socialist revolution would solve them. In particular, Mariátegui described the formation of a property regime that he called "gamonalismo," which concentrated land and power in the hands of a wealthy few. For Mariátegui, explanations of the "Indian problem" that failed to acknowledge how the once communal property of indigenous peoples had been appropriated by colonizers represented "sterile theoretical exercises."[22] And the most sterile of these analyses were ones that described the "Indian problem" as an "ethnic problem." These analyses, Mariátegui argued, were ones "nourished by the oldest repertoire of imperialist ideas," namely the "concept of the inferior races," which had long served "the white West" in its project of "expansion and conquest." Mariátegui also described the idea that indigenous emancipation would arise from race mixing "between the aboriginal race" and "white immigrants" as an "anti-sociological naivety."[23] "The degeneration of the Peruvian Indian," Mariátegui wrote, "is a cheap invention of the feudal legionnaires."[24] Though subsequent *indigenistas* drifted away from Mariátegui's revolutionary fervor, they continued to share his enthusiasm for preserving the distinctive features of Andean culture and with restoring the supposedly lost glory of the Inca race.

By the 1940s, Peruvian *indigenismo* began to shed its socialist roots and mutated into a movement oriented toward models of economic development that aligned with U.S. social science methods and political interests. This transformation can be observed in the trajectory of the Cusco-based archaeologist-historian Luis Valcárcel. In his early career, Valcárcel, much like Mariátegui, decried the influence of European racial thought and denounced "cultural mestizaje" as a colonial process that yielded deformities.[25] Valcárcel's ideas became widely known in Peru through his impassioned book *Tempestad en los Andes* (1927), which prophetically announced the resurgence of the Inca people thanks to the rise of "Indios nuevos"—schoolteachers, activists, and pastors—who stood poised to redeem their peoples after centuries of colonial oppression. By the mid-1940s, however, Valcárcel tempered his revolutionary fervor and became more of a reformer who promoted *indigenismo* through more conventional institutional methods. For instance, from 1945 to 1947, Valcárcel served as Peru's minister of education and facilitated a joint agreement between the Peruvian and Bolivian governments to coordinate efforts in indigenous education. The Peru-Bolivia agreement aimed to preserve Andean culture while recognizing that the region was made up of several dissimilar groups. Valcárcel, who rejected what he saw as Mexican *indigenismo's* emphasis on cultivating a mestizo identity, praised the agreement and argued that conserving "the cultural personality of indigenous groups" was crucial to their successful integration into Peruvian society. He also commended the agreement's goal of improving the standards of living in the highlands, which he believed to be an effective way of curbing the migration of indigenous peoples to cities.[26] As education minister, Valcárcel thus offered a conception of *indigenismo* that envisioned a process of modernization managed by indigenous leaders and where the preservation of the racial and cultural distinctiveness of the Andes served as a guiding tenet. In 1946, Valcárcel played an important role in incorporating the *indigenismo* movement into the state apparatus by becoming the founding director of the Instituto Indigenista Peruano (IIP).[27] As the IIP's first director, Valcárcel helped to articulate a new version of *indigenismo* that addressed the "Indian problem" through methods and concepts borrowed from U.S. applied anthropology.

SOCIAL MEDICINE AND INDIGENOUS DEVELOPMENT, 1940–1945

While Carlos Monge's views on the Puno-Tambopata project reflect the influence of Valcarcel's version of *indigenismo* from the 1940s, Métraux's endorsement of the resettlement project owed much more to the hygienic conception of Peru's interior regions articulated by Paz Soldán and his followers. Although Paz Soldán and others used the term "eugenics" at international forums, in the domestic context of Peru they used it much less. Indeed, when they looked to the Andean highlands and Amazon basin, social medicine practitioners did not typically frame their efforts in eugenic terms. Instead, these geographically distant regions prompted them to use notions of "social hygiene" and "colonization" to make sense of their seeming underdevelopment.[28] In an effort to develop these regions, practitioners of social medicine conducted state-sponsored "socio-medical" surveys of the Peruvian Amazon and the Andean highlands, which aimed to unleash these regions' economic potential by building public health infrastructure and roads. They also cautiously encouraged the migration of Andean highlanders to other regions with adequate public health support. Nonetheless, they often described the peoples from these regions as backward, malnourished, and unhygienic and proposed drastic changes to their environment as a means of improving their health and social circumstances. The "colonization" projects promoted by social medicine practitioners and their abiding concern with the regeneration of indigenous and rural populations thus demonstrate how the Lamarckian logic of Latin eugenics persisted and mutated during the 1940s.

A striking example of these colonization efforts can be seen in the work of the German physician Maxime Kuczynski-Godard, whom Métraux would later cite. From 1938 to 1944, Kuczynski-Godard led several sanitary campaigns to the Amazon region under the auspices of Paz Soldán's Institute of Social Medicine. Kuczynski-Godard, who was Jewish, had joined the ISM as "laboratory chief" in 1936 after fleeing from Nazi Germany because of racial persecution. Previously, he served as professor of pathology at the University of Berlin, where he had trained in medicine and anthropology. During the 1920s, Kuczynski-Godard led several medical expeditions in Siberia, Mongolia, and China concerned with studying the "ethnic pathology" of rural and indigenous populations.[29] During these expeditions, he developed an ethnographic approach and immersed himself in the communities he was studying to observe social conditions, perform anthropometric measurements, and photograph the physique and symptoms of individuals.[30] In Peru, Kuczynski-Godard expanded this anthropological approach through several "social hygiene" surveys of the Amazon sponsored by the ISM and the Peruvian government. Through these Amazonian expeditions, he honed the idea that building a public health infrastructure was crucial to the development of the Amazon region and its integration into the rest of the country. Echoing language later adopted by Métraux, Kuczynski-Godard wrote about sparking an "evolutionary process" that would liberate "Amazonian man" from poverty and servitude and "unleash" the capacities that lay dormant due to a harsh physical and social environment.[31] He also insisted that modernizing the Amazon would have to be more humane toward the region's indigenous inhabitants and avoid the carnage of the rubber industry.

Kuczynski-Godard also described the situation of Andean highlanders in similarly redemptive terms. By 1944, the Peruvian government lost interest in developing the Amazon region, and he accepted a new position within the Ministry of Health as a "technical advisor" of "socio-medical surveys in the highlands and tropical forest (montaña)."[32] In the following years, he conducted surveys in the highland regions of Puno, Cuzco, and Ayacucho. Kuczynski-Godard's Andean surveys characterized the region's indigenous inhabitants as poor, unhygienic, unhealthy, and malnourished. He also identified local practices such as coca chewing and a "magico-metaphysical" worldview as primitive coping mechanisms to a harsh environment and thus as obstacles to the region's potential development.[33] Given their poor state of health and propensity to infectious disease, Kuczynski-Godard also cautioned that the migratory patterns of these groups threatened to spark epidemics in other regions of the country.

These unflattering assessments of Andean highlanders were evident in his socio-medical study of mining communities in Puno. For this study, Kuczynski-Godard traveled to a hamlet next to the San Antonio de Esquilache mine, located 96 kilometers from the city of Puno on the shores of Lake Titicaca and situated at approximately 4,500 meters above sea level. His survey described the sanitary conditions of the indigenous laborers and how they were shaped by the socioeconomic conditions of the region, where people had struggled since colonial times to "exploit the riches in the ground."[34] Kuczynski-Godard paid little attention to the cultural beliefs of the miners and instead focused on the practical aspects of their daily life such as their wages, basic expenses, diet, and health. To assess their health, he considered the quality of their diet and whether it offered sufficient levels of protein and vitamins. He also gathered basic epidemiological data about the prevalence of physical conditions such as swollen gums and pyorrhea, lip fissures, scleritis, and spoon nail as well as respiratory diseases such as pneumonia, bronchitis, and tuberculosis. Based on these observations, the German concluded that the miners and their families were malnourished due to a lack of fresh fruit and vegetables. As other markers of nutritional deficiency, Kuczynski-Godard noted the short stature, premature aging, and delayed sexual development of the miners and their children. He also cautioned that the "Indian" miners had a "dangerously anti-hygienic" lifestyle: bathing infrequently, not washing their work clothes, which were often covered with lead dust, and not bothering to change into different clothes to sleep.[35]

Although he noted signs of ill health and poverty, Kuczynski-Godard also observed that indigenous groups had developed clever strategies to compensate for the trying environment of the highlands. For instance, he described a pattern of "migrations and oscillations" where men and youth from Puno traveled to warmer regions to cultivate and import fresh fruit and coffee and to cities to look for seasonal work. He also cautioned that these migrations posed a growing public health threat. When migrants returned, they often brought diseases such as tuberculosis, leishmaniasis, and intestinal parasites that were deteriorating the general health of Andean populations. "In view of the [migrant's] primitive ways of life and lack of hygienic control," Kuczynski-Godard wrote, "a situation has emerged that is full of risks and renders each migrant group a virtual vector of contagion."[36] The solutions to this challenging situation, he insisted, were straightforward. The Peruvian Ministry of Health could control the situation and secure the future prosperity and health of the nation if it increased the number of rural doctors in the region and trained more "subaltern sanitary personnel."[37] Echoing his proposals for the Amazonian region, Kuczynski-Godard suggested that the challenges posed by the Andean region's itinerant indigenous population was of a technical order that required the intervention of more medical experts and the creation of a more robust public health infrastructure.

Insofar as he was concerned with how to manage and improve Peru's indigenous groups from the Andes and Amazon regions, Kuczynski-Godard shared common interests with Peru's *indigenistas*. Yet by the end of the 1940s, Kuczynski-Godard and Paz Soldán made a concerted effort to differentiate their approach, which they viewed as objective and grounded in technical knowledge, from *indigenismo*. In 1948, Kuczynski-Godard and Paz Soldán published a scathing critique of *indigenismo* titled *Disección del Indigenismo Peruano*. Though they framed their assessment of *indigenismo* as an impartial and objective analysis, it is also quite evident that they viewed *indigenismo* as a rival and potentially dangerous movement. In his preface, Paz Soldán struck at the core of *indigenismo* by questioning the legitimacy of its central term. With stern disapproval, he described "Indigenismo" as a "lexically improper" and "indeterminate" term that had seduced many minds and sparked hopes that the "magical powers" of the state would make their ideological proposals for reform a reality.[38] Echoing Ralph Linton's work, Paz Soldán also described *indigenismo* as a "nativistic movement" characterized by "verbal exaltations that destroy the peaceful coexistence between men and peoples." He also decried *indigenismo's* "cosmotelluric fatality" and argued that the movement had rendered "indigenous" a magical word that was "hostile to everything non-indigenous."[39] By anchoring his perspective firmly within the vocabulary of medicine and science, Paz Soldán thus attempted to position social medicine as a necessary counterpoint to the affective excess of *indigenismo*.

Yet while offering social medicine as a tempering influence, Paz Soldán and Kuczynski-Godard could not help but frame their own approach in the eugenic language of "ethnic" improvement. Paz Soldán described social medicine as the study of "the laws of adaptation, endurance and perfection of the Ethnic groups" and thus as the discipline best suited to "guide the inquiring mind through the dark ideological labyrinth that is today "indigenismo Americano."[40] Like Latin eugenicists in Brazil and Mexico, both men also championed mestizaje as a crucial element of nation building and cited Peru's history of race mixing as a counterpoint to the alleged "dualistic orthodoxy" at the core of *indigenismo*. "The idea of Indigenismo," Kuczynski-Godard argued, "was born from antinomy," and its central category of "indigena" invited its followers to think of "two groups, and two dissatisfied human attitudes, each of which, from their cultural point of view, considers the other group as heterodox, and in the field of economics, as opposite."[41] Yet this binary did not withstand historical scrutiny, he argued. While "Indians" once existed as a well-defined "racial type," this purity had been eroded thanks to an intense process of "historical hybridism." Though in some cases indigenous villages had remained hostile to "major changes," most of Peru's "indigenous" inhabitants had "modified their human substance" and "perturbed the original structure of their way of life."[42]

Foreshadowing later debates about the Puno-Tambopata project, Kuczynski-Godard also used the example of Andean migration to illustrate this ongoing transformation of indigenous cultures. In his analysis of the "Indian problem" in the Andes, he identified a major lack of arable land in the Andes region as a major source of demographic anxiety. "Land hunger," Kuczynski-Godard argued, was the cause of Andean people's "unsettled and migrant" life. He also argued that the migrations provoked by land hunger opened new horizons for indigenous groups by prompting "assimilation" or "acculturation" toward "mestizo groups with more urban lifestyles."[43] He also warned that this migration was not without its risks. An itinerant lifestyle, he pointed out, often led to indigenous people's "self-destruction" because of the significant physical risks it entailed. Further, if left unchecked and unmanaged, land hunger might have catastrophic consequences for the nation.

SOCIAL MEDICINE AND "LAND HUNGER" IN THE ILO'S ANDEAN INDIAN PROGRAM, 1950–1952

By the 1950s, the political and economic fate of the Andes region became a matter of growing international concern. In 1952, Bolivia experienced a social revolution led by Víctor Paz Estenssoro, who introduced an agrarian reform program that redistributed land and led to the destruction of the country's landed oligarchy.[44] While Víctor Paz Estenssoro's government refused to pick sides in the growing ideological conflict between the United States and the Soviet Union and welcomed aid from both nations, the Peruvian dictatorship of Manuel Odría was closely aligned with the U.S. government and took strong pro-capitalist measures. These included outlawing the leftist APRA Party designed to encourage U.S. investment in the mining and petroleum sectors.[45] The differing fates of Andean countries and the potential for socialist revolution attracted close interest from the United States and from UN agencies, which resulted in the articulation of international development projects that doubled as communist containment strategies. These tensions can be observed in the the ILO's Andean Indian Program (AIP)—the broader regional development project of which the Puno-Tambopota resettlement project was one part.

The AIP ran between 1952 and 1972 and was one of the ILO's most ambitious endeavors of the Cold War era. It aimed to accelerate the economic development of Andean countries and assumed that the large indigenous population in the highlands posed an obstacle to this process. As a solution to this perceived social problem, the ILO and other UN agencies worked closely with Andean nations to create a series of "action bases" whose purpose was to transform the Aymara and Quechua inhabitants of the highlands into modern economic subjects.[46] Like proponents of social medicine, ILO officials viewed issues concerning the economic development of Andean highlanders as primarily technical issues and embraced nonbiological conceptions of race and indigeneity that emphasized the social dimensions of cultural change. In the mid-1950s at the height of the AIP's operation, Jef Rens, the ILO's deputy director, mused that the progress made during the project gave proof that "indigenous populations are perfectly capable of being transformed quickly into excellent producers and conscientious and active citizens."[47]

Ren's framing of the project owed much to the capitalist development strategy that typified the ILO in this era. In 1946, the ILO became part of the UN system and one of the key purveyors of the UN's technical assistance program designed to uplift the economies of "Third World" nations through tutelage from technical experts in industrialized countries. From 1948 to 1970, U.S. lawyer David Morse served as the ILO's director. During his tenure, Morse transformed the ILO into a development agency as a means of containing the spread of communism in the so-called Third World.[48] This project took shape primarily as an attempt to create the labor force necessary for worldwide capitalist development by establishing vocational training centers—managed

FIGURE 5.1 Building one of the AIP's community centers or "action bases" in Chucuito, Peru (photograph, n.d., ILO Multimedia Download Platform).

by technical experts from the Global North—in locations identified as problematic or potentially communist. The transformation of indigenous peasants into capitalist consumers and workers was also an important part of this anticipated prosperity.

From the earliest stages of the AIP, Kuczynski-Godard's theme of "land hunger" on the shores of Lake Titicaca became a focal point of the project and framed the field reports of the director of the AIP's first field mission—Ernest Beaglehole, an ethnopsychologist from New Zealand. It is not clear why Beaglehole was chosen to lead this mission, as he was not trained as a Latin Americanist and had not worked in the region. Yet by the late 1940s, he began to play a prominent role within UN agencies. As we saw in chapter 2, Beaglehole formed part of the committee of scientists that drafted the 1950 UNESCO Statement on Race. Two years later, Beaglehole became New Zealand's representative to the ILO's Committee of Experts on Indigenous Labor. And after the first meeting of this committee in La Paz in 1951, the committee chose Beaglehole to lead the field mission that initiated the Andean Indian Program.

Although he had no previous Latin American experience, Beaglehole had worked on questions of modernization and cultural change in indigenous groups in the United States and the South Pacific for two decades. Since the late 1930s, Beaglehole's research had focused on the psychological impacts of Western culture on Polynesian islanders and the Māori in New Zealand. Much of this work was influenced by culture and personality studies in anthropology and emphasized the psychological struggles and social disintegration that indigenous youth faced as they struggled to adapt to European ways of life. It also had a very practical bent and often proposed that the best policies for fostering modernization and addressing the challenges faced by indigenous communities were ones that encouraged quick and total assimilation to Western values. For instance, in the case of New Zealand, Beaglehole proposed the creation of adult education centers where Māori parents could learn "modern" ideas about child rearing and family values that would act as a catalyst in this process.[49] Beaglehole's field experience and applied anthropological experience coupled with New Zealand's growing international reputation for "racial harmony" and successful indigenous integration likely made Beaglehole a strong candidate to lead the mission in the eyes of his peers.[50]

FIGURE 5.2 "Presentó su Informe La Misión Indigenista Andina"; Beaglehole is fourth from left (*Noticias de La OIT*, no. 20, March 1953).

In the Andean context, however, Beaglehole ended up placing much less emphasis on questions of psychological adaptation and instead focused on the region's geographic extremes and the size and distribution of its indigenous population as obstacles to economic development. In his report, Beaglehole used dramatic language to describe a bleak situation in which geographic and colonial forces conspired to isolate indigenous communities and render them economically unproductive.

For instance, Beaglehole described the Andes region as "broken in places by mountainous knots which have produced a series of high-walled and isolated valleys within which the Indians of the region have lived for many centuries, forced to yield to their conquerors, whether Inca or Spaniard, their labour and often their lives, but never their integrity."[51] In another paragraph, Beaglehole wrote:

> The Andean region then is a region of violent contrasts—the high plateaus are cold and swept by knife-edge winds, bare, inhospitable, without coal, timber or oil. Here the Indian struggles against climatic aggression to gain hardly a living from agriculture and animal husbandry. The higher valleys are more sheltered and fertile: eucalyptus trees produce fuel and irrigated lands grow alfalfa—life though pleasanter than on the high plateaus is still a struggle between the Indian's desire to live and the small returns of an apparently over-populated and over-cultivated land. Lower still, on the slopes of the Andes are the *yungas* valleys, thinly populated, warm, tropical, fertile but often malaria-ridden and hazardous with tropical disease.[52]

And it was through this vivid depiction of "climatic aggression" and the delicate relationship between altitude, climate, and the economics of indigenous populations that Beaglehole outlined the core ambition of the ILO's Andean Indian Program—to redistribute and thereby "integrate" the Altiplano's indigenous populations without entirely dismantling their ways of life. For Beaglehole, the crux of the region's economic problems lay in an indigenous population that was overconcentrated in the inhospitable and unproductive higher plateaus while the fertile tropical valleys waited to be exploited. Beaglehole's framing of the problem echoed Kuczynski-Godard's work but also seems to have been influenced by input from officials from the Odría government. In comparison to Bolivia, Beaglehole's report described "booming" Peru as "further along the road towards industrialisation" and as offering an "attractive field for the investment of capital, both national and foreign."[53] He also noted that from the outset, officials from Peru's Department of Indigenous Affairs drew the mission's attention to the Puno region as a problem area due to the prevalence of large private estates devoted to animal husbandry, which were prompting a "spontaneous and unregulated exodus" of "Aymara Indians" westward to Arequipa and to coastal Lima. Peruvian authorities thus suggested that the AIP's most effective contribution to solving this problem would be "the study of migratory movements in the Puno area and the possibility of channelling [sic] such movements in a socially satisfactory direction," namely the Tambopata Valley, which the government had earmarked as a site for possible colonization by indigenous groups from the Altiplano.[54]

It was within the context of such questions about the region's poorly distributed indigenous population that the idea for the resettlement project in the Tambopata Valley began to take shape. Beaglehole estimated that the average population density of the Andean region was twenty inhabitants per square kilometer. Yet in some of the region's most "favoured areas" on the shores of Lake Titicaca, primarily inhabited by Aymara communities, the population density rose to as much as seventy-six inhabitants per square kilometer. According to Beaglehole, these alarming rates of population density were putting "acute" pressures on "land resources" and had given rise to a situation where only about a "half a hectare of land per person [was] devoted to agriculture." Echoing Kuczynski-Godard, Beaglehole observed that these acute population pressures had prompted the region's inhabitants to resort to "random and disorganized migratory movements throughout the region (often into the larger cities)." Although the sparsely populated and low-lying tropical valleys in the Amazon basin offered what seemed to Beaglehole like a logical destination for these migrants, he noted that the "prevalence of malaria and other tropical diseases" deterred them from coming in large numbers.[55]

Beaglehole's report did not offer an unqualified endorsement of resettlement but rather cautiously argued for further studies. Indeed, Beaglehole wrote that transferring groups from "congested districts of the altiplano to the undeveloped and potentially fertile lands at lowers latitudes" seemed like one of the "most tempting solutions to the Indian problem." Yet he also insisted on a "realistic attitude" toward the plan and stressed that such transference "must inevitably be costly" and riddled with "a number of very serious obstacles" that stand in the way of "any hope of success." Chief among these, he noted that Altiplano inhabitants had "their own specific physical constitution," which rendered their adaptation to lower altitudes "difficult and somewhat lengthy" and that "the whole subject has not yet been thoroughly investigated." He went on to say that past migrations of "altiplano Indians to the lower altitudes [had] been fatal to a high proportion of the migrants." "Indians themselves," Beaglehole continued, had thus adopted a "strong and well-founded prejudice against abandoning the altitudes to which they are accustomed."[56] Tempting as resettlement projects seemed, Beaglehole insisted that they should be considered only if the following conditions were met: "when further studies of Andean biology have been undertaken and completed; when the areas potentially suitable for colonization have been thoroughly surveyed; when satisfactory plans for the preparations of the areas to receive settlers have been worked out; when adequate capital for the execution of such plans has been made available; and when the potential settlers themselves have been persuaded to undertake the risks involved, and shown how best to protect themselves against them."[57]

CARLOS MONGE AND RACIAL ACCLIMATIZATION

Beaglehole's restrained endorsement of the resettlement project and call for further study of Andean biology suggests the influence of Carlos Monge. During the mission, Monge acted as a consultant and then wrote an appendix to Beaglehole's report. When Beaglehole arrived in 1952, Monge's reputation as an international authority on Andean biology and economic development was well established. Monge had also become a key figure within Peruvian *indigenismo* and in 1949 had succeeded Luis Valcárcel as director of the IIP.[58] During this early phase of the IIP, Valcárcel and Monge continued to insist on the importance of preserving Andean heritage yet increasingly articulated this vision through the idiom of North American cultural anthropology combined with Monge's accounts of Andean biology. As a result of this reorientation, IIP leaders strategically embraced modernization projects as long as they were designed to preserve crucial elements of Andean culture and biology, and the CPP that Carlos Monge codirected with Allan Holmberg from Cornell University served as the prime example.[59]

FIGURE 5.3 Ceremony of down payment on Vicos lands (photograph, in Allan R. Holmberg, Division of Rare and Manuscript Collections, Cornell University Library, Ithaca, NY).

In contrast to his enthusiastic support for the CPP, Monge offered an unflattering assessment of the Puno-Tambopata project. In an appendix to Beaglehole's report, Monge warned that the ILO project could have devastating consequences for the health of Andean highlanders. Recalling his decades-long concern with Andean biology, Monge wrote that "the Andean Indian is attuned to his climatic environment and, generally speaking, anything which tends to keep him there is the best means of ensuring his biological development."[60] He also opposed any project that would sever the ties between Andean Indians and their communities of origin and offered three "strong arguments" for why they would resist any "long-distance shifts":

> a biological "urge" to return to the place of origins—the plateau; a psychological aversion for the change, attributable to the distrust reaction which forms part of the Andean's acquired personality; and . . . the economic interest of the members of the community as joint owners of land which they do not want to lose and which they must cultivate.[61]

Echoing *indigenismo's* forceful claims from the interwar period, Monge wrote that for Andean Indians, "the community land is a permanent thing" while "the new settlement land is ephemeral to them."

Instead of rushing into resettlement, Monge suggested that the ILO follow indigenous people's already established migration and settlement practices. Rather than direct transfers from Puno to Tambopata, he proposed intermediary transfers to "medium altitude" areas (around 6,000–8,000 feet), which would allow Andean highlanders to "keep their vitality intact" and were already a "natural part" of "Andean population movements." Monge also described how "the Indians themselves" had already conducted successful settlement "experiments" that allowed migrants to regularly return to the highlands and maintain a connection to their communities of origin. He also noted that in Tingo Maria, migrants benefited from a state infrastructure including "suitable administrative personnel, government health services, a hospital, agricultural and stock-breeding experimental stations, colleges, etc."[62] These examples of highlanders' "instinctive migrations," Monge argued, "should be controlled along scientific and anthropological lines" and were likely to succeed as long as the "migrants do not leave the physical and mental environment to which they are accustomed."[63]

Monge also described the prospect of moving Andean highlanders outside of their climatic zone as highly fraught. The project of directly transferring "Indians" from the cold plateau of the "roof of the world" to the "torrid zone of the South American forests" was one with incredibly high stakes. He described it as a "biological experiment in which strange new climatic conditions invariably cause a morbid state—adaptation—culminating either in death by natural selection or in acclimatisation."[64]

As evidence of the potentially devastating consequences of such an experiment, Monge noted the "very high mortality rate from acute tuberculosis" among Bolivian and Peruvian soldiers, many of whom were indigenous people from the highlands, during the Chaco war. Monge also cautioned that such radical changes in climate were well known to cause "tropical fatigue" and reasoned that the only way to avoid it would be to carefully monitor the "hygiene, nutrition, and labour conditions" of the migrants. Accordingly, Monge argued that a resettlement project of this kind would have to be assisted by doctors with "special training in Andean physiology and tropical pathology" and anthropologists with training in "physiology and altitude biology."[65] Given the immense risk and logistical challenges, Monge concluded that "it is beyond question that the ideal solution would be not to remove people from their natural environment, as in the Peru-Cornell experiment."[66]

Though he articulated this concern with resettlement in response to Beaglehole's 1952 mission, Monge had been making similar claims for decades. Since the early 1930s, Monge had conducted research not only on the biological adaptations of Andean bodies but also on the social policies and procedures that the Inca Empire developed to cope with the "climatic aggression" of the Andes region. He had also positioned this research as relevant to the issues raised in the domain of eugenics and homiculture. For instance, in 1934, during a plenary session at the second Pan-American Eugenics and Homiculture Conference in Buenos Aires, Monge argued that the problem of biological adaptation to extreme environments was an ancient one that had vexed the Inca Empire. Drawing on textual evidence from the writings of Spanish chroniclers, Monge proposed that the "ancient communist empire of the Incas" had developed a sophisticated understanding of the interdependence between the biological and social realms, which was evidenced by their knowledge that "Highland" and "coastal" peoples suffered when transferred to unfamiliar climates. According to Monge, the Inca Empire had also developed sophisticated solutions to this biosocial problem, which was evidenced by their codification of "mitimaes"—an institution that he described as producing an "incessant interchange of populations that stimulated acclimatization to different altitudes and created a truly cosmopolitan man."[67]

Throughout his career, Monge continued to celebrate the ancient wisdom of the Inca Empire. In the years leading up to Beaglehole's mission, Monge expanded his arguments from the Pan American eugenics conference and argued for the superiority of Incan methods for dealing with the biological realities of different climates. Monge's case for Incan supremacy can be found at the core of his English-language book *Acclimatization in the Andes* (1948), which garnered significant international attention and was reviewed in prominent North American journals from multiple disciplines. In a chapter on "Racial Acclimatization," Monge again turned to colonial-era sources, which he interpreted as showing that the Incas had developed a "marvelous [and] intuitive comprehension of the problem [of climatic aggression]" and had confronted it with a "biological policy which took into account the age-old adaptation of the race to the altitude."[68] According to Monge, the Inca Empire's success owed much to its practice of "shifting" people around for the purposes of acclimatization and helped to explain how the Inca Empire had attained "such an extensive stage of development."[69] Monge drew one of the most striking examples of this policy from the chronicles of the Jesuit priest Bernabé Cobo. According to Cobo, the first thing that Incan kings would do after conquering a new province was to remove 6,000–7,000 families from the new territory and transfer them to other provinces under their rule. Incan rulers would also transfer "an equal number of people" into the newly conquered territory. When deciding where to transfer people, Incan rulers paid little attention to distance and were instead concerned with transferring people from similar climatic regions. Indeed, Cobo explained that they took every care to make sure that those transferred "should not move to just any region but more or less to one of the same air-temp[er] and qualities or very similar at least to the region they had left or in which they had grown up." It was by this means, argued Monge, that the Andean race had "since prehistoric times" responded "to the biological urge of acclimatization" and "succeeded in adapting itself to life by means of migrations within differing climatic zones."[70]

Monge's reverence for Incan antiquity and concern with racial acclimatization illustrates how different his views were from the typical expert described in the historiography of development. Most mid-twentieth-century development experts conceptualized the process of development (or modernization) as future oriented and nonbiological. Monge, by contrast, could not seem to fathom a conception of development that ignored the stark biological realities of racial acclimatization. Whereas many development experts took inspiration from ambitious infrastructural projects such as the Tennessee Valley Authority and did not see climate as much of an obstacle to modernization, Monge viewed climate as a potent force that tethered populations to circumscribed regions.[71]

Monge also departed from the perspective adopted in the race declarations of UN agencies. While documents such as the UNESCO Statements on Race confronted scientific racism by emphasizing the biological similarities between human populations and their cultural relativity, Monge insisted on seeing Andean highlanders as a distinct racial variety. Even before the AIP mission, Monge's willingness to confront the views of UN agencies had been apparent. In fact, between 1949 and 1950, Monge had been involved in a dispute with an international team of scientific experts who traveled to Peru as part of the UN Commission for the Study of the Coca Leaf. Through various reports and counterproposals, Monge challenged the work of the UN commission, which concluded that the indigenous practice of chewing coca leafs had negative psychological effects and reduced the economic productivity if indigenous workers. Prominent members of the UN commission had also publicly speculated that the coca chewing could lead to the "racial degeneration" of indigenous peoples from the Altiplano. By contrast, Monge insisted that the coca chewing was a benign physiological adaptation of "Andean Man" to a harsh climate and called for the creation of a national commission to study the issue.[72] Given Monge's seeming mistrust of UN agencies, it is not surprising that in the years after the Beaglehole mission, Monge's views became an irritant to the ILO's Puno-Tambopata resettlement scheme.

ALFRED MÉTRAUX: CULTURAL AND DEMOGRAPHIC PLASTICITY

Despite Monge's warnings, the ILO and the Peruvian government forged ahead with the Puno-Tambopata mission. As the next phase of the project, ILO officials approved another field mission to study the existing migration patterns of Andean highlanders. For this mission, ILO officials recruited someone with considerable Latin American experience:

Alfred Métraux. Métraux was persuaded to lead the Puno-Tambopata mission after meeting with Beaglehole, his former colleague at the Bishop Museum, in Paris in 1953. According to his diaries, Métraux read Beaglehole's "excellent" report to the ILO in Paris in early January 1953 and was particularly impressed with the report's account of the "Otavalo case" in Ecuador where "the Indians took charge of their situation and succeeded in creating some profitable industries."[73] A few weeks later, Métraux was invited to a meeting hosted by Alva Myrdal, then director of UNESCO's Department of Social Sciences (DSS), to meet the "members of the Andean Mission," which included Beaglehole and Enrique Sanchez de Lozada. During this meeting, Beaglehole expressed fondness for his time in Bolivia, and Lozada raved about the new "antiperonist and democratic" Bolivian government and mused that "the Bolivian army will no longer exist and the people will instead be armed."[74] By the end of 1953, Métraux prepared to travel to the Andes region to observe these exciting political developments himself.

From December 2, 1953, to January 28, 1954, Métraux traveled across Peru accompanied by the ILO economist Frank Bray and Dr. Luna Aguilera from Peru's Department of Indian Affairs and followed the migration routes that Andean highlanders had been using to travel to the Peruvian coast and the lower-lying tropical valleys of the interior. Yet the mission did not prove to be as exciting for Métraux as he had anticipated. In his personal correspondence during the survey, Métraux described feelings of loathing toward the Andes region and its inhabitants, which undoubtedly influenced his enthusiasm for development projects geared toward resettling Aymara groups from Lake Titicaca and remaking their culture. For instance, in correspondence with the photographer Pierre Verger, Métraux described his decision to lead this mission as an act of "masochism and imbecility." He also complained to Verger that he did not tolerate the high altitude well and that the "Aymara Indians" provoked in him a "violent physical aversion" and "nausea."[75] Even as he spearheaded UNESCO's antiracism campaigns, Métraux was unable to shake sentiments of loathing and antipathy toward indigenous people when they did not meet his expectations. Whereas once he espoused a sentimental posture oriented toward the salvage and preservation of indigenous groups, by this point in his trajectory Métraux had become an adept

LUNAR LANDSCAPE in the Andes has been scoured out by rain and wind. Buried in these mountains are great mineral riches which would have remained inaccessible were it not for the Indians. They alone are able to work at great altitudes, unaffected by the bleak desolation. (Photo Almasy.)

FIGURE 5.4 Lunar landscape (photograph, in "Awakening Continent: Latin America in New Perspective," *UNESCO Courier*, vol. 3, no. 2, 1955, p. 9. UNESCO Digital Library).

practitioner of a technical assistance discourse that masked uncomfortable sentiments.

After this trip, Métraux produced three reports for the ILO that described the demographic crisis of "land hunger" inducing the migrations, the state of the spontaneous settlements in Tampobata, and his proposals for how to accelerate migration to the valley. Unlike Monge's concern with the biological fragility of Andean Indians, Métraux's assessment was inspired by the notion of indigenous resilience he gained from his time in Hawai'i and thus emphasized the discontinuity between contemporary Andean groups and the ancient Incas and the need to transform Andean culture in the face of intense demographic pressures. Métraux based his arguments on the observations he made during his two-month trip as well as a careful reading of Peruvian social science literature. In his ILO report and other work on this topic, Métraux cited Kuczynski-Godard's work extensively and also relied heavily on the statistical data from the Peruvian demographer Alberto Arca Parró, whom he described as an "indisputable authority."[76]

In his reports, Métraux described Andean highlanders' increasing migration to the Peruvian coast as an alarming situation with two deep causes: "on the one hand, a demographic pressure accentuated by the misery of the rural masses, and, on the other hand, a nascent industrialization that attracts rural populations towards cities and mining centers."[77] Demonstrating his debt to Kuczynski-Godard's work, Métraux used "land hunger" to describe this situation and argued that it could have dire consequences for the region. Although the 1940 census suggested that Peru's national population density was 6 inhabitants per square kilometer, Métraux argued that 60 percent of the country "corresponds to [uninhabitable] deserts and mountains," which yields an "average density of 12.7 inhabitants per square kilometer." Like Beaglehole, Métraux also noted alarming demographic trends in the highlands where relative population density reached 20 inhabitants per square kilometer in many departments (such as Cuzco, Puno, Apurinac, Arequipa, Moquegua, Tacna, Ayacucho, and Ica) and climbed to an alarming rate of 80–100 inhabitants per square kilometer in certain districts of Lake Titicaca. The climate of these high-altitude regions meant that they were better suited to "cattle raising than agriculture," which led Métraux to conclude that economic "existence is precarious there" and to warn that the situation was approaching "overpopulation."[78]

Métraux also described a situation where Andean migrants were increasingly unwelcome in other regions. Before Paz Estenssoro's revolution in 1952, Aymara youth from Titicaca often migrated to the Bolivian Yungas where they could labor as plantation workers, rent plots of land for cultivating coffee or coca leaves, or even purchase their own plantations.[79] Yet after the revolution, Peruvian migrants became increasingly unwelcome as Bolivian indigenous workers organized unions as part of the revolutionary government's agrarian reforms and developed fierce nationalist sentiments. Although he favored the revolutionary measures of Paz Estenssoro's government, Métraux lamented the consequences they had for Aymara migrants from Peru who were excluded from this process and either "dispossessed of their lands or denied the possibility of exploiting them as they had in the past."[80] Because of these changes, Andean highlanders were forced to migrate mostly within Peru. Yet these migration routes not only forced highlanders to undertake long and perilous journeys, often by foot, they also created social and infrastructural problems that stoked racist fears of an "indigenous tide" sweeping the nation.

It was in this bleak context that Métraux viewed the Tambopata Valley as a promising place where migrants might find economic self-sufficiency and keep their culture intact. According to Métraux, the valley offered a place where migrants could cultivate coffee peacefully and "control their own destiny" with support from fellow migrants. "The colonization of Tambopata has been a success," Métraux wrote, "because the Indians were guided by other Indians who, thanks to their prior experience, had established a cultivation system adapted to the soil and climate." Newcomers to the valley also benefited from *el ayni*—an "old Aymara institution" or "system of mutual aid" through which individuals rely on the labor or money of friends and relatives as needed.[81] The valley "pioneers" also maintained close bonds with their families and land in the highlands where they often returned in the off-season to "breathe in the fresh air" during important festivals. In fact, most colonists participated in what Métraux described as a "dual economy" where they spent one part of the year laboring as "tropical farmers" and the rest of the year as villagers who cultivate "cold lands" or "lead a pastoral existence."[82]

In his subsequent reports, Métraux also directly confronted Monge's biological concerns. According to Métraux, the idea that tropical climates were potentially fatal to Andean Indians was widespread in both "intellectual and lay circles" and had been given additional scientific credence through Monge's work. Yet Métraux—who had already skillfully navigated the controversies provoked by the 1950 UNESCO Statement

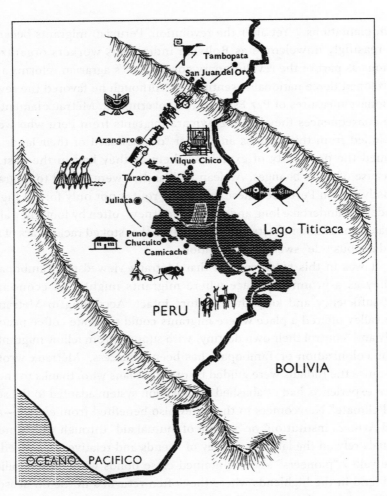

FIGURE 5.5 A map of the AIP's "action bases" in the Lake Titicaca region and the migration route from Puno to the Tambopata Valley (map, in *Programa Andino*. Geneva: ILO, 1961, p. 45).

on Race—questioned this seemingly deep-rooted belief. "Is this biological incompatibility as absolute as it appears and is the expansion of the Indians to the eastern slopes thus impossible?" Though he suggested that the question should be left to "medical experts," Métraux also insisted that anthropology could offer something to the debate. Europeans had also "succumbed to tropical illness" when adventuring in the Amazon jungle during the colonial period, yet such illness was not due to their biological makeup but rather something that could have been prevented with "better hygiene." Métraux also argued that anthropological studies, notably the work of Bernard Mishkin, had shown that the "penetration of Andean Indians into the Amazonian environment goes back to the prehistoric era and has continued uninterrupted until our current day." Aymara and Quechua groups, Métraux insisted, have settled in the Yunga valleys since ancient times, which led to a "Quechuization" of the eastern regions.[83]

In a further departure from Monge's work, Métraux also sided with the UN coca commission and cautioned against the economic dangers of coca chewing and intoxication. In an article for the *International Labour Review*, Métraux offered a bleaker outlook on the prospects of Andean modernization. "All those who have studied them [the Aymara and Quechua] closely," Métraux explained, "have observed the defiance and hostility of these groups not only towards white men and townsfolk but also towards the members of other Indian communities."[84] For Métraux, this "xenophobic and ultra-conservative" attitude was due to the "internal equilibrium that the communities have succeeded in creating" and which they try to maintain by resisting "anything that might disrupt it."[85] In addition to this xenophobia and conservatism, Métraux also characterized the Aymara "personality" as lacking a sense of method and order when it comes to work. Métraux based this argument on the work of anthropologist Harry Tschopik, who had characterized Aymara work habits as haphazard and lacking in efficiency because of the "frequent loafing and intervals of time out for coca" and "the degrees of alcohol consumed and degree of intoxication achieved."[86]

THE FINAL STAGES OF THE TAMBOPATA PROJECT

As plans for resettlement moved forward after Métraux's study, Monge became an irritant for project organizers. ILO officials—notably Jef Rens and Enrique Sanchez de Lozada, the latter part of the Beaglehole mission—believed Monge was an obstacle to the project and would "stop the program in its tracks if he could."[87] After Métraux's mission, Monge trumpeted the CPP as a successful model of how hacienda properties could be smoothly transitioned toward collective ownership by indigenous communities.[88] Monge also closely tracked public opinion concerning the project and informed his Cornell colleagues that major Peruvian newspapers had written glowing editorials praising the project's expropriation plans.[89] Tensions between Monge and the ILO escalated in 1957 when Lozada and Rens confronted Monge for allegedly sponsoring a press campaign that denounced their resettlement project as "pure and simple genocide"—a charge that Monge denied. Around this time, Lozada also warned ILO officials that Monge continued to argue that building roads into the lower-lying valleys would have no economic value and that the idea of mass resettlement to the Tambopata Valley was "illusory" and "dangerous."[90]

For all of the concern the Puno-Tambopata project generated, it also prompted other prominent Peruvian experts, like the demographer Alberto Arca Parró, to come out in support. In 1955, Arca Parró published a series of articles in the Lima newspaper *La Prensa* that challenged Monge's objections to the project and portrayed it as a viable solution to Peru's demographic problems. Arca Parró was an internationally renowned Peruvian demographer who served as director of the 1940 census and as representative to the UN Population Commission in 1947. During the 1930s, Arca Parró had been involved in the creation of the Peruvian Socialist Party and called for greater protection of the cultural and economic autonomy of indigenous communities. As census director in 1940, he endeavored to count Peru's indigenous population as accurately as possible and to modernize data-collecting practices through the use of new technologies such as IBM Hollerith calculators.[91] Using census data, Arca Parró identified problems associated with the distribution and racial composition of Peru's population. In his census report, Arca Parró argued that the Andes region was at risk of becoming overpopulated and observed that indigenous migration to Lima and to the coast was increasing rapidly. He also looked to the uninhabited parts of the Amazon as an area with untapped resources and a greater potential to absorb population growth than that of Peru's cities.[92]

In 1955 when he weighed in on the debates about the ILO project, Arca Parró leaned heavily on his census data. "Several years ago, when commenting on the results of the 1940 census, I pointed out that one of the most striking phenomena of Peru's demographic situation was the 'march on Lima.'" Since then, he cautioned, this march on Lima had only intensified as evidenced by the city's "rapid demographic and urban growth." Arca Parró also expected that the next census would reveal how "the Andean population is distributing itself within the national territory," how "overpopulated" the coast had become, and how the Andean population was poised to "populate the jungle." For him, the spread of the "Andean population" across the country was an entirely expected phenomenon given that the Sierra "is and will continue to be Peru's reserve of human potential" and because its "areas of cultivation" do not increase in function with its rapid demographic growth.[93]

From this lofty demographic standpoint, Arca Parró strove to offer an impartial perspective that did not seek to "defend or exalt" the Puno-Tambopata project. In a concession to Monge, he argued that the spontaneous migration that had been occurring for the past two decades lacked "adequate orientation" and that no mechanisms were in place to mitigate the adversity that comes with "the process of biological adaptation and cultural assimilation of migrant groups." He thus lauded Monge for urging caution and calling for careful studies of adaptation before pushing the project forward. Yet Arca Parró also argued that it would be "deplorable" to assume that colonization of tropical regions with migrants from the highlands was impossible. In a direct challenge to Monge's arguments, Arca Parró rejected the idea that the inhabitants of the highlands could only develop "socially, economically, and culturally" in their own environment. "No matter how powerful and respectable the biological premises that support this argument are," Arca Parró reasoned, "the volume and intensity of the process of internal migration is demonstrating precisely the opposite." Without endorsing the Puno-Tambopata outright, Arca Parró cautiously concluded that if migration were "directed by the state, in collaboration with international organizations," many of the current "errors and vices" could be avoided and become an effective pilot project to be emulated in the future.[94] From the statistical perspective he honed as a demographer, Arca Parró thus made the case that population dynamics and geography were better guides to managing Peru's economic future than the climatic and physiological determinism often implied in Monge's work.

Despite the growing opposition to his views, Monge maintained his convictions well into the early 1960s. In a 1962 symposium on "migration and integration" organized by Henry F. Dobyns and Mario C. Vázquez (two anthropologists previously involved with the CPP,

Monge again insisted on the need to consider the problem of Andean migration from a biological perspective.[95] Yet by this point he was swimming against the intellectual tide. Most of the social scientists who gave papers at the symposium focused almost exclusively on the economic, social, and demographic dimensions of the issue with little concern for the problems of biological adaptation. But this did not deter Monge from returning to the "biological" aspects of migration in his closing remarks, where he lamented that this perspective had been "generally forgotten when contemplating the problems of the populations of the highlands." In these closing remarks, Monge also described the "nomadism" of Andean highlanders as a necessary but hazardous feature of their way of life that took both a biological and psychological toll and inevitably prompted Andean migrants to return to their original "ecological habitat." He concluded by once again extolling the ancient Incas for having judiciously legislated to account for both the "sociological facts" and the "biogeographic" reality of the Andes.[96] Whereas the rest of the symposium participants seemed to follow the jurisdictional division between the biological and social typically seen as a consequence of the UNESCO Statements on Race, Monge stubbornly continued to blur these boundaries.

CONCLUSION

Métraux and Monge's debate concerning the Puno-Tambopata project demonstrates that interwar eugenic thought and Cold War development projects shared common epistemological terrain. Before the formalization of development practice at the level of UN agencies after World War II, eugenics offered Latin American and Peruvian medical experts like Paz Soldán and Monge an international forum through which to organize their biosocial projects of modernization and racial improvement. It is not, therefore, surprising that as eugenics fell out of favor at the international level, Latin American and Peruvian scientists easily pivoted toward economic development discourse as an alternative. In this new context, most experts dutifully conformed to the emerging consensus that insisted on holding the biological and the social as separate domains. Yet this did not prevent them from continuing to uphold long-standing racial hierarchies. By insisting on the biological specificity of Andean highlanders and viewing human diversity through the prism of distinct physiological types, Monge continued to defy this international consensus well into the 1960s while positioning himself as a defender of Andean peoples and a champion of their development. The history of the Puno-Tambopata project thus suggests that the historiographies of development and race science would mutually benefit from an approximation and that such a move might allow scholars to properly reckon with the ways in which eugenic and development discourse have intersected.

The Puno-Tambopata project also suggests that Peruvian racial thought cannot be easily subsumed within the category of "Latin race science." Key elements of what historians have called Latin eugenics— such as the celebration of race mixing, an emphasis on racial progress through sanitary and environmental reform, and an emphasis on a Latin identity defined in opposition to Nordic eugenic conceptions—can certainly be observed in the work of Peruvian experts like Paz Soldán and Kuczynski-Godard. Yet as Monge's work makes abundantly clear, Peruvian *indigenistas'* insistence on the biological specificity of the Andean race often stood in stark contrast to Latin eugenicists' concern with biological absorption and cultural assimilation. Even though Monge participated in eugenic conferences during the interwar period, the longer trajectory of his career suggests that his abiding concerns were vernacular in the sense that he firmly believed in the climatic and biological distinctiveness of the Andes region and thus the need to adapt scientific methods and social policies in recognition of this reality. Monge and other Peruvian *indigenistas,* in other words, were not invested in Latinity or concerned with improving a putative "Latin" race. They were Andeanists first and foremost. The regional divides within Peruvian racial discourse and strategic appeals to Andean racial distinctiveness thus render Peru and the Andes region an awkward fit within both the "Latin eugenics" and "racial conceptions in the Global South" frameworks.

The differing regimes of racial and developmental thought exemplified by Métraux and Monge thus suggest the need to more carefully situate Andean racial conceptions in relation to international trends. Indeed, the conceptual debates provoked by the Puno-Tambopata project not only elude the Latin race science framework but also frustrate "north-south" binaries. Though it has often been associated with

Boasian anthropology, the antiracist and non-typological regime of racial thought epitomized by Métraux also had southern oceanic roots as exemplified by Métraux's ties to Te Rangi Hiroa, Ernest Beaglehole, and the Bishop Museum. Similarly, Monge's ties to North American researchers and institutions suggest that he cannot be thought of as a provincial or southern intellectual in a strict sense. For all of his investment in conserving Andean distinctiveness, Monge was also cosmopolitan and maintained an international network of correspondents in North America and Europe. The streams of racial thought that Métraux and Monge were situated in demonstrate that the oscillation between the local, national, regional, and international scales has been an enduring feature of racial thought and perhaps explains its tactical mobility.

PART THREE

Engineering Racial Harmony and Decolonization, 1952–1961

CHAPTER SIX

A Brazilian Racial Dilemma

Modernization and UNESCO's Race Relations Studies in Brazil

Writing for the *UNESCO Courier* in 1952, Alfred Métraux explained that many people were puzzled and even alarmed by UNESCO's recently launched sociological inquiry of Brazilian race relations. Why, in the wake of World War II and the Holocaust, would UNESCO devote major resources toward studying a nation *without* significant race problems? Critics of the Brazilian studies not only questioned the need for such an inquiry but also warned that it risked making Brazilians conscious of the racial differences that they had long ignored. "Take care not to arouse by indiscreet questioning the sort of antagonisms which are always ready to flare up," Métraux's critics warned.[1] Rather than confront Brazil's racial issues, Métraux's critics suggested it would be better for Brazilians to continue to live unaware of racial tensions that may exist. Criticism of UNESCO's studies seemed guided by a tacit sentiment—if Brazil isn't broken, why try to fix it?

As a retort, Métraux dismissed these critiques as lacking in "careful thought" and offered a characteristically staunch defense of UNESCO's Brazilian studies. An enormous literature on race already existed in the United States and South Africa, Métraux explained, as well as "hundreds of organizations and public and private institutions" concerned with solving problems "arising from racial discrimination." What "new facts" or insights could UNESCO hope to add to these over-studied racial situations, Métraux asked? Rather than follow these well-trodden paths, UNESCO's campaign against racial prejudice would be better served by studying rare examples of "harmonious race relations." In a context where proponents of "racialism" proposed segregation as a means to counter the "moral and physical decadence" they believed to result from race mixing, Brazil and its reputation for racial harmony promised a potent empirical refutation of these unfounded dogmas. Yet Métraux also noted that the initial results of UNESCO's race relations studies in Brazil had revealed unexpected tensions. In the industrial centers of São Paulo and Rio de Janeiro, Métraux noted, rapid urbanization and competition between European immigrants and "large numbers of coloured [migrants]" from Brazil's northeastern region had brought "racialist feelings" to the surface. If left unchecked, simmering hostilities could destroy the nation's "racial democracy." Yet Métraux expressed a cautious optimism and described a "strong resistance" to this latent racialism "based on liberal and open-hearted traditions" that kept the nation from spiraling out of control.[2]

Métraux's defense of the Brazilian studies reveals the competing interests and intellectual frameworks that were often at play in UNESCO's race campaign. The cautious hope that Métraux placed in Brazil's "openhearted" traditions aligns with the liberal internationalist values that informed so much of UNESCO's work in the 1950s. Métraux's outlook also echoed the racial liberalism that had taken root in the U.S. social sciences during the 1930s and 1940s. In the decades preceding UNESCO's race campaign, prominent U.S. scholars and politicians adopted an approach to tackling racism that emphasized changing individual attitudes through social engineering and education.[3] This approach was exemplified by Swedish economist and sociologist Gunnar Myrdal's monumental study *An American Dilemma*, which famously framed "the Negro problem" as a moral issue located "in the heart of the [white] American" and arising from "the conflict between his moral valuations on various levels of consciousness and generality."[4] Framed in this way, the problem of racism in the United States came to be seen by many social scientists as a matter of changing white attitudes and feelings through state-backed educational initiatives. In part, this liberal framing of racism stemmed from the work of interwar sociologists from the University of Chicago who challenged the biological basis of race and instead insisted that race was a matter of

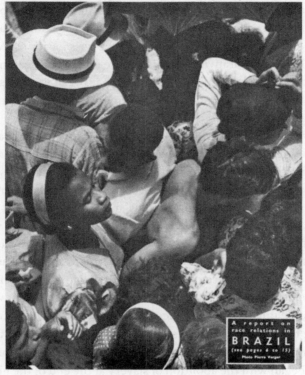

FIGURE 6.1 "A Report on Race Relations in Brazil" (cover image, *UNESCO Courier*, vol. 5, no. 8–9, August–September 1952).

consciousness, meaning the degree to which members of different racial groups attached meaning to racial difference. Instead of conceptualizing racism as something inherent to the structures and institutions of society, Chicago sociologists and social psychologists viewed racism, or what they more often referred to as "race prejudice," as a set of feelings and attitudes that took root in individuals typically in contexts where distinct social groups came together and competed over political and economic resources. Chicago sociologists thus endeavored to measure the degree of race prejudice in different sites and societies through a combination of surveys, questionnaires, and social statistics that sought to capture the perceived "social distance" between racial groups and the extent to which patterns of race relations conformed to "caste" or "class" stratification.[5] The project of race relations inquiry as it emerged in the work of Chicago sociologists, most notably Robert Park, was often premised on a comparative outlook, which assumed that race prejudice assumed a constant form that could be measured and compared across locations. And it was this liberal and comparativist framing of race relations research in North America that helped to cement the notion that Brazil was a distinctively harmonious society that could serve as a model for other countries battling with race prejudice.

And yet, as this chapter shows, rather than affirm the racial democracy thesis, the cycle of race relations studies that Métraux orchestrated in Brazil sparked the formation of an alternative and distinctively Brazilian approach to studying race relations. From the frame of U.S. racial liberalism, Brazil had appeared as an exemplar of racial harmony to many race relations researchers in Brazil and abroad in the 1930s and 1940s. Yet after the UNESCO-sponsored studies from the early 1950s, Brazilian researchers concentrated in Rio de Janeiro and São Paulo began to document the existence of a distinctly Brazilian racial dilemma that the sociologist Florestan Fernandes would describe as one in which the "white man" clung to the "prejudice of having no prejudice."[6] Yet instead of looking for psychological causes, Fernandes described it as a product of a historical process of economic development or "bourgeois revolution" that began with the abolition of slavery at the end of the nineteenth century yet systematically excluded Afro-Brazilians from participating in the nascent industrial economy of the 1950s. Fernandes and several other researchers who participated in the UNESCO-sponsored studies of Brazil's industrial centers thus came to see Brazilian race prejudice as an artifact of the slave trade that remained latent within the country's emerging industrial economy. In contrast to mainstream race relations researchers from North America who adopted individualist approaches consistent with racial liberalism during the 1950s, Brazil's urban sociologists adopted structural and economic analyses that highlighted inequitable and hierarchical distributions of wealth and power that correlated with skin color. They also began to conceptualize this inequitable structure, and the latent racial ideologies that supported it, as an important obstacle for Brazilian modernization and in this sense anticipated some of the themes that were later introduced by dependency theorists in the 1960s.[7]

In order to demonstrate how UNESCO's race relations studies in Brazil de-centered U.S. racial liberalism, this chapter begins by first tracing the influence of U.S. approaches to race prejudice during the early projects of UNESCO's Social Sciences Department (SSD). It then demonstrates how racial liberalism informed the work of U.S. sociologist Donald Pierson, whose work on race relations in Bahia during the 1940s affirmed the racial democracy thesis. The remaining parts of the chapter then demonstrate how Pierson's Bahia-centric analysis of Brazil's racial situation served as a starting point for UNESCO's cycle of race relations studies but was ultimately challenged and displaced by the urban studies of a new generation of sociologists including Florestan Fernandes, Roger Bastide, Luiz Aguiar Costa-Pinto, Fernando Henrique Cardoso, and Oracy Nogueira. By charting the waning fortunes of racial liberalism in Brazil, this chapter shows how UNESCO's internationalist ambition to identify a portable model of racial harmony unexpectedly yielded a thicker engagement with the afterlife of colonialism and slavery.

RACIAL INDIVIDUALISM AND INTERNATIONAL TENSIONS AT UNESCO

Before the 1950 and 1951 Statements on Race, UNESCO studies of race and racism were mostly subsumed within the SSD's broader cosmopolitan project of lessening international tensions and cultivating international harmony. In these early projects, SSD experts—mostly from North America—conceptualized race relations and racism as analogous to and hence not significantly different from other forms of intergroup tension such as conflicts between nations, genders, and cultures. Influenced by U.S. psychology and behavioral science, early SSD experts viewed racism and international conflict as problems arising from the personal prejudices of individuals. In this early period, SSD experts can thus be situated within the framework of "racial individualism," which historian Leah Gordon describes as one that "presented prejudice and discrimination as the root cause of racial conflict, focused on individuals in the study of race relations, and suggested that racial justice could be attained by changing white minds and protecting African American rights."[8] Yet when SSD experts began to plan the cycle of studies on race relations in Brazil in the early 1950s, they were confronted by circumstances that raised difficult questions for the North Atlantic paradigm of racial individualism. In a country supposedly free of racial tension, could race relations be properly understood through the frame of the psychological conceptions of race prejudice that informed UNESCO's early antiracist work?

Initially, most of the race relations work done by UNESCO social scientists occurred in the context of the Tensions Affecting International Understanding Project, which served as the flagship project of the SSD from 1947 to 1953. The kernel of the Tensions Project emerged from resolutions articulated at UNESCO's General Assembly in Mexico City in 1947. The resolutions instructed the director-general to promote research and action in five areas shaped by mid-century social psychology and culture and personality studies: to combat ignorance and prejudice by studying the distinctive character of different national cultures; to inquire into the conceptions and stereotypes that individuals from one nation hold of their own and other nations; to gather and promote research on methods for changing attitudes; to study the influences that predispose people toward either international understanding or aggressive nationalism; and to study the effects of population problems and modern technology on attitude formation.[9] By identifying individual attitudes and prejudices as the source of international conflict, the Tensions Project thus offered SSD experts a research agenda that conveniently avoided questions of geopolitical inequality and political economy and thus served to appease the national commissions of UNESCO at a time when imperial nations like the United Kingdom and France as well as the emerging U.S. empire held major sway.[10]

As the Tensions Project unfolded, it attracted prominent social scientists from elite North Atlantic institutions. Many of these experts hailed from departments and traditions that figure prominently in the conventional historiography of antiracism, notably Boasian anthropology, social psychology, and Chicago sociology. Perhaps the most influential participant in the Tensions Project was the Canadian social psychologist Otto Klineberg. Klineberg grew up in a Jewish-Canadian family in Montreal and earned a medical degree and a PhD in psychology at Columbia University in the 1920s. During his PhD studies, Klineberg trained in social psychology with Floyd Allport. As a graduate student, Klineberg also took several anthropology classes. These included a summer class with the linguistic anthropologist Edward Sapir on "Culture and Personality" and a departmental seminar taught by Franz Boas where he befriended budding young anthropologists like Ruth Benedict.[11] Klineberg's encounter with cultural anthropology had a profound effect on his PhD research and career trajectory. After conducting field research with children from Native American and African American communities, Klineberg completed a PhD dissertation in 1928 that demonstrated how their performance on intelligence tests was shaped by environmental and cultural factors.[12] On the basis of this research, Klineberg went on to become one of the most prominent critics against hereditary theories of intelligence that presumed the biological inferiority of non-European races. He developed this line of research in several key books, notably *Race Differences* (1935), *Negro Intelligence and Selective Migrations* (1935), and the UNESCO booklet *Race and Intelligence* (1951).[13]

After World War II, Klineberg also assumed a leading role in the Tensions Project and in the SSD. He served as director of the Tensions Project from 1948 to 1949 and later returned to the SSD for a second tour of duty as head of the newly formed Division of Applied Social Sciences where he continued with some of the Tensions Project research.[14] During his first SSD stint, Klineberg wrote a book titled *Tensions Affecting International Understanding* that offered an ambitious survey of social science research concerning international tensions. Two key objects that Klineberg linked together in this book were the study of "national stereotypes" and of "attitudes and their modification." And in both cases, Klineberg approached the issue of social tensions through methodologies that favored the individual as a unit of analysis, and of reform. For instance, Klineberg defined "national stereotypes" as "the pictures in people's heads referring to their own or to other national groups."[15] As an example of useful research from this domain, Klineberg described a pyschological study of U.S. magazines and how they portray minorities and foreigners through demeaning stereotypes, such as the "amusingly ignorant Negro, the Italian gangster, and the sly and shrewd Jew."

White Americans, by contrast, typically assumed the role of "approved" characters. In order to counteract these kinds of stereotypes, Klineberg stressed the importance of "developing an attitude toward individuals and groups . . . based upon the actual evidence obtainable, and not on the 'pictures in our heads.'" In a call to action, he also declared that "our educational attack must be against all forms of stereotyped thinking.[16]

As well as using education to combat negative racial stereotypes, Klineberg argued that psychology and the social sciences offered potent tools for measuring and modifying the attitudes that underpin racism and international tension. Drawing heavily from mid-century social psychology, Klineberg argued that "attitudes" were difficult to define yet could nevertheless be measured through methods such as observing how individuals choose companions in school or in recreation, through interviews and questionnaires, through "pictorial techniques" where reactions to persons represented in photographs serve to indicate attitude, and through projective techniques such as sentence completion and the construction of stories. Foreshadowing the SSD's later work, Klineberg also argued that psychologists, social scientists, and educators could play a role in changing individuals' negative attitudes toward other groups by both disseminating more accurate information about the "ways of life" of different groups and by prompting different groups to come into closer contact with one another.[17] Yet he also acknowledged that spreading information alone would not be sufficient for diminishing racist attitudes and also advocated for the creation of policies and incentives for members of different racial or national groups to voluntarily collaborate and cited studies on interracial housing in the U.S. Army as an example. Thus, even though the Tensions Project was not primarily conceptualized as antiracist, Klinberg's work demonstrates that much of the project was aligned with the racial individualism characteristic of mid-century antiracist discourse from the North Atlantic.

Though North American experts like Klineberg, held leading roles in the Tensions Project, the project also featured experts from Brazil and positioned the Lusophone nation as an exemplar of racial harmony. Klineberg himself had strong ties to Brazil, having spent two years (1945–1947) as a visiting professor in the Faculdade de Filosofia, Ciências e Letras at the Universidade de São Paulo (USP). During his time at the USP, Klineberg was tasked with establishing the institution's first

Department of Psychology, and he also worked toward institutionalizing psychology more broadly in Brazil by publishing introductory textbooks in Portuguese and by becoming a founding member of the Psychological Society of São Paulo.[18] Operating on the assumption that Brazil enjoyed less race prejudice than the United States and a pattern of greater race mixing, Klineberg also had plans (which he shared with Arthur Ramos) to conduct two major studies concerning race in Brazil: a first on differences in mental test results in schoolchildren classified according to "race (color) and economic class" and a second on "race relations" concerned with "patterns of friendliness and opposition between groups of different origin."[19] Though he was not able to complete these studies, when he published *Tensions Affecting International Understanding* Klineberg cited Brazil as an important example of how historical factors can produce unique "patterns of race relations." Echoing the Lusotropical narratives of the essayist and sociologist Gilberto Freyre, Klineberg described Brazil as enjoying a general pattern of "friendly" race relations that stemmed from Portuguese colonizers who had already had contact with a "colored" population—the "Moors"—and thus brought a friendly attitude that they extended to "Indian" and "Negro" women in Brazil.[20] A history of benign colonization coupled with the absence of civil war after slavery was abolished, argued Klineberg, went a long way toward explaining Brazil's relative lack of "agressive behavior" in the domain of race relations.

As the most internationally recognized proponent of this triumphalist version of Brazilian history, it is fitting that Freyre also played an important role in the development of the Tensions Project. In 1948, the Princeton University psychologist Hadley Cantril, who replaced Klineberg as director of the Tensions Project, invited Gilberto Freyre along with seven other prominent social scientists (including Gordon Allport, the Harvard social psychologist who specialized in the study of personality formation and prejudice, and the critical theorist Max Horkheimer) for a two-week stay at the UNESCO headquarters in Paris. During their stay, the group of social scientists discussed the causes of national aggression and how it might be dismantled and ended up producing a collectively written statement titled the "Causes of Tensions Which Make for War," which was published along with separate essays from the eight participants in a volume edited by Cantril.[21]

In his contribution to the volume, Freyre wrote an essay titled "Internationalizing Social Science" that called for greater international cooperation among social scientists and cited Latin America as an exemplar of harmonious social relations. After the conference took place, Cantril attempted to recruit Freyre for the position of director of the SSD, citing his "mellow wisdom" and international standing.[22] After Freyre declined the invitation, the SSD ended up hiring Arthur Ramos.

When Arthur Ramos was hired in 1949 as director of the SSD, his complaints about the excessive influence of U.S. psychology in the department's activities were thus well founded. In fact, the very idea of organizing an international conference of social scientists to discuss race prejudice was drawn straight from the playbook of the Tensions Project. And, indeed, SSD experts saw the 1949 conference in Paris that led to the publication of the 1950 Statement on Race as an extension of the Tensions Project. For instance, in a 1949 article describing the Tensions Project, Otto Klineberg cited the SSD's nascent campaign against race prejudice as one of the project's most important developments. Describing the SSD's plan to call together a group of experts to formulate a statement against racial prejudice, Klineberg noted a widespread need for "an educational offensive" geared toward the general public. While the "[North American] public generally believes that race differences are important" and "race prejudice is widespread," explained Klineberg, "scientists generally regard race as unimportant, and see no scientific justification for race prejudice." UNESCO's task was thus one of orchestrating an "educational offensive" that would reduce the gap between "popular and scientific knowledge in this respect."[23]

Given the prominence of Klineberg and racial individualism in the early stages of UNESCO's race campaign, it is not surprising that scholars have identified the UNESCO Statements on Race as potent symbols of mid-century U.S. racial liberalism alongside texts such as Gunnar Myrdal's *An American Dilemma* and the U.S. Supreme Court's *Brown v. Board of Education* ruling (in which Klineberg played a major role as an expert witness). The Tensions Project indeed shows how North Atlantic researchers like Klineberg often conceptualized dismantling racism as a technical problem to be fixed by engineering attitudes through psychological tools and placed much less importance on systemic and structural causes. Implicit in this approach was the notion that racial antagonism and intergroup tensions were the product of specific societal forces impinging upon individuals. Yet instead of dismantling social structures, adherents of racial individualism focused their energies on changing attitudes and promoting education designed to reduce personal prejudice. But as we will see in the rest of this chapter, though Brazil initially appeared to North Atlantic researchers to confirm this approach by offering an example of a place where individuals were largely free of racial antagonism, the research that UNESCO sponsored unsettled this view and prompted Brazilian researchers to turn to other ways of conceptualizing race and racism.

RACIAL INDIVIDUALISM IN BAHIA: DONALD PIERSON'S NEGROES IN BRAZIL

In the field of race relations, it was U.S. sociologist Donald Pierson's book *Negroes in Brazil: A Study of Race Contact at Bahia* (1942) that first articulated the view that Brazil was free of race prejudice. Accordingly, when SSD experts began to sketch out their plans for a cycle of race relations studies in Brazil in 1950, they cited Pierson's work as a major reference point. When *Negroes in Brazil* was published, its buoyant depiction of Brazil as a site of racial fusion had become commonplace in the minds of many Brazilian and North American observers and was most commonly associated with Gilberto Freyre's work.[24] Yet as a disciple of Robert Park and the Department of Sociology at the University of Chicago, Pierson's study offered a novel sociological framing of Freyre's Lusotropical narrative that was accompanied by numerous statistical tables depicting the spatial distribution of racial groups and their employment status. By focusing on this economic and geographic dimension, Pierson offered a variation of the racial individualism theme. Although he touched on themes such as attitude formation and race consciousness, Pierson's main conclusions in *Negroes in Brazil* concerned the social structure of Bahian society, which he viewed as a microcosm of Brazil as a whole. Pierson's main contention was that Bahian society was structured along "class" as opposed to "caste" lines. His study affirmed the racial democracy thesis by suggesting that interracial tensions did not impede upward social mobility and were ultimately manifestations of class prejudice as opposed to race prejudice.

In making this claim, Pierson drew a sharp contrast between the social structures of Brazil and the United States and thus exemplified the comparative approach to race relations favored by Chicago-trained sociologists who had initiated an ambitious transnational project concerned with mapping race relations in different parts of the world.

When Pierson was a PhD student in sociology, the idea of studying the Brazilian "racial situation" was one favored by many of the leading lights of Chicago's sociology department. After beginning his studies in 1934, the sociologist Herbert Blumer promptly called Pierson's attention to Brazil as "a locus for the study of racial and cultural contacts," and he was also encouraged to pursue this line of inquiry by Robert Redfield, Louis Wirth, and Ellsworth Faris. Having reviewed the available literature on Brazil in multiple languages, Pierson became increasingly enthused by the idea of conducting an "intimate and detailed" study of a racial situation that seemed distinct from both India where "the social order is based on caste" and the United States where "the Negro is now escaping from a caste status and becoming a racial minority similar to the Jew in Europe and elsewhere."[25] Pierson's dissertation project gained even greater traction in 1935 when Robert Park, the leading figure of the University of Chicago sociology department, returned from "an extended world-tour" during which he observed some of the most "important centers of racial and cultural contact, including Brazil" and gave his endorsement for Pierson's project.[26] With the help of the connections Park had established in Brazil, Pierson then spent two years (1935–1937) living and doing fieldwork in Salvador—"the old seaport of Bahia."[27]

When Robert Park wrote a glowing introduction for *Negroes in Brazil*, he framed Pierson's study and the field of race relations in expansive and comparative terms. Park argued that race relations at their core were fundamentally "ecological and biological" in nature and concerned with the territorial distributions of races and the changes that "inevitable miscegenation or interbreeding" brings about. Yet race relations studies had necessarily moved beyond this narrow focus. In the years leading up to Pierson's study, the field of race relations studies had expanded as had sociologists' growing awareness about the complexities of different racial situations. This growth and maturation of the field, Park argued, occurred in tandem with the global expansion of what he called "the racial horizon." "Wherever European economic expansion has brought

European peoples and the peoples or races of the world outside Europe into an association sufficiently intimate to produce a mixed-blood population," Park wrote, "the resulting situation is inevitably constituted a race problem." And as one of the world's most conspicuous "melting pots" distinguished by rapidly unfolding processes of "miscegenation and acculturation," Brazil offered a fertile site for a "comparative study of the problematic aspects of race and culture" at a time when "the structure of the existing world-order seems to be crumbling with the dissolution of the distances, physical and social, upon which that order seems to rest."[28] More importantly, however, Park argued that what made the Brazilian situation so interesting is that even though it had a "Negro population proportionally larger than the United Staes," it had no discernible race problem. In fact, after reading Pierson's manuscript, Park concluded that the main difference between the United States and Brazil was that "the people of Brazil have, somehow, regained that paradisal innocence, with respect to differences of race, which the people of the United States have somehow lost."[29]

Like many who commented on the racial situations of the United States and Brazil in this period, Park drew a stark dichotomy between these two nations. Yet, in practice, the generalizations and comparisons that Park and Pierson established between these two large countries were based on their experience with two specific regions: Bahia and the U.S. South. In drawing comparisons between these regions, Park and Pierson both advanced Park's comparative project and established connections between Brazilian scholars and government officials and scholars and institutions based in the U.S. South. When Pierson conducted field work in Bahia (1935–1937), Park retired from the University of Chicago and returned to the U.S. South. Before joining the University of Chicago sociology department, Park had spent seven years at the Tuskegee Institute—a historically black university in Alabama where he collaborated closely with Booker T. Washington.[30] In 1936, at the behest of the black sociologist Charles Johnson, his former student, Park left the University of Chicago and took up a position at another southern HBCU (historically Black college/university)—Fisk University in Nashville, Tennessee.[31] After finishing his Bahian fieldwork, Pierson joined Park at Fisk University where he worked on his book manuscript and taught a seminar with Park on "Race and Culture" in 1938. During this time, Park also arranged to have the anthropologist Ruth Landes, who was then a PhD student of Ruth Benedict's at Columbia, spend six months working with the Fisk library's African and African American collections before traveling to Bahia for her fieldwork.[32] Thus, when Pierson published *Negroes in Brazil*, he had already established strong institutional links between the U.S. South and Bahia.

In drawing a stark contrast between Brazil and the United States, Park and Pierson tapped into long-standing tropes concerning the lack of race prejudice in Brazil. Yet they also echoed the work of other Chicago race relations scholars who had come to interpret the race hierarchy of the U.S. South as a rigid caste society akin to the social system in India. The sociologists and social anthropologists who endorsed this interpretation of the U.S. South became known as the "modern caste school of race relations"—an epithet given to them by one of their fiercest critics: the Trinidadian American and Chicago-trained sociologist Oliver Cox.[33] Researchers in the caste school drew from an intellectual tradition that went back to nineteenth-century abolitionist movements and drew analogies between the Indian caste system and slave-holding societies, especially the United States.[34] The figurehead of the "caste school" was the social anthropologist W. Lloyd Warner, who served as professor of sociology and anthropology at the University of Chicago from 1935 to 1959. In an influential article titled "American Caste and Class," Warner introduced definitions of caste and class that were picked up by subsequent studies and echoed in Pierson's *Negroes in Brazil*.[35] In this article, Warner argued that the "Deep South" was unique insofar as it had accommodated two antithetical kinds of social stratification: a "caste system" and "class structure." According to Warner, "caste" and "class" refer to unequal social arrangements where "privileges, duties, obligations, opportunities, etc." are unequally distributed between hierarchically arranged groups. Yet whereas "class" systems allow for upward and downward mobility and intermarriage between members of different groups, "caste" systems maintain a rigid group boundary and only allow for mobility within a group. "Caste," argued Warner, functioned as a system where "marriage between two or more groups is not sanctioned and where there is no opportunity for members of the lower groups to rise into the upper groups or of members in the upper to fall into the lower ones."[36] In the context of the Deep South, Warner argued that the Jim Crow system had created a caste structure stratified by race (black and white), which also contained distinct class structures within it.

Pierson drafted *Negroes in Brazil* at Fisk University as the major studies of the "caste school" took shape. Though he did not cite these studies, the book's central arguments and concepts echo many of the themes of the "caste school." For instance, Pierson argued that the "Brazilian racial situation" was sufficiently distinct from that in "India" where the social order is "organized on the principle of caste." In contrast to India's caste system, Pierson instead drew an analogy between the racial situations of Brazil and Hawai'i, which he argued were examples of a "multiracial class society."[37] Echoing Gilberto Freyre, Pierson argued that the foundations of Brazil's multiracial class society lay in its benign colonial history, which allowed for an unprecedented "interpenetration of diverse races." As a result of this interpenetration, "the Brazilian Negro," like "the Brazilian Indian" before him, was inevitably disappearing through absorption into the "predominantly European population." This process of racial fusion coupled with the abolition of slavery dissolved racial boundaries and produced a competitive economic order in "which the individual finds his place on the basis of personal competence and individual achievement more than upon the basis of racial descent." Although Pierson observed the existence of a class system that corresponded with race, he argued that the representation of "blacks" and "mixed-bloods" throughout the "entire occupational scale" demonstrated the possibility of social mobility. "Since [. . .] the blacks, the mixed-bloods, and the whites do not constitute endogamous occupational groupings," Pierson reasoned, "the social structure is not that of caste."[38] As a corollary of this thesis, Pierson argued that discrimination toward darker-skinned people should not be mistaken for racism and should instead be interpreted as "class prejudice."

Although Pierson and his work received little attention in the United States, his analysis concerning the occupational distribution and mobility of racial groups in Brazilian society became an important benchmark for race relations research in Brazil.[39] As the first study to adopt the race relations framework for an analysis of Brazilian society, Pierson's *Negroes in Brazil* played a crucial role in prompting subsequent generations of researchers to abandon the concern with African survivals and acculturation and instead adopt a sociological approach concerned with the intersections between race and the economic stratification of Brazilian society. His interpretation of the Brazilian situation as a "multiracial class society" free of race prejudice also offered subsequent researchers a clear target. After his tenure at Fisk University, Pierson moved back to Brazil where he took up a position as professor of sociology and social anthropology at the Escola de Sociologia e Política in São Paulo, Brazil, from 1939 to 1959. From this position in Brazil's most populous city, Pierson played an important role in establishing São Paulo as a leading center for social scientific inquiry in the Americas.[40] And it would be the generation of students that Pierson helped to train in São Paulo in the 1940s and 1960s who would ultimately reject his interpretation of Brazilian society and what they would call the myth of Brazilian racial democracy.

RACIAL DETERIORATION: ALFRED MÉTRAUX AND THE SHIFT TO SÃO PAULO AND RIO DE JANEIRO

After a monthlong trip to Brazil at the end of 1950, Alfred Métraux wrote to Melville Herskovits and explained that he had been forced to rethink UNESCO's planned race relations studies. "Contrary to my previous plans," Métraux wrote, "Bahia will no longer be the focus of our project." Though he still planned to move ahead with a series of studies on rural communities in Bahia and social mobility in Salvador, Métraux explained that he now planned new studies concentrating on São Paulo and Rio de Janeiro where the racial situation was "rapidly deteriorating." By expanding the project's focus, Métraux expected to get "a picture of the racial situation in Brazil which will be close to reality and cover both the dark and the bright sides."[41]

Métraux further explained this change in approach in a letter to the Swedish economist Alva Myrdal (then director of the SSD). In this letter, Métraux explained that Bahia and the "favorable atmosphere" that permeated its race relations initially appeared the most appropriate site for addressing the resolution from the 1950 UNESCO General Conference, which asked the SSD to organize a pilot study in Brazil focused on "contacts between races or ethnic groups, with the aim of determining the economic, political, cultural and psychological factors, whether favourable or unfavourable to harmonious relations between races or ethnic groups."[42] Métraux also explained that a team of social scientists led by Charles Wagley was in Bahia conducting studies on social mobility and race, which had given him more reason to choose Bahia as the main site for the pilot investigation. Yet after traveling to Brazil and observing the issues firsthand, Métraux concluded that "the project as we had conceived it from a distance" risked yielding conclusions that applied only to "Bahia and not the whole of Brazil."[43] Métraux also cautioned that focusing narrowly on Bahia would expose UNESCO to legitimate criticism from sociologists from other regions in Brazil who "see the racial question from another angle."[44] For these reasons, Métraux "profoundly revised" the original program of study and established a new program of research in collaboration with Brazilian sociologists, which he asked Myrdal to approve.

Though brief by his standards, Métraux's trip to Brazil at the end of 1950 proved immensely consequential. The trip not only prompted Métraux to alter the scope of the Brazilian project but also paved the way for the emergence of a new and distinctly Brazilian approach to race relations research that differed from both racial individualism and the comparative studies favored by Chicago-trained sociologists. As we will see, this new approach emerged in São Paulo and Rio de Janeiro—Brazil's most populous and industrialized cities—and forced subsequent generations of scholars to reckon with the repercussions of slavery and the challenges they posed for modernizing Brazil. Yet when altering the scope of the project, Métraux and his collaborators in São Paulo and Rio first had to confront the racial individualism and racial democracy frameworks that held sway within UNESCO's ranks.

By confronting these frameworks, Métraux and his Brazilian collaborators would also begin drifting away from the strategy that had informed UNESCO's publication of the 1950 Statement on Race. As we saw in chapter 2, UN officials were interested in producing such a statement as part of a broader agenda focused on practical ways to diminish race prejudice, which represented an extension of themes from the Tensions Project and the framework of racial individualism. Thanks to Arthur Ramos's influence, the SSD had also begun to reorient its agenda to focus on issues concerning underdeveloped countries and the assimilation of Afro-descendent and indigenous groups into modern society. These two agendas converged during the fifth UNESCO General

Conference in Florence, Italy, which took place in May 1950. During this conference, SSD officials made the 1950 Statement on Race public for the first time. During the conference, representatives from UNESCO member states also approved the pilot study on race relations in Brazil.[45] The concurrence of these two initiatives at the General Conference was not coincidental. Indeed, according to Métraux, the pilot studies in Brazil—a country with a reputation for racial harmony and extensive race mixing—were meant to empirically refute the "racialist" claim that race mixing leads to the degeneration of populations and thus bolster the arguments of the 1950 Statement on Race. Yet it was this framing of Brazil as a straightforward exemplar and data mine that Métraux was forced to reckon with after his 1950 visit.

Even before his trip to Brazil, Métraux had begun to express doubts about received interpretations of Brazilian race relations and their over-reliance on Bahia as a representative site. When he joined the SSD in April 1950, he was appointed director of the newly created "Race Division" and soon given the task of drafting a plan of study for the pilot investigation in Brazil. The SSD also hired Ruy Coehlo—a Brazilian anthropologist and former student of Roger Bastide at the Universidade de São Paulo—to act as Métraux's assistant. In collaboration with Coehlo, Métraux reviewed the existing literature on race relations in Brazil and drafted a proposal for a project centered primarily on Bahia, which they sent to Otto Klinberg for commentary.

In this proposal, Métraux and Coehlo argued that existing studies on race in Brazil did not adequately address the topic of race relations. Métraux and Coehlo's proposal began by describing the "evolution of race studies in Brazil" and noted that "race problems" had typically not been considered problems *per se* in Brazil but rather as part of "general social problems."[46] The report noted that the existing research offered a detailed portrait of the ways of life of the "mixed populations of Brazil" and their material culture, folklore, religion, and social organization. Yet on the topic of race relations, Métraux and Coehlo argued that "much has been written, but much remains to be done."[47] Specifically, they argued that Pierson's *Negroes in Brazil* had made a start in the right direction but used questionable methods and thus arrived at unreliable conclusions. "It is doubtful," Métraux and Coehlo wrote, "whether the use of questionnaires, dealing directly with racial attitudes, gives an adequate picture of the situation."[48] Although such methods were used widely among race relations researchers in the United States, Métraux and Coehlo argued that for Brazil they would be of little use. Indeed, Métraux and Coehlo warned that given that in Brazil "it is considered disgraceful to have race prejudices . . . when [prejudices] do exist, [they] may assume covert and subtle forms . . . not revealed by the questionnaire technique."[49] To properly study race prejudice in Brazil and capture such covert attitudes, Métraux and Coehlo suggested that researchers would have to adopt other methods and techniques beyond questionnaires. For instance, they recommended the use of psychological techniques such as interviews and questionnaires based on attitude scales, pictorial techniques to gauge peoples' reactions to photographs of people of different races, and Rorschach tests and thematic apperception tests to study the range of personality types within groups.

Yet after Métraux's trip to Brazil at the end of 1950, these psychological approaches became less central to UNESCO's project. Once in Brazil, Métraux first traveled to Bahia where he met with Wagley's team and traveled to the field sites to observe the studies under way.[50] After spending several days in the northeast, Métraux traveled to Rio de Janeiro and met with Luiz Aguiar Costa-Pinto, the sociologist who had participated in the 1949 meeting on race and paid hommage to Arthur Ramos. Costa-Pinto expressed great interest in participating in the UNESCO studies and recommended Rio de Janeiro as a research site because of the structural changes that were taking place as a result of the industrialization transforming the city. Because the SSD had already displayed strong interest in studying the effects of industrialization, Métraux agreed that Rio would be a good research site and hired Costa-Pinto to conduct a study. By bringing Costa-Pinto aboard, Métraux thus gave one of the new generation of Brazilian sociologists the opportunity to shape the course of race relations studies for decades to come.

Yet it was when Métraux traveled to Brazil's most populous and rapidly growing city—São Paulo—that his vision for the UNESCO project began to radically change. In São Paulo, Métraux decided to recruit sociologists, anthropologists, and psychologists from two prominent centers at the Universidade de São Paulo: the philosophy faculty and the Escola Livre de Sociologia e Politica (Free School of Sociology and

Politics; ELSP). During the 1940s, the ELSP had established itself as the country's preeminent institution for graduate training in social science thanks to the service of foreign-trained sociologists including Donald Pierson (who acted as the center's director of graduate studies) and the French Durkheimian sociologist Roger Bastide.[51] Yet in order to avoid feelings of resentment and rivalry, Métraux followed the suggestion of his colleagues and established a committee of scholars from both the philosophy faculty and ELSP.[52] In his UNESCO report on the trip, Métraux also noted that Donald Pierson—"the best known specialist on race issues in Brazil"—had approved the project but would not be available to participate because he was conducting a study on race relations and industrialization in Brazil's São Francisco River valley.

During his visit to São Paulo, Métraux also gathered information on the state of race relations in the city. This firsthand information persuaded him that a well-managed sociological study might "bring to light little known facts and capture the beginnings of a process that could alter the possibly overly optimistic image of the racial problem in Brazil."[53] Not one to shy away from uncomfortable truths, Métraux cautioned in his report to the director-general that adding São Paulo as a research site could mean that the pilot studies would not respond to the original aims proposed during the 1950 General Conference:

> Conflicts and tensions are coming into being following from the rapid industrial development of São Paulo. This giant city presents us with a unique opportunity to learn the factors that incite racial antagonisms, which are otherwise latent or lacking in virulence. I don't ignore the fact that by organizing a study in São Paulo we risk arriving at conclusions that will not respond to the hopes of those that presented and voted for the aforementioned resolution, but it would be a treason to the scientific spirit that should inform our study if we brush aside newly arising problems in order to hold on to happy but outdated view of the situation. The Bahia study can only offer us an incomplete image of the racial question of Brazil.[54]

Métraux's decision to expand the study and steer it away from the aspirations of UNESCO officials marked an important turning point in UNESCO's race campaign and in the history of racial thought in

Brazil. At this point, thanks to the cheerful image of Brazilian race relations that scholars like Gilberto Freyre and Donald Pierson had crafted
during the 1930s and 1940s, the idea that Brazilian society offered compelling lessons on racial harmony was firmly established among both
French and U.S. scholars.[55] But as Métraux soon realized after his visits
to Rio and São Paulo, interwar scholars assembled this cheerful image
of Brazil as a country free of race prejudice on the basis of regional
studies of northeastern parts of the country and, above all, Bahia. The
geographic specificity of the racial democracy trope was also matched
by an anachronistic rendering of Bahian society as a place of heritage
and tradition where the best elements of Brazilian colonial society had
remained intact—a vision of Bahia that Anadelia Romo has aptly called
"Brazil's Living Museum."[56] By contrast, when Métraux traveled to Rio
and São Paulo—two cities that were rapidly transforming because of the
quickening forces of immigration, urbanization, and industrialization—
he was forced to grapple with evidence of racial tension that prompted
him to consider a more sober assessment of Brazilian society. Yet given
the preponderance of racial individualism and racial democracy in
the early stages of UNESCO's race campaign, the project of rendering
Brazil's racism visible internationally was an intellectual challenge that
prompted budding scholars in Rio and São Paulo to embrace new and
distinctive methods of inquiry that broke with received frameworks.

RACIAL DEMOCRACY AND MODERNIZATION: RACE
RELATIONS STUDIES IN BAHIA

The cycle of studies that UNESCO sponsored did not present a uniform vision of Brazilian society. While the studies that Métraux com-
missioned in Brazil's major urban centers would eventually challenge
received narratives, the studies conducted by Charles Wagley and his
team of Columbia University graduate students in Bahia reaffirmed the
status quo—a development that was welcomed by local politicians and
reformers who played an active role in shaping the scope and design of
these studies. Yet they also struck a cautionary note and warned against
the dangers of rapidly modernizing Bahian society and thereby eroding the
structure of racial harmony put in place during colonial times. The
Columbia-Bahia studies thus exemplified how mid-century experts so

often assembled antiracism in ways that responded to the concerns and interests of actors and forces at different scales including the cosmopolitan and internationalist ideals of UNESCO officials and the modernizing ambitions of local intellectuals and politicians.

Before Métraux and UNESCO became involved with the Bahia studies, Wagley and his collaborators envisioned the project as one primarily addressing modernization and social change. Instead of focusing on race relations, the original aim of the Bahian project was "to study the process of social and cultural change, especially the changes related to the appearances of different types of economy, of modern technology, of new ideology, and of concomitant complex administrative structures in a relatively non-technically developed society."[57] Within this frame, project participants emphasized the collection of demographic data concerning the composition of Bahia's population, migration patterns, and administrative divisions. They also emphasized the geographic diversity of Bahia State and divided the region into six "ecological zones": the Recôncavo, the Northeastern Sertão, the Southern Forest, the Central Plateau, the São Francisco Valley, and the Western Plateau. Wagley and his collaborators established these zones on the basis of data and bibliographic material collected by Thales de Azevedo—an anthropologist from Bahia who had originally trained as a physician at Bahia's School of Medicine and who served as codirector of the project. When fieldwork for the project began in 1950, Wagley and two of his graduate students (Marvin Harris and Ben Zimmerman) focused on three of these zones (Recôncavo, the Northeastern Sertão, and the Central Plateau). For each zone, Wagley, Harris, and Zimmerman chose one "traditional" community that tended to be "stable and conservative" and a second "progressive" community where "economic and social change seemed to be more [pronounced]." By comparing two contrasting communities across three different ecological zones, Wagley and his team thus hoped to establish a "picture of the dynamics and substantive changes underway in each of the three zones."[58]

This framing of the project reflected the political and economic changes that swept through Bahia and Brazil after the resignation of Getulio Vargas in 1945 and the return to democracy. In the decades after Vargas's resignation, Brazilian elites embraced a discourse of optimism and renewal and pushed through some of the most ambitious modernization policies of their country's history. At the outset of the 1950s, Brazil's economy was mainly geared toward the export of agricultural products, where coffee made up more than half of export revenues. In order to push the country forward, ruling politicians and leading economists sought to transform Brazil's economy from one reliant on agricultural exports to one more reliant on industrial manufacturing and more diverse commodities. The developmentalist ambitions of this period crystallized in the administration of Juscelino Kubitschek, who became president in 1955 and set his administration the ambitious goal of accomplishing fifty years worth of industrial modernization in five years.[59]

The Columbia-Bahia project's initial emphasis on issues concerning social change also reflects the influence of the progressive education reformer and Bahian Anísio Teixeira. An important institution builder within Brazil's educational system, Teixeira had studied with John Dewey at Columbia University in the 1920s and then spent most of the 1930s in Rio de Janeiro where he worked with Arthur Ramos and created the Institute for Educational Research in Rio's school system and later the Universidade do Distrito Federal in 1935. Because of increased persecution from the Vargas regime, Teixeira left Rio for Bahia and eventually, in 1948, became secretary of health and education for Bahia State.[60] During his spell as secretary of health and education, Teixeira oversaw a major overhaul of the state's education system that echoed the concern with racial uplift characteristic of previous generations of Bahian elites such as the Tropicalistas and Arthur Ramos. During this period, Bahia enjoyed an economic boom spurred by the discovery of petroleum and evidenced by the rapid growth in the population of Salvador, which doubled its population between 1940 and 1960.[61] The demographic and economic changes taking place prompted Teixeira to push through major changes to the school system with the overarching aim of better socializing children from "backward" groups and preparing them for the demands of modern civilization. A central pillar of Teixeira's school reform was the creation of a new kind of school, which he dubbed *escola-parque* (a "park-school"), that was premised on a curriculum that integrated basic education in reading, writing, and math with the cultivation of the social skills necessary for navigating modern society and the promotion of physical health through good nutrition and physical education. Teixeira argued that such ambitious reforms and new kinds of schools were necessary because Bahia's tropical environment was not favorable to modern civilization and because the state's population was made up of a combination of indigenous people, imported Africans, and Portuguese settlers with a tendency toward passivity. He also saw the various "premodern" traditions of Bahia's inhabitants as an obstacle to the state's modernization and chided social scientists like Arthur Ramos and Melville Herskovits for having brought so much attention to Afro-Brazilian traditions like Candomblé. By preparing children for the rigors of modern society and stripping them of their traditional customs, Teixeira also believed that the school system would neutralize the threat of social disruption during a period of significant societal change.[62]

When he first proposed the Bahia-Columbia project in 1948, Teixeira envisioned it as a study of the social differences between different regions in Bahia, which would shed light on how to adapt schools to suit local needs. With funding from the Division for the Development of Science of Bahia State and from Columbia University, Teixeira and Wagley assembled a research team made up of Wagley's aforementioned graduate students and several Brazilian social scientists: Eduardo Galvão, Luiz Aguiar Costa-Pinto, Gizella Valladares, and Thales de Azevedo. Initially, Teixeira and Wagley agreed that the objective of the community studies would be to examine processes of social change in the "folk" societies of Bahia's interior, which were undergoing change due to the introduction of radio, television, new roads, and schools. Wagley recommended examining several communities from different physical environments and at differing stages of progress in order to obtain a snapshot of how communities were transitioning from "premodern" to "modern." Yet after hearing that UNESCO intended to conduct a series of studies on Brazil's race relations, Wagley contacted his friend Alfred Métraux and recommended Bahia as an ideal place for these kinds of studies and suggested that he could oversee such studies because he had a team of researchers in place. Thus, as a result of the collaboration with UNESCO, the scope of the studies expanded beyond the question of social change and the process of modernization to encompass questions concerning Bahia's pattern of race relations.[63]

Like Donald Pierson, Wagley and his team drew their inspiration from the urban sociology and "community studies" approaches favored by many Chicago-trained sociologists. Adopting Wagley's concern that the studies cover the geographic as well as social diversity of Bahia, the studies focused on three separate sites: a traditional plantation community (Vila Recôncavo), which was studied by Harry W. Hutchinson; an old mining town in the Brazilian plateu (Minas Velhas), which was studied by Marvin Harris; and a typical community of the arid northeastern region (Monte Serrat), which was studied by Ben Zimmerman. When UNESCO published the results of these studies in a book called *Race and Class in Rural Brazil*, Wagley also included a chapter on race relations in the Amazonian community of Itá that was based on research that Wagley had conducted as part of the ethnographic surveys for the International Institute of the Hylean Amazon. The studies drew on community studies methods as well as psychological methods of attitude analysis including participant observation, interviews, questionnaires, surveys, and the use of statistical and historical documents to create a portrait of the daily life in the communities and the attitudes and preconceptions held between people of different races. Through interview and participation observation methods, the researchers carefully documented the criteria and categories that the community members from each site used to classify people's racial identity and collected information in the form of folktales and sayings to get a sense of what kinds of stereotypes people had about racial groups.

In addition to participant observation, the researchers developed survey and questionnaire techniques for gaining access to the attitudes that Bahians held about members of other racial groups. Hutchinson, for example, conducted a survey where he asked eighty-five community members of different racial and sexual backgrounds to examine eight photographs consisting of both males and females from each of the four locally acknowledged racial types (the *caboclo*, or Amerindian; the *preto*, or black; the *mulato*, or mixed; and the *branco*, or white).[64] No information was given about the social background of the people depicted in the photos, and the participants were asked to rank the photos by considering which they found most physically attractive, those they believed to be most wealthy, the best worker, the most honest, and the most religious. Marvin Harris showed to ninety-six individuals of different color, class,

FIGURE 6.2 Examples of the types of photographs used by Wagley's research team. Pierre Verger, (a) *Mulatto cowboy of the interior of Bahia* and (b) Cabocla of the interior of Bahia (photographs, in Charles Wagley, ed., *Race and Class in Rural Brazil*, 2nd ed., Race and Society. New York: UNESCO, 1963).

sex, and age three pairs of photographs corresponding with males and females from three "racial types" (Negro, mulatto, and white) and asked the participants to rank them using similar criteria (beauty, wealth, ability to work, honesty, religiosity).[65] Ben Zimmerman made use of a similar method by showing eight photographs (corresponding to men and women from four racial types) to one hundred participants and asking the participants to judge them on the basis of the same criteria but also asking the participants to indicate whether they would accept the man or woman as a neighbor, friend, guest, dance partner, brother-in-law, or sister-in-law.[66] The researchers believed that these tests made it possible to measure the attitudes of the various racial groups, the social distance between people of different color, and to evaluate to what extent physical appearance determined how individuals were allotted to different social and economic strata.[67]

The main conclusion that Wagley and his team drew from their observations and data was that "race" was of little importance in determining Bahian social structure and that Bahia could indeed be characterized as a "racial democracy." In the historical backdrop that the researchers provided for their studies, they suggested that Bahian society may have been formerly made up of castes, but that the extensive miscegenation that had taken place since the colonial era now rendered it a society of *classes*.[68] Yet although the studies concluded that "race" does not play a strong role in shaping Bahia's social structure, they did show Bahians to have strong inclinations toward making racial classifications based on phenotypic differences, as evidenced by their recognition of numerous racial types. Despite the evidence that racial type was an important and omnipresent element of Bahian social classification, Wagley and his team argued that it was only one of many factors including wealth, profession, and education that individuals used to determine social status. Thus, Wagley claimed that the "most crucial alignment in rural Brazilian society was that of social classes, and that racial type was generally but one criterion by which individuals were assigned to a social class."[69]

Wagley's studies also suggested that Bahia's harmonious relations between the races were the product of a premodern social structure in which there was little economic competition between social groups. Indeed, Wagley argued that Bahia's social patterns had not changed

APPENDIX II. Social Distance Tests

(a) Male Photographs

	Race				Class			Sex	
	N.	*W.*	*M.*	*C.*	*I*	*II*	*III*	*M.*	*F.*
	6	39	11	44	17	30	53	50	50

1. Would you accept this man as a neighbour?

Caboclo	2	18	4	24	7	14	27	26	22
Preto	2	5	2	11	2	6	12	9	11
Mulato	5	10	4	15	3	8	23	17	17
Branco	4	32	7	38	14	25	42	36	45

2. Would you accept this man as a friend?

Caboclo	2	18	7	25	9	16	27	30	22
Preto	3	4	3	11	0	5	16	9	12
Mulato	3	9	5	21	1	12	25	16	22
Branco	5	30	10	39	15	24	45	38	46

3. Would you invite this man to dinner?

Caboclo	3	17	5	26	7	14	30	27	24
Preto	4	7	2	18	3	6	22	16	15
Mulato	5	14	5	21	4	10	31	21	24
Branco	5	28	6	36	13	20	42	35	40

4. Would you let your daughter dance with him?

Caboclo	3	15	7	28	5	4	34	26	27
Preto	3	9	4	19	2	9	24	15	20
Mulato	5	16	8	24	2	15	36	25	28
Branco	4	29	9	35	11	21	45	34	43

5. Would you accept him as a brother-in-law?

Caboclo	3	12	8	20	3	11	29	19	24
Preto	4	6	5	8	1	6	16	12	11
Mulato	6	9	6	16	1	9	27	17	20
Branco	6	17	8	33	8	15	41	33	31

FIGURE 6.3 Example of the kind of questionnaire used by Wagley's team. Ben Zimmerman, "Appendix II. Social Distance Tests" (in "Race Relations in the Arid Sertão," in Charles Wagley, ed., *Race and Class in Rural Brazil*, 2nd ed. Race and Society. New York: UNESCO, 1963, pp. 114–15).

(b) Female Photographs

	Race				Class			Sexe	
	N.	*W.*	*M.*	*C.*	*I*	*II*	*III*	*M.*	*F.*
	6	39	11	44	17·	30	53	50	50

1. Would you accept this woman as a neighbour?

Cabocla	5	11	4	17	5	10	22	20	17
Preta	4	7	3	16	1	7	22	13	17
Mulata	6	22	6	25	4	20	35	28	31
Branca	6	29	8	36	7	27	45	40	39

2. Would you accept this woman as a friend?

Cabocla	5	14	4	20	5	11	28	22	22
Preta	5	12	5	11	3	10	20	12	21
Mulata	6	23	7	23	9	16	34	25	34
Branca	6	27	9	35	11	20	46	36	41

3. Would you invite this woman to dinner?

Cabocla	3	11	5	19	4	10	24	23	15
Preta	6	12	5	20	5	7	31	21	22
Mulata	6	22	7	27	6	17	39	32	30
Branca	6	27	9	35	12	18	47	38	39

4. Would you allow your son to dance with her?

Cabocla	5	20	7	25	9	14	34	30	27
Preta	6	21	7	28	9	15	38	30	32
Mulata	6	28	11	31	13	21	42	38	38
Branca	6	31	11	40	15	24	49	42	46

5. Would you accept her as a sister-in-law?

Cabocla	4	9	5	23	2	11	28	20	21
Preta	4	12	6	21	1	10	32	19	24
Mulata	5	17	8	30	3	18	39	28	32
Branca	6	26	10	34	11	19	46	35	41

FIGURE 6.3 *Continued*

since the abolition of slavery—a phenomenon he attributed to a relative lack of modernization in the region:

> Thus, the picture of relations between social classes and racial groups which emerges from our studies of these rural communities is on the whole pre-industrial and pre-urban (in the sense of the twentieth century urban complex). The patterns of class and race relations in these communities approximate to those which had taken form in Brazil in the last century out of the unique Brazilian past.[70]

Although Wagley asserted that Bahia was largely shaped by practices and traditions stemming from the nineteenth century and argued that many of these traditions should be preserved and strengthened, he also suggested that "social barriers are likewise uncovered which prevent the full utilization of the human resources of these rural communities and which must be erased if Brazil is to become the great democracy it promises to be."[71]

Wagley's studies thus situated Bahia's amicable race relations within a narrative that suggested if Brazil was to prosper, it would need to both preserve its traditional culture and modernize through technological and industrial change. Indeed, Wagley concluded *Race and Class in Rural Brazil* by predicting that Brazil would continue to develop economically and to expand its educational system thus allowing "the people of darker skin colour" to improve their economic situation and assume their rightful place among a growing middle class. For Wagley, the notion that Brazil was free from racial prejudice and lacked a caste-like system like the United States led him to believe that "there are no serious racial barriers to social and economic advancement, and as opportunities increase, large numbers of people will rise in the social system."[72] Given the meritocratic and democratic nature of Brazilian society, Wagley explained that Brazilians should have no reason to fear change or make a connection between modernization, industrialization, and racism:

> Both Brazilians and foreign observers have the impression that Western attitudes and concepts of racism are entering Brazil along with industrial and technological improvements. But there is no

inherent relationship between Western industrialism and technology and Western racism, no necessary connexion [*sic*] between the widespread improvement of social conditions and the development, through competition, of tensions and discrimination between racial or minority groups. Aware of the dangers and pitfalls and taking care to avoid them, Brazil may enjoy the benefits of technological change, and of greater rewards for its underdeveloped potentialities, without losing its rich heritage of racial democracy.[73]

Wagley and his research team thus saw Bahia as teeming with economic and cultural potential. Like the discourses of cultural change that were discussed in chapter 2, Wagley suggested that the best way forward for Bahia would be to establish a careful equilibrium between "tradition" and "modernity." Indeed, for Wagley, Bahia's potential lay in the fact that its "traditional" culture was already characterized by democratic and prejudice-free relations and could thus be easily grafted onto a more competitive and industrialized way of life. Wagley's optimistic assessment of Bahia's future is thus emblematic of the kinds of redemptive aspirations that provoked the interest of UNESCO and the international social science community. As described by Wagley, Bahian society represented an ideal blend of tradition and modernity wherein colonial-era social relations were redeemed as a social order that could serve as a template for industrialized nations whose race relations were comparatively less democratic.

Thus, by downplaying the significance of "race" within Bahia's race relations, and by denying any necessary link between racism and modernization, Wagley's UNESCO-supported studies furthered the race-as-caste analogy that had become central to the American interpretation of the function of racial ideology. On the basis of an implicit and at times explicit contrast with the race relations of the United States, Wagley and his team of researchers concluded that there was little evidence of race prejudice and that "race" in fact had little significance in Bahian society. Yet there is a notable tension in Wagley's conclusions. On the one hand, his research team argued for the minimal relevance of race within Bahia society, while on the other hand his research team introduced extensive evidence suggesting that

Bahian social relations were racially coded and that Bahians possessed intricate and elaborate racial classifications that often furthered racial stereotypes. However, because these ways of conceiving "race" did not conform to the segregationist model of the U.S. South that had been established as a paradigm of antagonistic and problematic racial relations, Wagley's research team attached little significance to the palpable racial ideologies that they observed in Bahia. Because racial tensions tended to align with class divisions, and because Wagley and his colleagues tended to believe that class differences could be reduced through modernization, they arrived at the conclusion that "race" did not appear to pose a major social problem. Thus, whereas the race-as-caste metaphor proved useful in the American context for describing the U.S. system of race relations, in Bahia it had the effect of obscuring North American researchers' understanding of the ways that "race" continued to inform and shape Bahian society and Brazilian society more broadly.

DOCUMENTING RACE PREJUDICE IN SÃO PAULO AND RIO DE JANEIRO

In contrast to the Bahian studies, the studies conducted in Rio de Janeiro and São Paulo developed forceful counter-narratives to the "racial democracy" thesis by showing how racial tensions and anti-black prejudice rose as Afro-Brazilians became upwardly mobile. Through these studies, scholars developed influential antiracist conceptual frameworks that insisted on the geographic specificity of race relations and described how latent colonial structures within Brazilian society encouraged individuals to deny the existence of race prejudice. Many of the key claims from these studies were the product of Rio and São Paulo sociologists' engagement with the work of emerging Afro-Brazilian activists and associations like the Teatro Experimental do Negro (TEN). These associations emerged during the 1930s and 1940s amidst the social upheaval of the Estado Novo regime of Getulio Vargas and called for Afro-Brazilians to assert and reclaim a distinctive black consciousness as a means of countering race prejudice and overcoming the inheritance of slavery.[74] In their UNESCO-sponsored studies of race relations, Rio and São Paulo sociologists argued that the race prejudice that they documented through economic and census data merely corroborated the race prejudice that Brazil's Afro-Brazilian groups had long experienced. Through their critical engagement with received Brazilian frameworks and ties to Afro-Brazilian groups, this new generation of race relations scholars, who came of age during the Cold War, articulated a novel analysis of race relations that broke with the mainstream approaches that circulated in North Atlantic channels.

From the very outset, the leading researchers of the São Paulo and Rio studies declared their intent to unsettle existing race relations frameworks and the portrait of Brazilian society that had emerged form the work of scholars who studied Bahia. Before the race relations project began in earnest, the lead investigators of each regional study sent a project proposal to Métraux's office in Paris. For the Bahian studies, Charles Wagley sent a four-page "tentative plan of research" and budget that briefly outlined the different sites of study, the personnel, and the need for a social psychologist on the team. By contrast, Roger Bastide and Florestan Fernandes—the main architects of the São Paulo studies— sent Métraux a lengthy twenty-five-page position paper titled "Racial Prejudice in São Paulo."[75] In this methodological and conceptual essay, which was later published as an appendix in their book on race relations in São Paulo, Bastide and Fernandes offered a critique of the "prejudice" concept as defined by American race relations researchers and stressed the need to adapt and refine sociological concepts and categories to suit local realities. They also argued that a traditional "research plan" would not suffice for their study and instead focused on building a theoretical framework from which they would draw a guiding hypothesis for the project. Through this document, Bastide and Fernandes signaled their intent to break with existing approaches to race in Brazil.[76]

In charting a new direction, Bastide and Fernandes took aim at the "prejudice" concept that served as the backbone of U.S. racial individualism. Though they conceded that the prejudice concept was necessary for their envisioned research, Bastide and Fernandes insisted that its use by U.S. social psychologists and sociologists like Gordon Allport and Arnold M. Rose was too abstract and "imprecise." They also argued that race relations studies from the United States—notably those of the "caste" school of race relations and the work of black sociologists like E. Franklin Frazier, Horace R. Cayton Jr., and S. C. Drake—were of limited use for São Paulo because they described instances of "interracial adjustment" on the basis of "segregation" and on a combination of "[caste] and class" regimes. If the concept of race prejudice was to have any use for their study of race relations in São Paulo, the two sociologists insisted, it would have to be adapted to suit the distinct social dynamics at play in the rapidly growing city and would have to be given greater scientific rigor. In this vein, they also worried that UNESCO's race relations project lacked a solid scientific foundation insofar as it was motivated by "extra-scientific" goals; namely, the international organization's desire for practical knowledge that could be immediately put to use for the purpose of the "social re-education of adults" and its basic project of "bringing races together."

To give the prejudice concept a more rigorous foundation, Bastide and Fernandes eschewed the psychological framing of U.S. scholars and instead conceptualized it through an expansive structural functionalist framework inspired by the work of Émile Durkheim. Quoting from Durkheim's *Les Règles de la Méthode Sociologique* (1947), they argued that the origins of prejudice, like any other social process, has to "be sought in the constitution of the internal social environment." As such, a proper sociological study of prejudice requires an examination of how it emerges as a social phenomenon shaped by the "moral and material conditions" of "human co-existence." They also insisted on defining "race" in a strictly sociological sense—as nonbiological and distinct from the "race" of physical anthropology—and argued that race prejudice is conditioned by specific societal norms, rules, and codes that establish "limits to the fluctuation of individual behavior." So instead of approaching race prejudice as an individually held attitude, they conceptualized it as a phenomenon that arises from and inheres in social relations and societal norms.

Citing the work of Robert Park, they also described race prejudice as emerging from the vertical social structures created by economic inequality. "It seems that racial prejudice," they argued, "tends to develop as a natural consequence of the intermittent or continuous contact of people or groups of people belonging to different 'races,' whenever conditions of economic and social inequality contrast racial marks with notable discrepancies in regards to occupations, wealth, standard of living, social status and education."[77] And they also insisted on treating race prejudice as a fluid phenomenon that changes over time. "Racial prejudice necessarily reflects all the fluctuations and transformations of importance that operate in given social situations," Bastide and Fernandes concluded, and explained that "its meaning and function change continuously, in the course of the evolution of societies." By insisting that race prejudice can only be understood in historical and socioeconomic terms, Bastide and Fernandes thus departed from the methodological individualism that took root in North Atlantic studies of race relations.

This historical and sociological approach directly informed how Bastide and Fernandes conceptualized Brazilian race prejudice. While many had interpreted Brazil's extensive race mixing as marking an absence of race prejudice, Bastide and Fernandes instead argued that race prejudice was often latent and manifest in indirect ways that were rarely externalized. This coded and veiled form of racism, they argued, reflects its origins during slavery where it became a "condition for the accommodation of whites and blacks." A consequence of this accommodation is that "racial prejudice did not regularly reach ostentatious or violent expressions during the period of slavery and later remained latent in the behavior of whites." Nor did they see the abolition of slavery as a major rupture point in Brazilian racism's trajectory. Even after abolition, they noted, racial prejudice served to undermine Afro-Brazilians' newly won status as "citizens" yet was not openly denounced until much later and even so only in urban centers. And it was precisely this shift toward an open confrontation with Brazil's enduring racial ideologies that they believed was unfolding in São Paulo. "The integrity of the old racial ideology, with the disappearance of the slave social order and with the competition of blacks in the free labor market," they speculated, "is being shaken more or less deeply thanks to the change in the contact situation produced by immigration, population and the development of social classes."[78]

It was precisely this kind of historical and geographically specific analysis that they saw as a major omission of Donald Pierson's Bahian studies. Like many "specialists with a rigorous scientific formation," they argued, Pierson had been "disoriented" by Brazil's bewildering pattern of race relations, which defied "global comprehension." What Pierson failed to recognize is how race relations in Brazil conformed to latent racial ideologies that were remnants of the colonial social order. After the abolition of slavery in 1888, Bastide and Fernandes maintained, "the old racial ideology did not immediately collapse nor did it lose the function it held in the social order of slavery."[79] "Any change in the social status of the Negro," Bastide and Fernandes explained, "was merely legal."[80] Pierson's conclusion that only class prejudice existed in Brazil thus neglected how class correlates with skin color and failed to see the continuities with the social order of slavery. Pierson's failure to see how latent racial ideologies continued to shape the Brazilian social order was further compounded by his erroneous assumption that Bahia served as a microcosm of Brazilian society writ large.

With Bastide and Fernandes in charge, the studies of race relations in São Paulo adopted methods and practices that stood in stark contrast to the Bahian studies and reached quite different conclusions. Most strikingly, Bastide and Fernandes made a concerted effort to collaborate with Afro-Brazilian groups in significant parts of the research study. For instance, they assembled a research team that included leaders from prominent Afro-Brazilian associations in São Paulo and "white students from the Faculty of Philosophy" whom they asked to go through "a kind of psycho-analytical confession on the racial problem." In addition to doing interviews, life histories, and questionnaires with Afro-Brazilians in "slums and factories," Bastide and Fernandes also set up three discussion groups that met "once a week or once a fortnight." The first of these groups was composed of Afro-Brazilians from "every social stratum," the second group was made up of leaders delegated by Afro-Brazilian associations, and a third group was made up of "colored women." As Fernandes would later observe, the study was an important first attempt to assess race relations by "working cooperatively and by utilizing systematic, empirical research as a technique for raising 'social consciousness.'"[81]

The links that Bastide and Fernandes made with Afro-Brazilian associations reflects the growing influence of Afro-Brazilian activist groups in the postwar era. After the end of Getulio Vargas's authoritarian Estado Novo regime in 1945, activist groups emerged and began to promote the civil rights of Brazil's Afro-descended communities. Perhaps the most prominent of these groups was the Teatro Experimental do Negro, led by the scholar and artist Abdias do Nascimento, which was formed in 1944 with the goal of eliminating "color prejudice" and enhancing Afro-Brazilians' possibilities for education. In May 1949, Nascimento and other leading figures organized the First National Negro Congress—an event designed to foster research on black issues that was attended by leading social scientists including Gilberto Freyre, Donald Pierson, and Roger Bastide.

In *Brancos e Negros em São Paulo*, the book that emerged from their studies, Bastide and Fernandes offered extensive evidence for the existence of anti-black racism in São Paulo. Although São Paulo had become South America's leading industrial center with plenty of job opportunities, the existence of what they called "latent" anti-black stereotypes thwarted the upward social and economic mobility of Afro-Brazilians.[82] But instead of attributing anti-black racism to individual attitudes, Bastide and Fernandes argued that it was a product of São Paulo's unique social structure, which had been assembled during the city's transformation from a "traditional city" to "tentacular metropolis" in less than half a century. Because of this meteoric development, Sao Paulo's social structure became one where the remainders of a patriarchal "slave-owning society" coexisted with all the innovations of a modern "capitalist society." In this context, the "color prejudice" whose previous function had been to justify the enslavement of Afro-Brazilians was now being used to maintain a "class society" stratified along racial lines.[83]

By documenting the persistence of race prejudice in São Paulo, *Brancos e Negros* also corroborated the personal experience of Afro-Brazilians, which had been erased by "racial democracy" narratives. For instance, the young Marxist sociologist Oracy Nogueira, who contributed a chapter to the study, argued that race relations scholars had generally underestimated the degree of race prejudice in Brazilian society, and North American researchers had been unable to see it. According to Nogueira, the UNESCO-sponsored studies in Rio and São Paulo represented the first time that social scientists corroborated the racism that Afro-Brazilians had long experienced. Thus, "for the first time," Nogueira wrote, "the testimony of social scientists frankly admits and corroborates the contention of the non-white population of Brazil based on their own experience."[84]

Race Prejudice in Rio de Janeiro

Like the São Paulo studies, the study conducted by the sociologist Luiz Aguiar Costa-Pinto in Rio de Janeiro broke from existing approaches to race relations. As we saw in chapter 2, Costa-Pinto trained with Arthur Ramos and saw himself as the successor of Ramos and Nina Rodrigues.[85] Yet while Costa-Pinto paid tribute to his Bahian mentors, he also sought to render their work a relic of the past. In the introduction to *O Negro no Rio de Janeiro*, the book that emerged from his UNESCO-sponsored study, Costa-Pinto drew a sharp distinction between his "realistic" and sociological approach to the "Negro" and the idealizing approach of his predecessors. Previous studies, he argued, treated Afro-Brazilians as a "spectacle" and took the alterity of the "Negro" as their primary concern. This tendency to exoticize and create an "abstraction" out of Afro-Brazilians was evident in the themes chosen by anthropological studies, namely how Afro-Brazilians differ from Europeans, and the survival of African culture, folklore, and religion. Costa-Pinto also interpreted this emphasis on cultural difference as a remnant of social relations from the slave-holding era. The tradition of research concerned with "o Negro no Brasil," he insisted, was crafted from the standpoint of "ruling" whites and was thus structured by relations of dominance. Instead of offering clear insights on race relations, anthropological studies of acculturation, cultural contact, and accommodation obscured the real nature of "concrete relations." To break with this tradition, Costa-Pinto steered clear of the "comfortable refuge" of a purely descriptive monograph or, alluding to Freyre, of a literary essay "full of insinuations but lacking in analysis." Instead, he studied the Brazilian "Negro" from a strictly sociological perspective, which he described as one that entails examining relations between diverse ethnic groups within "social structures that are historically concrete."[86]

Like Fernandes and Bastide, Costa-Pinto also challenged the "racial democracy" narrative and insisted on the geographic specificity of race relations. Every country and region, he maintained, develops its own distinct pattern of race relations marked by specific social and historical circumstances. Accordingly, he argued that it was "entirely arbitrary" to take one country's race relations—usually the United States—as a reference point and then judge other concrete situations by whether they conform to the particularities of the North American model. Because of this comparativist tendency, Costa-Pinto observed, most race relations studies end with the predictable conclusion that "everything is ok because it's not as bad as the Deep South."[87] In this vein, he lauded Métraux's decision to expand the original focus of the race relations studies beyond Bahia and argued that UNESCO would stand to lose scientific and intellectual credibility by continuing to present the "traditional" pattern of race relations to the world through folkloric monographs and literary essays. He also wrote that it was high time to fix this "bias" in Brazilian thought, which had been nurtured by a "false conception of national pride" and "mental inertia" and then "exported all over the world."[88]

Through his sociological study of Rio's race relations, Costa-Pinto also introduced new concepts for analyzing the distinctiveness of the Brazilian situation. Like his São Paulo colleagues, Costa-Pinto interpreted race prejudice as a concealed phenomenon that reflected underlying social tensions produced by rapid social change. Social conflict is inherent to the "developmental process," he suggested, and assumes various forms and gradations that bubble to the surface as "tension discharges." In Brazil, behind the dogma and appearance of racial democracy, lay "feelings of bitterness and obvious uneasiness." To capture this subterranean racism, Costa-Pinto adopted the term "cryptomelanism," which he defined as "the fear to admit or the desire to hide the importance which is, in fact, attached to the question of race and of color" and functioned as a "refusal to face up to reality."[89] "What seems typical of the Brazilian situation," Costa-Pinto wrote, "is not the complete lack of prejudice, but the absence of violence in the forms of discrimination shown to the Negro." In contrast to overt and extreme forms of anti-black violence in other countries, Costa-Pinto described Brazilian racism as composed of accumulated daily micro-aggressions that leave black Brazilians with a feeling that they might be exposed to injustice or discrimination at any moment. "It is this insecurity which produces in many Negroes a serious state of anxiety, of fear and instability," Costa Pinto observed. And this constant threat of injustice is "sufficient to put him [the Brazilian "Negro"] on the defensive and to give him a feeling of victimization."[90]

Costa-Pinto's assessment of the Brazilian racial situation was also shaped by his engagement with the work of scholars and activists from Afro-Brazilian associations. For instance, Costa-Pinto wrote that the papers given at the First Brazilian Negro Congress in 1949 were of immense value to him as he drafted his manuscript. He also used articles, advertisements, speeches, conferences, essays, and chronicles published from the black press for the purpose of observing the "social life" of the "Negro" in all sectors and within the structural framework of the city.

Yet although he tracked the arguments and work done by Afro-Brazilian associations quite closely, Costa-Pinto also made an effort to maintain sociological distance and, perhaps reflecting his distaste for community studies, did not make his own ethnographic observations. In fact, at the outset of his study, he collaborated with the Afro-Brazilian anthropologist Edison Carneiro, who had played an important role in organizing the Second Afro-Brazilian Congress in Bahia in 1937. Costa-Pinto tasked Carneiro with observing the Samba schools and Macumbas (Afro-Brazilian religious groups) of Rio and with interviewing prominent black leaders. As such, Costa-Pinto's thoughts on Afro-Brazilian associations were based on Carneiro's notes. Although Costa-Pinto had planned to coauthor *O Negro no Rio de Janeiro* with Carneiro, this didn't work out and he instead relied on Carneiro's notes for relevant sections. Ultimately, Costa-Pinto's writing suggests an ambivalent attitude toward Afro-Brazilian associations. Although Costa-Pinto sympathized with Afro-Brazilian associations and engaged with their arguments in his analysis, he also maintained a sense of scientific detachment and insisted on subsuming their narratives within his sociological framework.

Costa-Pinto's complicated relation to Afro-Brazilian associations is apparent in his discussion of the Teatro Experimental do Negro and the politics of negritude. In a chapter devoted to the "new type" of associations, Costa-Pinto described TEN as an exemplar of the second wave of Afro-Brazilian associations that emerged during the Estado Novo regime. According to Costa-Pinto, the emergence of this new black elite represented a major break with the "traditional" black associations that emerged after the abolition of slavery in 1888. In harsh terms, Costa-Pinto described traditional Afro-Brazilian associations as led by an "evolved" black elite whose main concern was to escape their blackness and to promote the upward social mobility of Afro-Brazilians through an embrace of all things "Nordic, European, Aryan, and Classical"—a process Costa-Pinto described alternately as "Whitening" or "Aryanization." The goal of these traditional associations, Costa-Pinto argued, was to produce "negroes with white souls."[91] By contrast, Costa-Pinto described the new black elite typified by TEN more favorably. "The new black elite," Costa-Pinto wrote, "intend to rise as black elites, without ceasing to be black, black more than ever, avowedly and proudly black, unapologetic for their blackness."[92]

Yet although Costa-Pinto seemed to welcome TEN's departure from a politics of whitening, he also admonished the group's leaders for logical inconsistencies in their antiracist tactics. Although TEN had enjoyed great success that transformed it into a "movement" as a opposed to a mere association, Costa-Pinto argued that it remained beholden to an ideal of training "black people in the styles of the predominantly white and socially ruling classes of Brazilian society." He also suggested that some of their core principles were marked by elitism and paternalism toward the "colored masses." For instance, he noted that Abdias do Nascimento had argued that the black masses suffer from "low cultural levels" and a "pre-logical" mentality and that they were thus impervious to ideas, concepts, and "literate techniques." Costa-Pinto thus interpreted Nascimento and the TEN's emphasis on the dramatic arts as reflecting an assumption that Brazil's black masses can only be reached by "manipulating" their penchant for "all that is mystical and fantastic," "their artistic and musical tendencies," and "sentimentality . . . and taste for recreation."[93]

Costa-Pinto also flagged one of TEN's most successful initiatives— their "Queen of the Mulattas" beauty contests—as a striking example of their flawed antiracist approach. As the sociologist noted, TEN organizers endorsed beauty contests because they believed they would "promote the social valuation of the eugenic and aesthetic riches" of black and mestizo women and serve as a means of countering internalized racial repression. By encouraging the valuation of women of color, TEN organizers hoped that the beauty contests would prompt a "mass redemption [desrecalcamento em massa]." Yet in Costa-Pinto's estimation, this line of thinking was based on faulty assumptions. Women of color's repression, argued Costa-Pinto, did not stem from a "lack of aesthetic valuation" on the part of white men. In fact, Costa-Pinto noted that many white men clung to the stereotype that women of color had "exceptional qualities" for "extramarital sexual relations." Women of color were thus subject to an excess of "purely carnal valuation" on the one hand and, on the other, an inverse devaluation on the "economic, social, and intellectual plane." Instead of serving as instruments of "mass redemption," Costa-Pinto argued that "ebony beauty contests" served to assuage the insecurities of the "colored man" for whom "the white woman is relatively much more inaccessible than the colored woman to the white man." By demonstrating that "colored women" possess "fully desirable aesthetic and eugenic qualities," the black beauty awards thus sought to convince the "colored man" that he need not be "frustrated" by his limited ability to choose.[94]

Ultimately, Costa-Pinto concluded that the ideology of "negritude" espoused by TEN and other new black associations constituted a kind of "inverted racism." In the final chapter to *O Negro no Rio de Janeiro* where he offered a description of the main features of the Brazilian racial situation, Costa-Pinto described the ideology of negritude as the most sophisticated form of "race consciousness" to have emerged in Brazil. Negritude, argued Costa-Pinto, had become a potent symbol of "the aspirations of a black elite fighting for social advancement." Yet Costa-Pinto concluded that this race consciousness was one that had been imposed on Afro-Brazilians rather than freely chosen. He thus interpreted negritude as an ideology that whites had imposed on the "conscience of the petit-bourgeois black" who wished nothing more than to forget his racial status. He expanded this analysis in a remarkable passage:

> Seen this way—and this is the dialectic of things—the idea of blackness, above all, represents the flowering in the heads of a black elite of a seed that was planted there by the attitudes of whites. In other words, just as it may be repeated here once again that there is no black problem—for the problem is the white man who has false ideas about black people and acts in accordance with these false ideas—one could also say, conversely, that the idea of blackness is not black—it is white, it is the inverted reflection, in blacks' heads, of the idea that whites have about it. It is the result of awareness (also in false terms, by the way) of the resistance that the white makes to the social rise of the black. It is, in short, racism in reverse.[95]

Costa-Pinto's assessment of negritude as a false consciousness thus reveals a strict adherence to sociological truth that was shared by so many mid-century race relations experts. From the perspective of the strictly materialist and structuralist sociological frame that he was keen to promote, race could only be seen as an illusion, wanting in material reality. Accordingly, Costa-Pinto argued that negritude represents an ideology that renders race a "mystical entity" and an "extrapolation" of daily lived experience "to the plane of the fantastic." He thus concluded that negritude reflects an "overvaluation of race" that makes it appear like a "cosmovision" and "philosophy of life" but is in fact a "defense mechanism managed by an aggressive elite, a rationalization of concrete social tensions, in which dominant and ethnically different groups, using race as a criterion of discrimination against some, lead the discriminated ones to make the cause of their problem the mystical flag of their redemption."[96]

At stake in Costa-Pinto's dismissal of negritude was the political authority of the social sciences within Brazil's burgeoning developmental state. As part of their project of professionalizing the social sciences and establishing their authority, the new generation of scholars who led the UNESCO race relations studies not only sought to challenge prior interpretations of Brazil's racial situation but also offered interpretive frameworks that aimed to shed light on the obstacles facing Brazil on its path toward economic development and modernization. From this lofty standpoint, racial identities appeared as by-products or epiphenomena of broader and more fundamental economic processes. Accordingly, these lofty structural frames afforded this new generation of sociologists a new means of enacting scientific detachment and playing the role of commentators on the success of Afro-Brazilian groups. Thus, although they offered incisive critiques of "racial democracy" narratives and Afro-Brazilian studies, they also could not escape the trappings of scientific elitism and became highly invested in seeing everything through a structural lens. Costa-Pinto made this standpoint clear in a *UNESCO Courier* article where he summarized his main conclusions from *O Negro no Rio de Janeiro* and argued that Rio's racial tensions are a subset of wider social tensions produced by rapid industrialization and modernization. Yet unlike Wagley, who concluded that there was no need to fear modernization, Costa-Pinto was less upbeat. Rapid industrialization would not necessarily free Brazil from its history and could instead result "in [either] a crisis or a solution according to the nature of Brazil's social evolution in the coming years."[97]

CONCLUSION: RACE RELATIONS AND MODERNIZATION

The frame of racial individualism that became predominant in U.S. race relations during the Cold War gained little traction in Brazil. As we saw in this chapter, the generation of Brazilian sociologists who came of age in the 1950s challenged existing interpretations of Brazilian race relations, which had been articulated during the strident nationalism of the Vargas era and celebrated Brazil's lack of race prejudice on the basis of studies conducted in Bahia. In an effort to move away from this framing, urban sociologists like Luiz Aguiar Costa-Pinto and Florestan Fernandes adopted a structural and historical analysis of Brazilian race relations that identified latent and veiled forms of racism that they interpreted as remainders from the colonial social order that impeded Brazil's path toward modernization. At stake in their interpretations of Brazilian society was the question whether Brazil could transcend the traditional agrarian economy and social structure of the colonial era and become a fully modern and industrialized society. For Brazil's urban sociologists, race relations served as one part of a broader project concerned with discerning and anticipating Brazil's path toward economic development and modernization. Insofar as they foregrounded structural and economic dimensions, Brazil's urban sociologists thus developed frames of analysis that paralleled the work of influential development economists, notably Raúl Prebisch and Celso Furtado, who joined forces at the offices of the UN Economic Commission of Latin America (CEPAL) in Santiago, Chile.[98]

Another sign of Brazilian urban sociologists' discordance with the Cold War liberalism of the United States can be seen after 1964 following the U.S.-backed military coup d'etat of Brazil that led to the rise of a military dictatorship that lasted for twenty-one years. In contrast to Gilberto Freyre, whose Lusotropical ideology was warmly received by Brazil's military regime and informed its foreign policy in Africa, many of Brazil's urban sociologists were forced to go abroad during the dictatorship.[99] For instance, in 1970, just six years after

Florestan Fernandes had become full professor at the Universidade de São Paulo (USP), the military regime dismissed him from his position and forced him into retirement. Fernandes accepted a position at the University of Toronto in the early 1970s and eventually returned to Brazil in 1973 but was forbidden to teach by the military regime.[100] A similar fate befell Fernando Henrique Cardoso—Florestan Fernandes's student who contributed one of the chapters to *Brancos e Negros em São Paulo*. Like Fernandes, Cardoso was also forcibly retired from the USP in 1969, and he relocated to Santiago, Chile, where he joined the economists at CEPAL (the United Nations Economic Commission for Latin America). During his time at CEPAL, Cardoso and the Chilean sociologist Enzo Faletto coauthored *Dependency and Development in Latin America*, one of the landmark texts of dependency theory. In Rio de Janeiro, the military regime also forced Luiz Aguiar Costa-Pinto to retire from his position, which prompted Costa-Pinto to relocate to Canada in 1960 where he spent the rest of his career.

Rather than an extension of racial liberalism, the UNESCO-sponsored studies of race relations in Brazil's urban centers mark the beginning of an alternative tradition that was broadly aligned with structural and socioeconomic perspectives. Instead of the predominant concern with cultural assimilation, biological absorption, and the erasure of racial difference typically associated with race conceptions in the Southern Hemisphere, Brazilian sociologists conceptualized the project of confronting racism as one that entailed major changes to the socioeconomic structure of Brazilian society. From this perspective, antiracism did not appear as straightforward as in UNESCO's education projects and did not yield tidy solutions. Indeed, rather than offer prescriptive solutions to Brazil's racial situation, scholars like Costa-Pinto and Fernandes instead concentrated on confronting the myth that Brazil was free of race prejudice and on documenting the societal challenges that racial tensions continued to pose. In order to render Brazilian race prejudice visible, Brazil's urban scholars amassed and interpreted significant amounts of demographic, ethnographic, and census data. Yet they also challenged the universality and portability of frameworks and concepts from North Atlantic race relations research and sought to shift Afro-Brazilian studies away from their traditional concern with culture, religion, and folklore and instead engage with the sociological realities of Brazil's black population. Central to the scientific antiracism they articulated was an insistence on recognizing the historical and geographic specificity of racism and a critical rebuke of comparative studies of race relations. UNESCO's urban studies thus challenged the institution's core assumption that racism was an international issue requiring a universal response. As Costa-Pinto's tense relations with Abdias do Nascimento and TEN suggest, Brazil's race relations scholars also wrestled with the issue of how to draw the boundaries around what constitutes an objective and rational understanding of racial issues and whether race relations should be an apolitical domain of inquiry. As we will see in the following and concluding chapter, these political and epistemological challenges were ones that would continue to vex race relations scholars throughout the world during the Cold War and would ultimately pose fundamental challenges to the field's coherence as a domain of inquiry.

CHAPTER SEVEN

A White World Perspective and the Collapse of Global Race Relations Inquiry

Describing the 1955 conference in Bandung, Indonesia, that sparked the beginning of the nonaligned movement, the African American novelist Richard Wright wrote that it was the "kind of meeting that no anthropologist, no sociologist, no political scientist would ever have dreamed of staging."[1] For Wright, the conference represented a meeting of nations that had nothing in common except a feeling of being rejected from the Western world. The meeting could only have been organized, Wright claimed, by "brown, black, and yellow men who had long been made agonizingly self-conscious, under the rigors of colonial rule, of their race and their religion." As such, Wright insisted that the meeting was not motivated by ideology as many claimed but rather by deeply internalized sentiments. "The agenda and subject matter had been written for centuries in the blood and bones of the participants," Wright wrote.[2]

Wright did not name any specific figures or institutions when describing the contrasting worldviews of social scientists and the anticolonial leaders of Africa and Asia. Yet ten months before the Bandung meeting, a group of U.S.-based social scientists staged a conference called "Race Relations in World Perspective" in Honolulu, Hawai'i, that lasted from June 28 to July 23, 1954. The Honolulu conference was organized by three white U.S. sociologists who were graduates of the Chicago school of urban sociology: Andrew Lind from the University of Hawai'i, Everett Hughes from the University of Chicago, and Herbert Blumer from the University of California, Berkley. When writing from Bandung, Wright may have had social scientific experts like them in mind. As a young writer in Chicago during the interwar period, Wright developed a close association with Chicago sociology and befriended many of its leading lights including Louis Wirth, Robert Park, Horace R. Cayton Jr., and E. Franklin Frazier.[3] Wright learned important sociological lessons from these relationships. He would have certainly been familiar with the ideas of objectivity held by Chicago sociologists—ones that implored social scientists to refrain from attempting to transform society without first developing a firm grasp of its empirical reality.[4] It is quite plausible, therefore, that when Wright described the inability of the social sciences to fathom an event like Bandung, he may have recalled his own interactions with Chicago sociologists.

Though Wright did not name them in his account, the organizers of the Honolulu conference did indeed adopt a dismissive attitude toward Bandung and feared Wright's romantic vision of a gathering of all the darker and nonaligned nations and peoples.[5] For instance, when Andrew Lind—the chief architect of the meeting—assembled the Honolulu conference papers into an edited collection and wrote the introduction, he described the meeting as a necessary countermeasure to the geopolitical threat posed by Bandung. In his opening sentences, Lind described the Bandung "gathering" as creating an "urgent" demand for rigorous and systematic knowledge like the kind that emerged from the Honolulu meeting. Though he said little about what actually occurred or what was said during the Bandung meeting, Lind did raise concerns that such a meeting threatened the stability of the current world order. "The mere fact of calling a conclave predominantly of colored people, from which the former colonizing powers of the West were excluded," Lind wrote, "is itself of great moment to the future peace of the entire world."[6] In preparation for the Honolulu meeting, Lind, Hughes, and Blumer assiduously followed their Chicago training and envisioned the conference as a sober and objective counterpoint to the anticolonial movements emerging in Africa and Asia, which risked unleashing, in their minds, a global tide of anti-white antagonism. Accordingly, they designed the Honolulu conference to serve as a forum where race relations would be discussed in a spirit of free inquiry and dispassionate analysis and convened forty participants and numerous observers representing different areas of regional expertise including West Africa, the Middle East, the South Pacific, South Asia and East Asia, and South America. In assembling a conference of this lofty scale, the organizers also attempted to lay the foundations for a "world perspective" on race relations and to craft a general and portable theory of race relations that could be used across different regions. They also hoped to join the recent trend of creating international societies of social science disciplines—such as the International Sociological Association—and to create an international society of race relations. Yet for all of Lind, Hughes, and Blumer's drive and optimism, these objectives went unrealized.

What do the differences between these two meetings tell us about the differing ways that scholars and political activists responded to racism during the wave of African and Asian decolonization from the 1950s and 1960s? In her recent book, *Worldmaking After Empire*, Adom Getachew offers a compelling revisionist history of decolonization that helps frame the gaping chasm between the Honolulu and Bandung meetings. Getachew's account interprets the imaginaries of mid-century anticolonial intellectuals and political leaders like Kwame Nkrumah and W. E. B. Du Bois as concerned with a "thoroughgoing reinvention of the legal, political, and economic structures of the international order."[7] Getachew aptly calls this form of politics *worldmaking* and contrasts it with the narrow nationalist concerns often attributed to anticolonial leaders in the historiography of decolonization. Framed this way, the worldmaking ambitions of Bandung participants stand in sharp contrast with the decidedly less revolutionary project of crafting a "world perspective" favored by the organizers of the Honolulu conference. The worldmaking practices of Bandung leaders were concerned with creating a new world order, which involved an insistence of the self-determination of decolonized nations and a restructuring of the international economic order. By contrast, this chapter argues, the project of crafting a "world perspective" favored by Honolulu organizers was concerned with preserving a geopolitical order that emerged from European imperialism and with turning race relations inquiry into an ostensibly neutral and cosmopolitan field of study.

As this chapter shows, the "world perspective" of the Honolulu race relations conference was closely aligned with the liberal internationalist standpoints that prevailed in institutions like UNESCO during the 1950s. Like UNESCO in this time period, the Honolulu conference struggled to accommodate more radical decolonizing alternatives. The kind of institution-building practices envisioned for the meeting were commonplace in the social sciences of the North Atlantic region in this period. For instance, as Perrin Selcer has shown, in the 1950s U.S. social scientists associated with UNESCO attempted to create an international system of knowledge production that would serve as a foundation for "democratic governance of a world community." By creating a network of international scholarly associations, experts associated with UNESCO's Social Sciences Department (SSD) crafted a framework that attempted to create "unity in diversity" by bringing together national and regional perspectives through a shared technical language. As such, SSD experts played a leading role in creating several international associations including the IEA (International Economic Association), ISA (Sociology), IPSA (Political Science), and ICLA (Comparative Law).[8] The envisioned purpose of these international associations was to "construct a synoptic view of the world community" by coordinating the perspectives of scholars from distinct national cultures—a project that Selcer suggestively calls the "view from everywhere." By making the creation of an international society for race relations a key outcome of the Honolulu meeting, Lind, Blume, and Hughes sought to extend the "view from everywhere" into the domain of race relations inquiry. Yet, as the meeting unfolded, many participants in the Honolulu meeting began questioning the core objectives of the event and argued that race relations inquiry had often been framed from the perspective of white elites. Thus, as much as Honolulu organizers attempted to frame their project as strictly "descriptive" and non-interventionist, the tensions that emerged during the conference increasingly revealed how it was in fact concerned with preserving an established world order and with undermining attempts to imagine the world in other ways. The debates that took place during the conference thus revealed how the "world perspective" desired by conference organizers favored the standpoint of white Chicago trained sociologists.

Lind, Hughes, and Blumer's ambition to transform race relations into a cosmopolitan science faced considerable obstacles. As we saw in the previous chapter, the project of fashioning a politically neutral and universal science of race relations in the mold of Chicago sociol-ogy was one that had already been questioned by Brazilian scholars in response to UNESCO's cycle of race relations studies. In contrast to Chicago-trained scholars like Donald Pierson who made frequent com-parisons between U.S. and Brazilian patterns of race relations, the new wave of Brazil-based researchers who participated in UNESCO's studies of race relations in São Paulo and Rio de Janeiro—including Florestan Fernandes, Roger Bastide, and Luiz Aguilar Costa-Pinto—challenged the portability of Chicago sociological methods and insisted on the geographic and historical specificity of race relations. At stake in Brazil-ian debates over how to define the parameters of race relations inquiry and how to conceptualize race prejudice were important and unresolved questions about the degree to which Brazilian social relations remained tethered by the slaveholding era and whether the Lusophone nation could escape its condition of dependent capitalism. In the United States during the 1950s, the paradigm of racial individualism took root and framed rac-ism as a matter of individually held prejudice. By contrast, Brazil's urban sociologists increasingly turned to structural economic perspectives in order to foreground inequitable distributions of power and wealth that could thrive in conditions of seeming racial harmony. Similar tensions and a similar shift in perspective would emerge during the course of the Honolulu conference and its aftermath. By tracking the tensions that arose during and after the Race Relations in World Perspective confer-ence, this chapter examines how conceptual frictions like those that arose in Brazil also troubled internationalist projects concerned with scaling race relations up from the regional and national to the "world" domain.

HAWAI'IAN RACIAL HARMONY AND BUILDING A WORLDVIEW OF RACE RELATIONS

In the prospectus for the Race Relations in World Perspective confer-ence, Andrew Lind extolled Hawai'i's "harmonious" social relations and geographic location as fitting traits for a conference concerned with global order. Echoing the cheerful portraits of Brazil penned by Park and Pierson, his Chicago colleagues, Lind wrote that Hawai'i enjoyed "a world-wide reputation for its multi-racial population" and "for the ease with which its numerous immigrant stocks are adjusting to each other and to the emerging Hawaiian-American culture."[9] Like Brazil, Lind mused that Hawai'i served as a unique observatory for the "fusing of ethnic groups" and a model of "effective community living on an interracial basis." Yet Lind also highlighted Hawai'i's unique geographic location in the heart of the Pacific as one of its distinguishing features and reasons for its lack of race prejudice. Indeed, Lind attributed the island's supposed racial harmony to its "mid-oceanic detachment from the major points of racial tension" and to a "well-established tradition of tolerance and objectivity" on racial matters. This combination of geographic and racial detachment, Lind ventured, rendered Hawai'i an ideally situated place to sponsor a "new, bold venture in interracial understanding."[10]

Though Lind penned this prospectus in the early 1950s, he and his colleagues at the University of Hawai'i's Department of Sociology and the Romanzo Adams Social Research Laboratory had been making similar arguments about Hawai'i as a racial paradise since the 1920s when they first began to conceptualize Hawai'i as a "racial laboratory." Much like Gilberto Freyre's Lusotropical narrative, Lind and Hawai'i sociologists' buoyant portrait of the island was one that celebrated and sanitized its settler colonial history and relied on stark contrasts with regions like the U.S. South where racism was described as more violent and overt. As Maile Arvin has argued, during the interwar period this conception of Hawai'i as a racial paradise and racial laboratory paradoxically flourished despite ample evidence to the contrary, such as the highly publicized scandal known as the "Massie Affair," which was prompted by the allegations of Thalia Massie, the white wife of a white naval officer who claimed to have been assaulted and raped by a group of Native Hawai'ian and Asian American men in 1931.[11] Although hatred toward Native Hawai'ians and Asian American settlers was commonplace during this period, Chicago-trained scholars like Romanzo Adams and Andrew Lind portrayed the island as a cosmopolitan exemplar of racial harmony on the basis of sociological studies of interracial marriage and human biological studies of race mixing. Like Donald Pierson in Brazil, Adams and Lind viewed Hawai'ian society through the frame of Park's assimilation cycle, and they upheld Hawai'i's high rates of interracial marriage as compelling evidence for little social distance and prejudice between the island's racial groups and interpreted the anthropometric studies of race mixing conducted by human biologists Harry Shapiro and Louis Sullivan as providing further evidence for the emergence of a distinct "Neo-Hawai'ian" racial type.[12] By portraying the island as both a racial laboratory and "melting pot," Adams and Lind aimed to show that the island's inhabitants could assimilate American ways of life and thus joined a chorus of political, religious, and business leaders advocating for Hawai'ian statehood. Similarly, they portrayed Hawai'i's unique location in the middle of the Pacific Ocean as offering the geographic conditions for the formation of a distinctive cosmopolitan culture that fused together the best elements of East and West.

Given Hawai'i's racial and geographic distinctiveness, Lind and his collaborators also saw it as a place uniquely suited for the development of race relations research and an appropriate site to launch a new phase of race relations research that would be even more cosmopolitan in its approach. Conference organizers and their colleagues also saw the meeting as a crucial moment in a trajectory of race relations that they traced back to Chicago. The centrality of the Chicago tradition within the framing of the conference is not surprising. The conference steering committee was made up entirely of members from the Chicago sociology department and alumni who had helped to establish the Chicago approach at other prominent West Coast institutions: Herbert Blumer (University of California), Robert W. Redfield (University of Chicago), Everett Hughes (University of Chicago), and Andrew Lind (University of Hawai'i).

In the opening session of the conference, the presidents of the University of Hawai'i, the University of Chicago, and the University of California gave opening remarks, which were followed by opening statements from Everett Hughes, Herbert Blumer, and Andrew Lind. In these opening remarks, speakers tethered the history of race relations to the University of Chicago and stressed the importance of adopting the ecumenical and dispassionate approach to research that became the calling card for Chicago-trained sociologists during the conference proceedings and during future endeavors. For instance, in his opening remarks, Everett Hughes situated the Honolulu conference as the culmination of a trajectory of race relations research whose important landmarks included the 1911 Universal Race Congress and the 1928 meeting of the American Sociology Society on the theme of "race and culture contacts."

Hughes also spoke of how race relations research had taken root in different parts of the world and had become an increasingly specialized domain studied from the vantage of various disciplines. The "factor of race has entered into economics, politics, human geography, anthropology, and sociology," Hughes explained. Hughes also described how this specialization had produced diverging strands of race relations inquiry in England, the United States, Holland, France, and South Africa that had yielded a voluminous literature. Accordingly, Hughes hoped that the Honolulu conference would bring coherence to this situation and would have the effect of reducing "ethnocentrisms in the study of race relations and of dispelling the mistaken idea that the particular racial situations with which participants are familiar are unique."[13]

In his opening remarks, Herbert Blumer described the conference's goals and made a similar pitch for the important role it was poised to play in bringing coherence to race relations research and to institutionalizing it along Chicago lines. The lofty scale at play in Blumer's remarks echoed the "view from everywhere" promoted by UNESCO as well as other ambitious data-gathering projects of this period, which Rebecca Lemov has called the "Big Social Science Era."[14] The first goal of the conference, Blumer argued, would be the "sharing of knowledge and perspectives" and the "expression of orientations which have been shaped by regional and scholarly backgrounds of participants." This collective pooling of information and orientations, Blumer posited, would then provide the scaffolding for the second goal of the conference—"[the] development of a body of analytical knowledge which would take the form of generalizations and insights which can be applied to a wide variety of areas." In order to achieve these objectives, Blumer insisted on the need to adopt a narrow "scholarly" approach as opposed to one "of advocacy or rejection of particular racial policies." He also urged participants to engage in "open and free" discussion with the proviso that "matters of existing racial issues and policies be examined dispassionately with the intention of discovering the factors involved and their effects." "It is hoped that personal sensitivities will not interfere with open and free discussion," Blumer petitioned. Pointing to future directions, Blumer also hoped that the conference participants might also develop some "type of organization, preferably informal, which would lead to a continued contact and exchange of research and experience with reference to the matters which will be discussed during these meetings." Through these kinds of initiatives, Blumer concluded, the conference could potentially "capitalize" on the "impetus given to the development of a world view of race relations."[15]

Hughes's and Blumer's remarks, which were also echoed by Lind, reveal precisely the sort of project they had in mind and how it contrasted with the *worldmaking* politics of anticolonial leaders. While anticolonial leaders sought to disrupt and unsettle the existing world order to create a new one, Chicago and Honolulu sociologists privileged notions of "stability" and "equilibrium" in their analyses, which were derived from their reading of plant and animal ecology. Accordingly, their work did not anticipate or concern itself with anticolonial revolution and instead focused on describing how societies emerged out of and established a new balance in the wake of European colonialism. In fact, through the frame of "human ecology," Chicago sociologists approached intergroup tensions and conflicts as benign natural processes leading to equilibrium. For instance, as Christine Manganaro has shown, in Andrew Lind's landmark sociological study of Hawai'i, *An Island Community*, the paradigmatic concept and focal point was the notion of *succession*, which he borrowed from plant ecology. In plant ecology, interwar experts had used succession to describe the process whereby a plant community successively gives way to another until an equilibrium is reached. In the field of race relations, Chicago-trained sociologists like Lind adopted plant ecologists' conception of succession in order to naturalize the process whereby a social hierarchy is created.[16] In *An Island Community*, Lind used the lens of "succession" to craft what Manganaro calls "a developmental economic history" that imagined a process of modernity unfolding in Hawai'i according to three stages: "(1) 'stone age,' (2) 'feudal' society through the (3) 'era of agricultural expansion' culminating with the era of (4) 'stabilization,' 1900–1936."[17] According to Lind, a crucial element in this process of modernization was the competition between Hawai'i's different immigrant groups, which he conceptualized in ecological terms as invaders in a new land whose position would be determined by their ability to "compete" and rise in social status. Though he admitted that "race prejudice" existed in Hawai'i, Lind downplayed its significance by describing it as a predictable, natural, and ultimately temporary product of groups competing with one another. As Hawai'i continued its economic development and became fully integrated with the United States and the global economy, "race prejudice" would inevitably fade. Lind also argued that American colonization had helped to stabilize Hawai'i and other colonies after their balance was disturbed by European contact and predicted that Hawai'i's future would be peaceful and prosperous given the progress of global capitalism on the islands.[18]

Yet by the early 1950s when Lind spearheaded the organization of the Honolulu conference, he and his Chicago-trained colleagues began to see the emerging nationalist movements in Asia and Africa coupled with the spread of communism as alarming trends that gave new salience to the issue of race and threatened the stability of the existing global order. In his prospectus for the conference, Lind wrote that "expanding nationalism in the contracting colonial world, revolutionary movements, and international alignments" had become intertwined with "racial considerations" and thus transformed race relations into a problem of "world-wide proportions." Given this new context, Lind argued that a "world perspective comparable to the scope of the problem is urgently needed."[19]

Yet for all of his efforts to frame this emerging worldwide problem evenhandedly, Lind also made it clear that it was the continued dominance of the white Western world that was under threat. In contrast to Hawai'i's mid-oceanic detachment and interracial harmony, Lind warned that the awakening of nationalistic aspirations throughout the colonial world, particularly in "South East Asia and Africa," were "invariably accompanied by a rising flood of anti-white antagonism and a corresponding suspicion of the intentions of American and western European powers." According to Lind, this increasingly hostile context had also turned the "so-called colonial powers, including the United States" into the "principal objects of this antagonism and suspicion." Given such dangers, Lind concluded that nothing less than "the security of the emerging world order" was at stake during the Honolulu conference and that this security would require "skill and perspicacity" for dealing with "the complicated problems of race relations in the so-called 'backward areas' of the world."[20] Making matters worse, Lind also described how the Western antagonism that was growing in the decolonizing world was being opportunistically stoked by the Soviet

Union. Indeed, Lind suggested that through "propaganda and maneuvering," Soviet Russia and other communists were exploiting "racial resentments" in "backward areas" by fostering "the view that racial discrimination and "exploitation of suppressed races by the dominant" are inevitable and ineradicable under capitalism." "In some colonial areas," Lind wrote, "[the Soviets] have succeeded in identifying communism in the minds of many 'colored' inhabitants of those areas with nationalism and with the movement for political independence from 'white oppressors.'" Like so many mid-century race relations researchers, Lind thus framed the political movements of non-Europeans as laboring under a false racial consciousness.

IS THERE SUCH A SUBJECT AS RACE RELATIONS?

As Lind and his colleagues' opening remarks suggest, the organizers of the Honolulu conference were concerned with systematizing and institutionalizing race relations inquiry in the service of maintaining the existing global order. As such, the conference organizers conceptualized race relations as a technical and ostensibly apolitical domain of inquiry. They also viewed race relations as a field concerned with phenomena at multiple geographic scales and thus requiring the creation of a coordinated international infrastructure. Yet they also recognized that the boundaries of race relations research were potentially limitless and would have to be carefully drawn. As Andrew Lind put it, the conference organizers fully embraced Robert Park's "catholic conception" of race relations and joined him in acknowledging that the interaction of people and races in the modern world is so "vast and irresistible" that it assumes the character of a "cosmic process."[21] Though potentially infinite in its scope, Lind and his collaborators also placed faith in conventional scientific methodologies of accumulating data, comparing across different sites, and generating abstract theories that capture the observed phenomena at its highest level of generalization.

Lind's awareness of the potentially cosmic scale of their project is evident in the conference design. In assembling a roster of participants and planning the agenda for the meeting, Lind and his colleagues struggled to balance competing forces: a desire for interdisciplinary and geographic breadth and the competing demand for abstracting general principles; a commitment to a strictly scientific approach with the "action-oriented" demands of some of the participants; and a recognition of the need for scholarly exchange with the competing demand for communicating to the public. Such competing and even contradictory forces were often constitutive of the kinds of antiracist projects that liberal social scientists assembled within international institutions. In an effort to strike this balance, Lind and his colleagues reviewed a list of more than 200 names of people in light of the conference objectives. In deciding who to invite, they made a "serious attempt" to secure "qualified observers and research scholars form the more distinctive areas of race relations around the world." Yet instead of focusing on their qualifications as social scientists, they considered their general "understanding of what happens when races meet" and chose scholars with field experience over those whose knowledge was "exclusively theoretical."[22] They ended up with a roster of forty participants, which included a wide range of professional representation including representatives from fields such as economics, political science, public administration, journalism, geography, education, history, philosophy, social psychology, and, primarily, sociology and anthropology. The official roster also included some notable "observers," namely John Bell Condliffe, a New Zealand economist who later wrote a biography of Te Rangi Hīroa, and Alfred Métraux, who served as a UNESCO representative.

In the spirit of the "view from everywhere," Lind and his collaborators also made every effort to cultivate an atmosphere that would allow for a genuine "meeting of minds and interstimulation" among the keenest observers of race relations around the world. For instance, they put together a handbook for conference participants, which included a cordial welcome and "Aloha" from Lind, a full list of participants, as well as a thirty-six-page "Who's who" document that included a brief biography for each scholar along with a list of relevant publications.[23]

Yet for all of their efforts to create a calm and collegial environment, once the conference began, significant differences among the participants began to emerge. As tensions rose to the surface, participants began to question some of the organizers' basic premises and assumptions and the degree to which these truly reflected the "world perspective" to which they aspired. Participants also raised difficult questions about the coherence and portability of race relations as a domain of inquiry.

These fault lines began to emerge during the second plenary session of the conference, which was titled "A Frame of Reference for the Study of Race Relations." During this session, presenters raised probing questions about the coherence of race relations as a field and whether it was amenable to the sort of generalization to which the organizers aspired. For instance, in his presentation titled "The Concept of Racial Adjustment," the Japanese American sociologist Jitsuichi Masuoka, who grew up in Hawai'i and worked at Fisk University, a historically black university, observed that the scientific study of race relations is often predicated on the assumption of an "orderly life."[24] As a result, Masuoka argued, race relations have generally been been analyzed as "systems" unto themselves. Masuoka thus interpreted race relations as something more than fragmentary contacts between people; race relations were the "consequence of an emerging social order in which racial groups are involved." Yet, in what would become a common theme during the proceedings, Masuoka argued that the stability that had conventionally underpinned race relations was becoming ever more "dynamic" because of expanding communication methods in society.[25]

Like Masuoka, the British social anthropologist Kenneth Little (and author of UNESCO's *Race and Society* pamphlet) questioned the internal coherence of race relations in his presentation. In his paper titled "Race Relations in Commonwealth Affairs," Little questioned whether "a subject of race relations as such" could be said to exist. Little based his argument on the observation that race is but one of many factors, such as age and sex, that differentiate members within a society. Little also pointed out that any local system of race relations is almost always "determined by wider systems of social relations." As an example, Little, who specialized in the study of British West Africa, described the differences between race relations in West Africa and Central Africa. According to Little, in West African countries such as Nigeria and the Gold Coast (now Benin), black Africans fully participate in the legislature and thus no "color bar" could be said to exist. However, in Central Africa, where there was a greater number of European settlers, a distinct color bar could be observed. The difference between these two regions, argued Little, reflects broader institutional and economic pressures. In Central Africa, he insisted, financial and industrial institutions (both locally and globally) had become dependent on cheap African labor in farms and mines. Commercial institutions were thus invested in maintaining existing racial hierarchies and were ready to resist any "reactionary" trend that might disturb this stability. By highlighting the economic forces that underpin social orders, Little thus questioned the degree to which race relations represented an independent domain of inquiry.[26]

The final presentation of the session, by Herbert Blumer, the sociologist who played a pivotal role in developing the field of symbolic interactionism, also questioned whether race relations as a concept was losing its analytic purchase. In his paper titled "Shifting Character of Race Relations," Blumer argued that a 200-year period of "relative stability in racial orders" produced by European colonialism was coming to an end. In ominous terms that echoed the historical sensibilities of many mid-century race experts like Arthur Ramos, Robert Park, and Alfred Métraux, Blumer argued that "the character of modern civilization" was undermining the stability of these racial orders. Indeed, Blumer mused that the forces of modern civilization—including technological change, increasing urbanization, the multiplication of communication methods, and the formation of specialized interest groups—were eroding traditional racial barriers and introducing an ever greater degree of "fluidity" in race relations. Whereas during the previous regime of stability, scholars could have conceivably devised a general scheme for analyzing race relations throughout the world, the fluidity of modern society had rendered it next to impossible to formulate a coherent "body of theory."[27]

In the discussion that followed the presentations, the participants amplified the critical questions raised by the panelists. According to conference notes, two questions at the heart of the discussion were "Is it possible to analyze race relations as separate and distinctive phenomena?" and "If race relations exist what are their central features and how might they be understood?"[28] Participants also raised the point that in many places, "state interference" and "political ideologies" were prompting "fundamental changes in race relations." As an example of this, participants described a recent "succession" of racial ideologies in the Soviet Union—from internationalism, to the "elder brother theory," to the latest idea of the development of "brotherhood" between Europeans and Asians. In response, some participants contended that the growing influence of ideologies render the preoccupation with comparative "patterns" of race relations "outmoded and inadequate."

Given the increasing "fluidity of the social scene," they argued, other forces such as "capital investment" were becoming more significant than "racial systems." As a retort to these points, other participants wondered whether a frame of reference designed to "clarify the factor of race" was nonetheless necessary.[29] By the second day of the conference, the participants had thus already begun to question the very foundations of race relations as a domain of inquiry. The liberal framework of "racial individualism" that Brazilian race relations scholars challenged through UNESCO-sponsored studies also began to unravel at the Honolulu meeting.

It was at this point in the proceedings that the Indian scholar Kodanda Rao made one of the conference's most trenchant interventions. Unlike many of the other participants whose primary affiliations were academic institutions, Rao joined the conference in his capacity as a member of the Indian Servants Society—a nationalist association created by moderate liberal upper-caste elites who promoted Indian self-determination through constitutional reform and educational initiatives such as mobile libraries, night-school classes for workers, and hygiene and public health education.[30] According to the "technical experience" section of his biography in the conference handbook, Rao visited, lectured on, and wrote about issues concerning the Indian diaspora in East Africa (Kenya, Uganda, Tanganykia, and Zanzibar), the British settler colonies (Canada, Australia, and New Zealand), Asia (Fiji, Malaya, Japan, China, Siam, French Indo-China, Dutch East Indies, and Ceylon), and the Caribbean (Trinidad, British Guiana, Dutch Guiana). He had also previously served as a Carnegie scholar in Yale's Department of Race Relations (1934–1935) and participated in the Conference on Education in Pacific Countries in Honolulu (1936) that had been jointly organized by Yale and the University of Hawai'i.[31] In his remarks, Rao offered the provocation that the most salient factor in contemporary discussions of race was not the technological advances cited by Blumer and others but rather the anxiety of "Whites over their loss of power over Non-Whites." In a candid departure from the preceding lines of analysis, Rao suggested that it was not sociological or economic factors that were the most salient influence on race relations but rather political ones. Accordingly, Rao suggested that the logic that governed race relations in colonial situations could be boiled down to a basic formulation:

"White men may rule Whites, as in England; Whites may rule Non-whites, as the British over Indians; Non-whites may rule Non-whites, as in China; but Non-Whites may not rule Whites."[32] Rao's intervention prompted others to pose even sharper questions. "Is this conference primarily concerned with the ways Whites can deal with their anxieties?" asked one participant. "What are other people in the world thinking about race?" asked another. The next day, during the discussion after the third plenary session, which was titled "Types of Racial Frontiers" and included presentations by Masuoka (again), Edgar Tristram Thompson, and Albert Hourani, participants again raised questions about whether the era of "White dominance on racial frontiers of the world" was coming to an end.[33] This theme continued over the discussions the following days, which considered whether the encroaching influence of the Soviet Union in Asia and Africa was extending the era of white dominance and whether militant and messianic movements in places like West Africa could be described as adopting a form of counter-racism directed toward their former white rulers. Thus, while Lind and his fellow Chicago-trained sociologists introduced abstract diagnoses concerned with how processes of modernization threatened to undermine existing racial orders, Rao's intervention and the contributions of Africanists centered on issues concerning the lasting effects of white colonial rule. For all the conference organizers' concern with epistemic unity and coherence, significant divides began to emerge as participants began to call out the implicit whiteness embedded in the Chicago framework.

By the fourth plenary session, the conference organizers felt compelled to revisit the intended goals of the conference and to reconsider their working methods. Members of the steering committee asked participants to think about how to adjust the existing arrangements so as to better meet the conference objectives. Albert Hourani, the Lebanese-British historian of the Middle East, then presented the conference participants with three proposals to consider.[34] Hourani's proposals included: (1) a modest plan to tweak the proceedings by having fewer papers during the plenary sessions and more discussion time; (2) a radical plan to scrap the plenary sessions and establish independent "commissions" that would meet and focus on specific topics such as "the development of White predominance and subsequent adjustments" or "the ideology of race"; and (3) a compromise plan with shorter plenary sessions supplemented by "voluntary commission meetings."[35] After vigorous debate, the participants voted to go with the compromise plan, which they kept in place for the remainder of the conference.

Although this compromise plan seemed to appease some of the tensions that had emerged, it seems as though the differences were never fully reconciled. After the conference, Hourani wrote a short reflection where he emphasized that the conference raised more questions than it provided answers. According to Hourani, the participants were not able to develop compelling answers to the fundamental questions that some participants raised about the boundaries and coherence of race relations as a field: "Is there such a subject as 'race relations'?" "Does the concept of 'race' suffice to define an intelligible field of study, divided off from other fields by reasonably clear boundaries, and possessing an essential, not merely accidental unity?"[36] Rather than come to any sort of agreement on these questions, Hourani explained that the conference participants instead developed opposing views. Whereas some argued that a "science of 'race relations' *was* possible," particularly if races were thought of as "historic communities," others maintained the view that such a definition of race was arbitrary and insufficient for the development of a distinct science of race relations because it could potentially apply to all communities. Hourani also wryly observed that some conference participants "seemed to assume that American experience applied to all 'racial situations'" and conjectured that this was possibly due to a "lack of an adequate analysis of the purpose of conflicts."[37] Offering a lukewarm assessment of the meeting, Hourani concluded that although the "conference discussed many interesting things, it was never able to decide what it ought to be discussing."[38]

INTERNATIONALIZING RACE RELATIONS

Although some participants, like Hourani, left the near monthlong conference unsure of what they had accomplished, other participants saw the meeting as an opportunity to build toward an international society for the study of race relations and to craft a research trajectory for the field. With an eye to planning for the future, members of the conference steering committee reserved the final plenary session of the conference, which took place on July 23, for discussing proposals for future research, for appraisals of the conference, and for discussing the creation of an international society. Both the discussion of future research and the discussion of the potential international society revealed many of the themes and fault lines that emerged during the conference and complicated the development of a "view from everywhere." They also revealed a growing reflexivity that prompted race relations scholars to begin stiudying many of the practices, figures, and situations that had been taken for granted by many of the antiracist scholars encountered in previous chapters. For instance, one of the proposed themes for future research was the study of "Western Racial Ideas." As key elements of this research program, the proposal listed topics such as "a historical study of the idea of White superiority seen as a product of changing ideas in western civilization and of the expansion of Europe"; the role of "racial ideas in various imperial systems"; and the "image of the 'black man' in Western Civilization."[39] The envisioned research thus broke from the comparative study of racial situations in different regions that had typified race relations inquiry since the interwar period and instead gestured toward a genealogical approach not unlike the most recent historiography of race science.

In addition to considering proposals for future research, the conference participants also attempted to take stock of the key takeaways from the conference and how they might inform future directions of the field. In their appraisals of the conference, participants notably observed that the major themes of the conference reflected a shift in racial thought from the preceding decades. One shift noted was a move away from the paradigm of "racial individualism." As evidence of this shift, participants observed that "the matter of the relation between race and culture" received much less attention than it would have twenty years ago.[40] In a similar vein, they also observed that "the matter of prejudice," which would have likely been discussed at length ten years ago, was not mentioned. Instead of the emphasis on "culture" and "prejudice" that had previously been central to the field, participants remarked that their discussions during the preceding three weeks had placed major emphasis on "power structure and race relations, racial stratification, and nationalist movements."[41] In charting future directions, some participants suggested that the field needed to much more carefully examine the hypothesis that had been implicit during most of their conversations: "that if the power structure in a racial situation is known, there is no need to examine individual attitudes." Although they had little to say about whether they had been successful in defining what might constitute a "world perspective," they certainly seemed more resolute about the growing importance of structural approaches to the analysis of racial situations.

As well as moving away from notions of culture and prejudice, participants noted how the conference had displaced the centrality of Chicago sociology. As one speaker remarked, the "nucleus" of the conference was undoubtedly formed by "those who were of the 'Park School,'" and the framing of the conference thus owed much to concepts from this school. And yet, according to this speaker, as the sessions unfolded "the Conference developed a life of its own" and "new emphases" emerged that displaced the conventional Chicago approach and raised fresh "theoretical considerations for future consideration."[42] In a summary of the conference published in *Social Process*, Lind also described a similar phenomenon. Whereas the conference had been organized around different facets of what Lind called simply "race relations theory," he conceded that "some of the most stimulating and profitable sessions" were those "devoted to the examination of conceptions emanating from the conferees." As an example of these stimulating moments, Lind noted that the European participants were responsible for injecting "certain ideological concepts such as negritude, the 'elder brother' theory of race relations, and race as an idée force [*sic*]" into the conversation. He also noted that the value of these concepts in race relations had "not previously impressed very many American scholars" and, similarly, that many of the concepts developed by the American conferees "such as the race relations frontier, race making situations, and racial types" were eagerly examined and "criticized by non-American scholars."[43]

One of the outcomes that Lind and his collaborators envisioned for the conference was the creation of some entity that would give race relations a lasting international presence. As Lind wrote, "the entire conception of the conference, the preparations for it, and the development within the conference itself, all pointed quite naturally to the formation of some organization to give permanent expression to its ideas with respect to race relations research."[44] To this end, the conference organizers devoted a significant portion of the final session to discussing a proposal for an "International Organization in the Field of Race Relations." The proposal circulated to the conferees adopted the very same Chicago framework that had informed so much of the conference. According to the proposal, the proposed organization was to be "scientific and scholarly in character and thus concerned with the objective study of race relations." Presumably as a way of clarifying what it meant by the "objective" study of race relations, the proposal explained that the organization would not "espouse racial policies, nor agitate against given racial policies." Instead of wading into the choppy waters of policy and advocacy, the envisioned organization was to focus on the much more pedestrian task of coordinating research among international scholars and thus providing a framework for the growth of the field. For instance, the proposal mentioned that the organization could keep the members of the conference and those with similar interests in "closer and more continuous contact with another" and also perform other coordinating functions such as serving as a "clearinghouse" for the exchange and dissemination of news and research updates, preparing relevant bibliographies, and suggesting problems for local research and assisting members with their local studies. Like many other recently created social science organizations, the proposal also envisioned cooperating with UNESCO. "While maintaining complete autonomy and independence [the organization] could affiliate with UNESCO," the proposal explained.[45]

Yet during the discussion of the proposal, some conference participants again pushed back on the narrow scholarly framing and suggested expanding the scope of the proposed organization. Several speakers insisted on casting a wide net when recruiting for the organization by seeking not only scholars from around the world but also experts from "government agencies and non-academic organizations" engaged in activities concerned with the study of race relations. One of the key functions of the organization, these speakers suggested, could be to "open up channels of communication" between academic and non-academic experts concerned with race relations.[46] According to Melvin Conant, the director of the Pacific and Asian Affairs Council who later wrote a book summary of the conference proceedings, several participants also campaigned for the organization to be "action" oriented presumably meaning one that might advocate for or against specific policies. Ultimately, this more expansive approach gained little traction. According to Conant, the majority of the participants were primarily concerned with encouraging further research into the problems discussed and felt confident "that if the studies were sufficiently precise and objective they would be of important aid to other scholars, administrators, observers, and politicians."[47] In the end, the participants agreed to create a nongovernmental and independent body that was given the name International Society for the Scientific Study of Race Relations (ISSSRR). During the meeting, the conference participants elected the African American sociologist E. Franklin Frazier as chairman of the organization, Quintin A. Whyte (director of the South African Institute of Race Relations) as vice chairman, and William O. Brown (director of the African Research and Studies program at Boston University) as secretary-treasurer. The conferees also elected four additional members of the conference, Georges Balandier, Albert Hourani, John A. Barnes, and Andrew Lind, to the executive committee and tasked them with expanding the membership of the organization to include "qualified students and administrators in the field of race relations on as wide a geographic basis as possible."[48]

THE ISSSRR: A SHORT-LIVED EXPERIMENT

The makeup of the elected officials for the nascent ISSSRR reflects the tensions and faultiness that emerged during the conference itself. By electing a prominent cohort of Europe-based experts (Balandier, Barnes, and Hourani), two of whom were Africanists and one a Middle East specialist, the Honolulu conferees appeared to acknowledge the over-representation of Chicago-trained sociologists in the field and took a significant step toward de-centering their influence. The election of Quintin A. Whyte from the South African Institute of Race Relations seemed to reflect a similar desire. By electing E. Franklin Frazier as president, the conferees also made a bold choice that suggested a desire to change the whiteness of the field. Although he trained at Chicago with Robert Park, Frazier became the most prominent black sociologist of his generation and was the chair of the sociology department at Howard University, the leading HBCU in the United States. He had also built his career on challenging the status quo and had never adhered to the predominantly liberal politics favored by most Chicago-trained sociologists. Thus, by electing Frazier as president and an executive committee composed of European scholars, the Honolulu conferees appeared to wager that they could simultaneously address the overrepresentation of white and U.S.-trained scholars of race relations in the field and thus open up the field to what we might now call "global" and "diverse" approaches.

For Frazier, the stakes of his appointment were considerably high. Unlike the organizers of the Honolulu conference, whose main concern was crafting a "world perspective" on race that would preserve the status quo, Frazier's political sensibilities were much more aligned with the "worldmaking" politics of the leaders of the nonaligned movement. After Frazier completed his doctorate with Robert Park in Chicago in the early 1930s, he also departed from Chicago sociologists' liberal politics and was a committed socialist for most of his life. Much like Richard Wright and W. E. B. Du Bois, he also associated with communist and Pan-Africanist movements during crucial moments of his trajectory. For instance, in 1934 he gave Nancy Cunard, the British writer and supporter of the U.S. Communist Party, permission to reprint his article "The Pathology of Race Prejudice" in her anthology *Negro*.[49] In 1936, Frazier joined the editorial staff of the newly founded journal of Marxist thought and analysis, *Science and Society*, and the following year he gave a keynote address at the May Day rally in Washington, D.C., organized by the Communist Party where he called for the creation of a workers' movement that would unite black and white workers in a single struggle.[50] During the 1950s, when he was investigated by the FBI and the Justice Department for his previous involvement in black popular front activism, Frazier stuck to his socialist principles and produced a scathing critique of the liberal aspirations of the black middle class, titled *The Black Bourgeoisie*.[51] In this study, which many consider to be his most important work, Frazier drew comparisons between the alienation faced by the black middle class in the United States and the new bourgeoisie of decolonized nations. During this mature phase of his career, Frazier also developed an admiration for Kwame Nkrumah, the Ghanaian anticolonial leader and first president of Ghana. Frazier viewed Nkrumah as a role model for black Americans, and such was his enthusiasm for the Ghanaian independence movement that he eventually donated his library to Ghana.[52]

In the months following the Honolulu meeting, Frazier and the executive committee made good on their promise and began recruiting potential members to the society. Frazier, who had served as the head of UNESCO's Applied Social Sciences Division in the early 1950s, was quick to contact Alfred Métraux, who had attended the Honolulu conference as a UNESCO observer, with an invitation to become a founding member of the society and to solicit his advice on other potential members.[53] Métraux gladly accepted Frazier's invitation and recommended several scholars associated with the SSD including: Alva Myrdal, Otto Klineberg, Roger Bastide, and Henri Vallois (the director of the Musée de l'Homme). Métraux also promised Frazier that he would do "whatever is possible to help you in placing the Society on a firm basis."[54] Not one to rest on his laurels, Métraux promptly took the initiative and arranged a meeting of the society's executive committee using unspent UNESCO funds left from the 1954 budget. Métraux sent an invitation to Balandier, Hourani, and Barnes (who resided in London and Paris) and proposed that they meet in Paris to begin discussions on the society's prospectus and goals. A few days before the meeting, Métraux sent Frazier a letter describing his plans. He also emphasized the importance of UNESCO's role in the creation of the proposed society: "I was right when I insisted in Honolulu despite strong opposition that UNESCO should be brought in some way otherwise no support could come from the outside."[55]

On December 17, 1954, the executive committee met at UNESCO's headquarters in Paris and soon after, Métraux sent a letter to Alva Myrdal describing "two very fruitful meetings."[56] Métraux explained how he, Barnes, and Balandier had concluded that "the International Society for the Scientific Study of Race Relations could not come into existence without the aid of Unesco," and that their conversations focused "on ways and means of helping the society get started."[57] He also shared with Myrdal that one of the committee's best ideas was for the ISSSRR to publish an international yearbook on race relations, which would "not be a dull bibliography" but rather an "exposé" of the trends and ideas that have come forward on the research made during the year. Echoing the aspiration for a "world perspective" voiced at the Honolulu meetings, Métraux explained that the yearbook could also include "regional coverings on activities and trends" and that it should be "something interesting and lively, and serve the purpose of advancing science and not only of informing people of what is going on."[58]

In his haste to get the ball rolling on the creation of the ISSSRR with UNESCO support, Métraux solicited little by way of feedback from Frazier, the elected president of the society, and did not seem to consider how Frazier may have felt about his lack of involvement. Not surprisingly, Frazier was not impressed with Métraux's actions. In a curt response to Métraux, cosigned by William O. Brown, Frazier shared that he was "surprised and puzzled" by Métraux's decision to gather the members of the provisional executive committee for a meeting at the UNESCO headquarters.[59] Frazier reminded Métraux that "it was the consensus of the Honolulu meeting that the organization of the International Society for the Scientific Study of Race Relations should not be set up within the Unesco framework" and claimed that it was "strange" that Métraux arranged to have the executive committee meet in Paris without "communicating with the officers who were duly elected to assume responsibility together with the Executive Committee for all matters pertaining to the organization of the new society."[60] Frazier, speaking on behalf of the society's officers, suggested, "we would appreciate an explanation of this extraordinary action on your part." Further, Frazier described Métraux's statement that "no support could come from the outside unless Unesco [sic] was brought in" as "baffling" and suggested that "since it may be interpreted either as a threat or a statement of belief," Métraux's formulation might not "communicate exactly" his intended message.[61]

Métraux and the executive committee members who met in Paris were both puzzled and offended by Frazier's letter. In a letter to Métraux, Barnes claimed to be "surprised" by the exchange and offered to send Métraux the report of the meeting that he had written so that he could pass it on to Frazier and "clarify the purpose of our Paris meeting, and of the nature of our discussions."[62] In a dramatic reply to Barnes, Métraux wrote that he was also puzzled by Frazier's letter and that he was "so hurt about it" that he had resolved to "never again have anything to do with the whole business."[63] Métraux explained that he thought he was doing the society a service by organizing the meeting and that the reactions of Frazier and Brown were such that he would "not want to deal with the Society any more" and declared that he would only renew contact if "obliged to do so as a Unesco [*sic*] official."[64] Finally, Métraux expressed regret that an "initiative which was meant in all innocence to be a gesture of good will should have been so maliciously interpreted."[65]

In his reply to Frazier, Métraux poured on the disappointment. In no uncertain terms, he explained that he would usually express "surprise" in response to a letter of the sort that Frazier had written, but that on this occasion his response was rather one of "pained astonishment."[66] Métraux explained that it was certainly not the first time in his career "that an initiative taken in all good faith and with the desire to be helpful has been misconstrued," but also that he never expected that his initiative of bringing the executive committee together "would result in a letter such as the one you and Dr. Brown have seen fit to write to me."[67] In his defense, Métraux argued that he had "no apology to make," and that "the facts are simple and clear, no wicked or torturous intentions were meant."[68] Appealing to Frazier's experience as a "former Unesco official," Métraux explained that there is usually some money left over at the end of the budgetary year and that it was the availability of these funds that allowed him to arrange a meeting with Balandier, Barnes, and Hourani. Métraux also explained, "Unesco's cooperation was envisaged" in their discussions but that "it was purely on a hypothetical basis and would have been subordinated to official action of the Society."[69] Further clarifying his position, Métraux insisted that "there was no intention of controlling the Society or including it within Unesco's [*sic*] framework" and explained that he had informed UNESCO's director-general of "the attitude adopted by the Honolulu conference."[70] Métraux concluded his letter by stating that he found himself "at a loss" to understand how Frazier could have interpreted his "innocuous remark" in his letter as a threat, and that he could no longer cooperate with the society as a "private person" and would only continue to do so in his capacity as a UNESCO official.[71]

Following Métraux's letter, Georges Balandier also sent a letter to Frazier seeking to clarify the executive committee's position and how they viewed UNESCO's involvement. Balandier repeated many of the same arguments as Métraux, and added that "it would be a pity if the Society, hardly formed, should get divided on account of such things" and also explained to Frazier that he considered his actions to have "damaged the Society" and that he felt "certain that I shall have to resign as a member of the Executive Committee of the Society."[72]

Later that year, Frazier sent an apologetic letter to Alva Myrdal pledging his loyalty to the "Division of Social Sciences" and attempting to clarify the situation.[73] Frazier began his letter by thanking Myrdal for a previous letter where she "gave a clear statement concerning what would be the relationship of a society set up by UNESCO [i.e., the proposed international race relations society] and the UNESCO organization" and then sought to clarify his own position.[74] "Let me make it clear," Frazier explained, "that although my statement of the situation was not true it was nevertheless the way the people who attended the conference at Honolulu felt about such an arrangement with UNESCO."[75] Seeking to mend bridges, Frazier assured Myrdal that he maintained "a deep sense of loyalty to UNESCO and especially the program of the Division of the Social Sciences."[76] In a further pledge of loyalty, Frazier explained that he had always made an effort to speak about UNESCO and the "important work which it is doing in the social sciences."[77]

Once the dust settled following the ISSSRR's executive committee meeting at UNESCO headquarters in Paris, Frazier and the elected officers struggled to bring the society to fruition. Though Métraux had committed to using UNESCO resources to help the society, after his terse exchange with Frazier he lost enthusiasm for the project. Frazier, for his part, attempted to find other sources of institutional support with little success. In 1957, a proposal that Frazier submitted to the Ford Foundation for a grant to establish a headquarters for the society was rejected on the grounds that it was contrary to the foundation's policy to sponsor professional societies. In a letter to the ISSSRR membership that year, Frazier explained that despite this setback, he was planning to submit funding requests to two other foundations. He also made a plea to the society's 175 members for patience and understanding. Frazier wrote that he and William Brown (the secretary-treasurer) were "keenly aware" that three years had passed since they had been tasked with creating the society. Though they accepted responsibility for not having made progress during this time, Frazier and Brown also asked the members of the society to "appreciate the difficulties under which we have worked in carrying out your mandate." As examples of these difficulties, Frazier noted the "lack of time" available to them due to the "heavy responsibilities" they had to their universities and to the "programs of African Studies which we are attempting to establish." He also noted that their universities had absorbed most of the "expenses incident to secretarial work." Frazier also pledged that they would continue to do everything possible to "keep the Society alive."[78]

Despite Frazier and Brown's best efforts, the ISSSRR did not materialize. After his initial show of enthusiasm, Alfred Métraux did little more to assist Frazier and Brown with creating the society and instead began incorporating some of the doomed society's proposed functions into UNESCO's program. For instance, after UNESCO's director-general passed a resolution at the 1956 General Conference in New Delhi to "take all appropriate measures toward the elimination of racial prejudice" including "gathering and disseminating documentation on race relations, including trend reports on current research in this field," Métraux promptly began maneuvering to make this happen. With the goal of performing some of the "clearinghouse" services envisioned for the ISSSRR, Métraux coordinated with the editor of UNESCO's *International Social Science Bulletin* to publish a special issue the following year consisting of a series of articles on the "status of the studies undertaken in various parts of the world, in the field of race relation [*sic*]."[79] To this end, Métraux promptly signed contracts with two prominent race relations scholars: the sociologist Anthony H. Richmond from the University of Edinburgh, who agreed to write a review article on trends in race relations in the United Kingdom, and Herbert Blumer, who agreed to write a similar article relating to the United States. Métraux also attempted to recruit scholars to write similar pieces relating to South Africa, Brazil, Hawai'i, New Zealand, and Southeast Asia. Curiously, in a letter to T. H. Marshall, who was then acting ahead of the SSD, Métraux also explained that he would not approach "the Russians" at first because they already had a study on race relations for UNESCO in the works and because he suspected that no previous research had been carried out in the field for the "good reason" that "officially, there is no race antagonism in Soviet Russia."[80]

After working tirelessly to recruit scholars and suffering a few set-backs, Métraux saw the special issue of the *International Social Science Bulletin* come to fruition in 1958. He also positioned it as the first of two special issues dedicated to reviewing trends in race relations across different regions—a goal that revealed the lingering ambition of crafting a "world perspective" central to the Honolulu conference. The foreword to the special issue, which had an anonymous author, made it clear to the journal readers that UNESCO had stepped in to pick up the slack left by the failed ISSSRR. The foreword described how the wording of the UNESCO resolution from the New Delhi conference that called for trend reports on the state of the race relations research had been influenced by the central functions that the conferees of the Honolulu conference had envisioned for the ISSSRR. "The chief function of that non-governmental organization would have been to serve as a docu-mentation centre, responsible for the preparation of bibliographies and directories of current research" the foreword explained.[81] "As the society was never set up," the foreword continued, "Unesco felt called upon to undertake some of the activities for which it would normally have been responsible."

The articles from the 1958 special issue served as a fitting reminder of the challenges the Honolulu conferees encountered in attempting to articulate a singular "world perspective" on race relations. On the one hand, the articles confirmed that the conceptual frameworks of Chicago sociology and U.S. social psychology exerted a major influence in many places. Yet the articles also showed that these U.S. frameworks were not hegemonic and were rarely cited in some contexts. The articles also testified to the fact that, as an object of study, "race relations" did not necessarily cohere in different regions. The issue consisted of four articles: one on Britain by Anthony H. Richmond, one on East Africa by the London School of Economics social anthropologist Barbara E. Ward, one on the Federal Republic of Germany by Kripal S. Sodhi from the Free University of Berlin, and one on the United States by Herbert Blumer. Not surprisingly, the articles on Britain and the United States were the ones where the influence of Chicago sociology and the racial individualism paradigm were most evident and where the race rela-tions framework seemed to have greatest purchase. Yet these articles showed that even in these contexts, scholars had been prompted to question the portability of U.S. frameworks in response to local pressures and realities. For instance, in his article on Britain, Richmond described how Robert Park's concepts of assimilation, accommodation, conflict, and adjustment had not necessarily helped British researchers to understand the situation of ethnic minorities in Britain. Instead of Park's conceptual scheme, Richmond argued that the theoretical framework used by Israeli sociologist Shmuel Noah Eisenstadt to study the "immigration of Jews in Palestine" in his book *The Absorption of Immigrants* was better suited for understand the situation of Britain's ethnic minorities.[82]

In the survey of East Africa, Barbara Ward noted that "race relations" had not yet developed into a distinctive domain of social inquiry. and would thus refrain from treating "race relations as a separate subject."[83] Ward argued that this was necessary because "inter-racial contacts" were so pervasive in East Africa that existing studies tended to treat them as a "by-product" of other social relations rather than singling them out for "separate analysis." Ward also noted how, in Uganda, the phrase "race relations" did not have a simple recognizable meaning. According to the sociological sense of the phrase, argued Ward, the phrase "race relations in East Africa" could be taken to mean social interactions between individuals or groups typically labeled "African," "Asian," or "European" and identified by distance physical traits such as "skin colour and hair form." Yet the bureaucratic categories and on-the-ground realities painted a much more complex picture. "In official census returns," Ward pointed out, "a separate so-called 'racial' category exists for people originating from the Portuguese possession of Goa, in the Indian sub-continent, and another for 'Arabs.'" The census also used the term "coloured" to refer to individuals of mixed parentage. The phrase "race relations" could also easily refer to interactions between these groups listed on the census and to the social interactions between "groups and individuals from different African 'tribes' and territories." Ward thus suggested that "the obvious and most widely discussed" groupings of African, Asian, and European were in fact crude approximations of other identities and that individuals from these broad categories rarely, if ever, "formed a single co-acting 'group.'"[84] Echoing arguments made by Africanists during the Honolulu conference, such as Kenneth Little, Ward thus questioned whether "race relations" was an adequate framework for describing the social complexities and realities of East African nations.

In 1961, the *International Social Science Journal*, the successor to the *International Social Science Bulletin*, published a second special issue on "recent research in race relations," which reflected the culmination of Métraux's efforts. The anonymous foreword to the special issue proved the enduring influence of the Honolulu conference and particularly its suspicion of anticolonial politics and activist-oriented research. According to the foreword's author, the articles in the special issue covered the regions of "North Africa, Tropical Africa, South Africa, South-east Asia and Latin America." As a selling point for the special issue and the field of race relations more broadly, the author celebrated that some regions, like Africa, constitute "real laboratories" where "the relations between groups can be studied from a great many angles." The author also described such regions as offering fertile and complex terrain for the social scientist due to the "variety of situations and historical backgrounds" and "the multiplicity of recent developments," which had the effect of producing "a whole range of those 'variables' which the social scientists envy their colleagues in the natural sciences." Yet the author also issued a note of caution. The articles from the special issue demonstrated that "the race problem" can never be considered in isolation and cannot thus "be dissociated from the development of industrialization and urbanization which affects so many countries that have recently become independent and others aspiring to independence."[85] And in this context of decolonization and rapid development, the notion of race was "on the way to becoming a weapon in the armoury of the élites seeking to assert themselves *vis-à-vis* the West, which they have chosen even as they have rejected it." As an example of such weaponization of race, the author cited the example of "the vague concept of 'négritude,'" which was often "set against the pride and self-assurance of the Whites." Echoing the sorts of criticisms Métraux levied at Peruvian *indigenistas*, the author described negritude as an "awakening to the consciousness of a past for which a mystic value is claimed and of a future inevitably envisaged as radiant" and as part and parcel with "outbreaks of nationalism in racist guise."[86] In a concluding sentence that captured the ethos of many trained Chicago sociologists, the foreword's author reached the conclusion that "the purpose of the social sciences is to help us to live in peace, by bringing to our awareness human problems viewed with objectivity and detachment." From this perspective, the author arrived at the modest conclusion that the issue could be considered a "useful source of information."[87]

CONCLUSION

According to historian Henry Yu, one of the lasting contributions of Chicago sociologists was the creation of a new social scientific outlook on race and culture that centered the concerns of enlightened white experts from the United States. In contrast to U.S. eugenicists from the 1920s who called for an end to large-scale immigration due to the supposed biological incompatibility of non-white races, Chicago sociologists viewed themselves as enlightened cosmopolitans who opposed racism and occupied the forefront of academic thinking about race. Like Arthur Ramos, Chicago sociologists reformed the study of racial differences by taking the mental states of individuals as their primary object of study and thereby questioned the causal force of racial phenotypes. For Chicago sociologists, argues Yu, "it was not physical difference or incompatibility that led to racial conflict and antagonism" but rather "the awareness of physical difference that was the problem, particularly among one of the social groups that such an awareness created and reinforced: whites."[88] Yet during the Honolulu meeting in 1954, the sizable congregation of experts convened by Chicago sociology's leading lights began to challenge the implicit whiteness and Americanness of race relations inquiry and thereby questioned the field's intellectual coherence. In contrast to Chicago and UNESCO experts' assumption that a better understanding of race relations would come from aggregating more regional and disciplinary "perspectives," non-U.S. participants at the Honolulu conference like Albert Hourani and Kodenda Rao began to question the suitability of race relations frameworks beyond the United States and to ponder the degree to which they obscured rather than illuminated the operation of power at a global scale. Like Brazil's urban sociologists who became dissatisfied with U.S. conceptions of race prejudice, Honolulu participants began to question Chicago sociology's emphasis on individual attitudes and instead called for a greater engagement with "power structures" as well as action-oriented research.

Although the influence of Chicago sociology on U.S. analyses of racism waned after the 1950s, it was not until the 1990s that U.S. sociologists began to more widely embrace structural interpretations of racism.[89] Yet as the last two chapters have shown, structural accounts of racism had been articulated in Brazil and other areas beyond the North Atlantic in the 1950s.

UNESCO and Chicago sociologists' failed attempt to create an international society for race relations inquiry after the Honolulu conference also demonstrates the limits and historical specificity of this project. The tensions that quickly emerged between E. Franklin Frazier, Alfred Métraux, and the European members of the nascent society's leadership team illustrates how experts from differing regions and institutions could not seem to reach agreement on the purpose of race relations inquiry beyond the pedestrian task of stockpiling information. What Chicago-trained sociologists saw as a relatively straightforward and apolitical task of scaling race relations inquiry toward a "world perspective" proved to be a highly contentious endeavor that exposed the colonial foundations of the project. The tensions from the Honolulu conference and the failed attempt to create an international society for race relations inquiry thus not only reveal the implicit whiteness of Chicago studies of race but also the limits to the empiricism that was so firmly ingrained in many of UNESCO's race projects from the 1950s. Although participants from the Honolulu conference frequently noted that the era of "white dominance" seemed to be coming to an end, Chicago sociology's conceptual toolkit seemed to offer little guidance for imaging a different future.

Conclusion

"Racism Continues to Haunt the World"

In April of 1963, the same month that he ended his life, Alfred Métraux published his last article for the *UNESCO Courier.* Titled "Does Life End at Sixty?," Métraux's revealing final words offered a stark contrast to "An Indian Girl with a Lesson for Humanity," the article he had published more than a decade earlier. "An Indian Girl" looked to the future and buoyantly celebrated indigenous assimilation, third world development, and the expansion of modern civilization. By contrast, "Does Life End at Sixty?" ruminated on the West's failings while praising the wisdom of "savage" societies and their treatment of the elderly. In the past century, Métraux observed, the average life span in industrialized countries had increased by "leaps and bounds" giving rise to an ever growing percentage of "old people in our population." Having turned sixty just months before and forced to retire as a matter of UNESCO policy, Métraux interpreted this changing demographic structure of modern civilization as one that posed profound dilemmas. In a not-so-veiled reference to his own situation, Métraux asked "how can we reconcile the very understandable impatience of new generations with the enforced retirement of men and women who still have many years of life ahead of them and who still feel energetic enough to remain active far beyond the time limits imposed on them by society[?]" More pointedly, Métraux asked "how can we find occupations for these people whose physiological age does not correspond to their real age?"[1]

Métraux's final thoughts in the *UNESCO Courier* offer a striking reference point from which to historicize the race campaign he served. For much of his postwar career in the UN system, Métraux pragmatically championed Western assimilation and economic development as horizons of possibility opened up by dismantling scientific racism. His career as an antiracist technical assistance expert thus entailed a reluctant acceptance that the indigenous communities he studied could not survive without conforming to Western civilization. By the 1960s when he wrote his final article for the *UNESCO Courier*, such noblesse oblige seemed increasingly out of touch with the political forces that would be unleashed during this decade. By this point, as we saw in chapter 7, the "Third World" and nonaligned politics initiated at the Bandung conference posed a forceful counterpoint to the bipolar politics of the Cold War. Anticolonial revolutionaries in Asia, Africa, and the Caribbean toppled imperial rule and demanded global political equality and a new international economic order. Inspired by such movements, indigenous peoples in settler nations and elsewhere called out assimilationist policies and began to assert their historical treaty rights and to demand recognition of their land and territorial sovereignty.[2] By 1963, the paternalistic antiracism that Métraux had championed for most of his career seemed out of date. As he contemplated whether life ended at sixty, new and world-making movements were being born.

Métraux's about face coincided with a resurgence of racism, which prompted UNESCO to reassess the strategies behind its race campaign. In the year after Métraux's death, UNESCO officials became increasingly unsettled by a resurgence in racism and began to reevaluate the organization's race campaign and to replace narrow approaches to racism that emphasized individual prejudice and cultural difference. This reassessment occurred just as nineteen newly formed African states joined UNESCO's ranks. It also followed the organization's recognition of the reawakening of anti-Semitism in Europe and the intensification of racial segregation in South Africa's apartheid state: two topics that figured prominently in the October 1960 issue of the *UNESCO Courier*. In light of these sociopolitical trends as well as developments made in biological studies of human diversity, participants from UNESCO's twelfth General Conference in Paris (1962) called on the organization to update its Statements on Race and recognized the importance of addressing the "social" and "biological" dimensions of racism simulta-neously. Participants at this conference asserted that "the growth of population genetics as a discipline, and the very high rate of fossil dis-coveries" had made it necessary to revise the 1951 Statement on Race. They also insisted that it had become necessary to "enlarge [the 1951 statement's] scope" by including previously missing points dealing with "the nature and forms of racial prejudice in inter-racial situations."[3] In response to this proposal, UNESCO convened two meetings: one

FIGURE C.1 "Racism!" (cover of the *UNESCO Courier*, October 1960).

FIGURE C.2 Stamps against racism (image, *UNESCO Courier*, November 1971, p. 23).

that took place in Moscow in 1964 and brought together twenty-three biologists, geneticists, and physical anthropologists and a second that took place in Paris in 1967 and assembled a group of eighteen experts representing the fields of genetics, social sciences, history, law, and philosophy. The groups of experts who participated in these meetings produced two statements: the 1964 Proposals on the Biological Aspects of Race and the 1967 Statement on Race and Racial Prejudice. To a greater extent than the committees that drafted the previous statements, the committee at the Moscow meeting and at the Paris meeting included experts who hailed from or had considerable experience in regions of the Global South.[4]

The 1964 and 1967 statements and especially the latter moved decisively beyond the frame of "racial individualism" that informed the liberal social sciences of the United States during the Cold War. From the outset of the 1967 meeting, participants expressed a desire to do more than educate and instead called for a statement with direct policy outcomes that could be brought "to the attention of the governments of Unesco's Member States" during the next General Conference. The participants in the 1967 meeting also demonstrated a keen awareness of the protean nature of racist discourse and the need to tackle racism on multiple fronts. They acknowledged that the 1964 Moscow propositions served as a necessary point of departure and argued that it continued to be necessary to publish "widely the truth on biological aspects of 'race'" given that racists continued to use biological arguments in favor of racial inequality. Yet they also observed that racists could easily pivot and adapt. The report noted that when they could no longer use a "biological justification," racists "resorted to other excuses—divine purpose, cultural differences, differing levels of 'achievements,' etc" and emphasized that it "was this latter type of argument which was now the most often used."[5] Going further still, another participant raised concerns about building antiracism from a strictly scientific foundation. "To build anti-racism solely on the basis of scientific findings was to build it on a very fragile structure," this participant asserted. For this participant, "the present findings of modern science" suggested the possibility that the races could one day "be found to be 'unequal' in this or that particular sphere." For this reason, this participant implored the committee to base its use of notions such as human unity and equality on something "more fundamental." The "human rights aspect," suggested this participant, should be stressed in the committee's 1967 statement.[6]

Given the tenor of the deliberations, it is not surprising that the final 1967 statement was much bolder than its predecessors. Although the statement briefly reaffirmed the claims of biological human unity from the previous statements, it eschewed the detached tone of the previous statements, and the bulk of its attention focused on directly confronting racism and its historical and structural causes. This show of intent was evident in the opening paragraph, which declared that "racism continues to haunt the world" and to act as a "particularly striking obstacle to the recognition of equal [human] dignity."[7] The statement also reiterated the point made during the meeting that when the false biological basis of racism is exposed, "racism finds ever new stratagems for justifying the inequality of groups." It also cautioned against an antiracist strategy focused only on overt forms of racism and thus called attention to the ways racism can endure without naming race. "Many of the problems which racism presents in the world today," the statement declared, "do not arise merely from its open manifestations, but from the activities of those who discriminate on racial grounds but are unwilling to acknowledge it." The statement also repudiated the notion that racism is an inevitable facet of human experience. Instead, it insisted that "racism has historical roots" and was "not evident for long periods in world history." In fact, the statement offered a precise genealogy of racism that identified its origins in "conditions of conquest," notably "the case of Indians in the New World" and the "justification of Negro slavery and its aftermath of racial inequality in the West."[8]

Further separating itself from previous statements and from much of the work done in North America under the race relations rubric in the 1950s, the 1967 statement invoked a broad set of structural causes to illustrate what racism is and how it might be dismantled. The task of undermining racism, the statement suggested, required more than biologists exposing its fallacies. "It is also necessary that psychologists and sociologists should demonstrate its causes," the statement declared. And here the statement asserted that "the social structure" always acted as an "important [causal] factor" albeit with "great individual variation in racialistic behaviour" depending on the "personality of the individuals and their personal circumstances." Similarly, instead of dwelling on agreed biological facts, the statement listed a set of conclusions that the committee of experts agreed on about "the social causes of race prejudice." One notable conclusion illustrates the scope of the structural analysis adopted by the committee:

> Social and economic causes of racial prejudice are particularly
> observed in settler societies wherein are found conditions of great

disparity of power and property, in certain urban areas where there have emerged ghettoes in which individuals are deprived of equal access to employment, housing, political participation, education and the administration justice and in many societies where social and economic tasks which are deemed to be contrary to the ethics or beneath the dignity of its members are assigned to a group of different origins who are derided, blamed and punished for taking on these tasks.[9]

In response to this structural and even systemic conception of racism, the 1967 committee also included a series of specific proposals for "changing those social situations which give rise to prejudice." And here the committee emphasized that any important changes in social structure that might lead to "the elimination of racial prejudice" would require "decisions of a political nature" as well as the mobilization of "certain agencies of enlightenment" such as education, mass media, and the law. Offering concrete examples of what such changes might look like, the committee include proposals such as ensuring that school curricula contain scientific understandings about race and human unity, that school and other educational resources be made fully available "to all parts of the population with neither restriction nor discrimination," and that societies take "corrective measures" in cases where "certain groups have a lower average education and economic standing" due to "historical reasons."[10] Such measures, the statement insisted, should do everything possible to ensure that "the limitations of poor environments are not passed on to the children." As a further measure, the committee recommended that governments and other organizations give special attention to "improving the housing situations and work opportunities available to victims of racism" and argued that this would "not only counteract the effects of racism" but also offer a "positive way of modifying racist attitudes and behaviour."[11]

The 1967 meeting in Paris represents the last time that UNESCO convened a committee of scientific experts for the purpose of drafting a "statement" on race. Yet it was certainly not UNESCO's last initiative against racism. In fact, the organization produced its most comprehensive policy instrument against racism a decade later when it published the 1978 Declaration on Race and Race Prejudice. Although the 1978 declaration incorporated many of the basic principles from the four previous UNESCO Statements on Race and cited these statements as important antecedents, UNESCO officials engineered the declaration with even loftier goals in mind and assembled a very different set of experts to write it. In this regard, the 1978 declaration was much more aligned with the "worldmaking" ambitions of anticolonial leaders than with the Honolulu conference project of crafting a "world perspective." While the previous Statements on Race had been primarily educational in their ambition, UNESCO officials conceptualized the 1978 declaration as a much more impactful measure that would be ratified by UNESCO's member states and provide them with a framework and set of standards for taking concrete actions against racism. The declaration was also designed to complement a recent outpouring of international legislation against racism within the UN system.[12]

On the face of it, UNESCO's 1978 declaration recapitulated and expanded the core claims of the previous statements such as the importance of recognizing the unity of the human species, the rejection of biological determinism, and the insistence on conceptualizing material differences between groups as the product of nonbiological forces. Aligning itself with previous UNESCO Statements on Race, the declaration stated that "the differences between the achievements of the different peoples are entirely attributable to geographical, historical, political, economic, social and cultural factors. Such differences can in no case serve as a pretext for any rank-ordered classification of nations or peoples."[13] Yet despite such conceptual continuity, the 1978 declaration invoked decolonial standpoints that were far less visible in the previous statements. Whereas the previous statements often sought to distance themselves from a conceptual past marked by biological determinism, the 1978 declaration positioned itself as participating in historical changes that were creating a new world order. For instance, the declaration's preamble stated that it was "mindful of the process of decolonization and other historical changes which have led most of the peoples formerly under foreign rule to recover their sovereignty." It also invoked the process of decolonization as one that will create "new opportunities [for] eradicating the scourge of racism and [for] putting an end to its odious manifestations in all aspects of social and political life, both nationally and internationally."[14] Moving away from the racial individualism that featured prominently in many of UNESCO's antiracist publications from the 1950s, the 1978 declaration also conceptualized racism as an ideology that is incommensurate with principles of state sovereignty and with an international order based on human rights. To this effect, the third article of the declaration stated:

> Any distinction, exclusion, restriction or preference based on race, colour, ethnic or national origin or religious intolerance motivated by racist considerations, which destroys or compromises the sovereign equality of States and the right of peoples to self-determination, or which limits in an arbitrary or discriminatory manner the right of every human being and group to full development is incompatible with the requirements of an international order which is just and guarantees respect for human rights; the right to full development implies equal access to the means of personal and collective advancement and fulfilment in a climate of respect for the values of civilizations and cultures, both national and world-wide.[15]

Through the emphasis they placed on the systemic basis of racism and the right to self-determination, the 1967 statement and the 1978 declaration eclipsed the comparatively narrower focus of the 1950 and 1951 statements. Yet, as previous chapters have shown, the 1967 and 1978 documents were not without precedent. They in fact epitomize many of the southern currents of racial thinking that this book has described. For instance, the 1967 statement's analysis of the social, historical, and economic causes of race prejudice bears a striking resemblance to the arguments made by the cohort of Marxist-inspired sociologists like Luiz Aguiar Costa-Pinto and Florestan Fernandes who developed novel analyses of the Brazilian racial situation in the 1950s. Similarly, the structural and environmental transformations the 1967 statement proposed, such as improving housing conditions and transforming educational institutions, evoke the Bahia Tropicalista school of medicine from the late nineteenth century as well as the ambitions of Te Rangi Hīroa and the Young Maori Party from the early twentieth century. Rather than attempt to forge a new "world perspective" on race prejudice, the 1967 statement and the 1978 declaration put forward proposals that were much more concerned with discerning and reinventing the political and economic structures of the international order—the sort of "worldmaking" project that Adom Getachew associates with anticolonial leaders.

Given these historical trends, it is not surprising then that in his last UNESCO article from 1963, Métraux began to make an about-face. Instead of accepting the inevitable spread of Western "civilization," he began to question its fundamental premises. He lamented how poorly modern societies treat their elderly and how much better "primitive" societies fare in this regard. "Primitive" societies, explained Métraux, held their elderly in high esteem, kept them tightly integrated within their communities, and granted them positions of great power and privilege. In groups as geographically distant as the "Eskimoes" of the Arctic, the "Crow Indians" of the Great Plains, the Kayapo of Brazil, the "Akambas" of Kenya, Aboriginal Australians and the "Navahoe Indians," elderly people were treated as "living archives" whose duty it was to transmit their group's myths, traditions, and healing practices to future generations. In many cases, elderly people also exerted great influence and prestige to an extent that Métraux argued it would be no exaggeration to call these societies "gerontocracies." Looking back at modern society from the standpoint of "primitive societies," what became clear to Métraux is that "technological progress" had come at an immense cost. "We certainly live longer than 'savages,' but we have paid a high price for this privilege," Métraux declared.[16]

Thoughts on the toll of aging also permeated Métraux's personal correspondence in his last days. In one of his last letters, he complained to his close friend and self-declared "twin," the photographer Pierre Verger, of the physical and psychic burdens of turning sixty. Whereas Verger, born the same year as Métraux, relished turning sixty in his adopted home of Ibadan, Nigeria, where he enjoyed a "benedictine" life of writing and solitude, Métraux complained that his sixties had started poorly. "All this winter," Métraux wrote, "I was plagued by sciatica which made me lame and insomniac, not to mention the morale damaged by a foretaste of 'old age' where these ailments will become chronic."[17]

For Métraux, it seems, this state of ill health is not one he was willing to bear. On Thursday, April 11, 1963, just a few months after being appointed the secretary-general of France's Societé des Americanistes,

Métraux took his own life. Even in death he remained committed to the rigorous empiricism so often associated with his work. As his friend and colleague Claude Lévi-Strauss explained, Métraux committed suicide by overdosing on barbiturates and recorded the stages of his intoxication in a notebook for as long as he was able.

In tributes following his death, Métraux's friends and colleagues praised his career in ways that evoked the wisdom of indigenous elders—the "living archives" Métraux described fondly in his last article for UNESCO—as well as the trope of vanishing races that has such deep roots in European colonial conquest. Lévi-Strauss described his death as a "crushing loss" and lamented that Americanists had lost a deep well of accumulated knowledge that sprang from Métraux's extensive field-work, much of which had only been partially published. Lévi-Strauss's tribute also framed Métraux as one of the last portals to vanishing indigenous worlds. "His disappearance therefore marks," Lévi-Strauss explained, "the definitive extinction of many indigenous populations that he had once known, still in possession of their traditional culture and which, alas, for many years had practically no longer existed except in his knowledge."[18]

While Lévi-Strauss's tribute highlighted the loss of indigenous knowledge, others, such as Métraux's student Pierre Clastres, empha-sized an abandonment of self. For Clastres, Métraux's body of work was notable for its epistemological and ethical rigor which was evident in how Métraux adhered to the strictest methodological standards and chastised all forms of ethnological amateurism. Métraux represented what Clastres called a "militant de objectivité" and explained how it had often led students to misinterpret his writing as betraying a cold and indifferent attitude toward indigenous people. Yet beneath this stern commitment to objectivity lay what Clastres called a "rigorous ethic" characterized by a sensibility toward "the Other" so extreme it could lead to a "total oblivion of self." "The Indians were never pure objects of observation for Métraux," Clastres noted and insisted that the contrary was actually the case—"in the field, [Métraux] attempted to reduce that irreducible distance that nevertheless separates a White and an Indian." Métraux's attitude toward "the Indians" could thus be described as a "loyalty to friendship," and it could be measured by the fact that he was willing to risk his life for their benefit.

To illustrate this last point, Clastres offered a personal anecdote. In 1963, during his first-ever field trip to Paraguay and Brazil, Clastres traveled to the Gran Chaco and visited the Chulupi at Métraux's suggestion. The Chulupi's territory straddled the Pilcomayo River, which marks the border between Paraguay and Argentina. During his visit, Chulupi members told Clastres the story of a soldier that saved their lives during the Gran Chaco war some thirty years before his visit. Back then, the mysterious soldier emerged from the river fully clothed and sopping wet to warn the Chulupi that an Argentine military garrison stationed across the river planned to raid their camp at dawn and massacre them. Given that soldiers were typically concerned with killing them rather than saving their lives, the Chulupi were unable to fathom why a soldier would disclose such plans. Yet, out of caution, they heeded the soldier's warning and abandoned their camp before dawn. As they left they were able to observe, from a safe distance, the arrival of heavily armed Argentine troops. When they related this story to Clastres in 1963, the Chulupi still could not make sense of the soldier's motives. Yet Clastres was convinced that they were mistaken. Their mysterious protector was not an Argentine soldier but none other than Alfred Métraux. Clastres's certainty stemmed from his last meeting with Métraux. On the day before traveling to South America, just a few weeks before Métraux's death, Clastres spent the day with his mentor and found him in an uncharacteristically relaxed and unguarded mood. During their conversations, Métraux told Clastres a story that echoed the Chulupi's. During his early fieldwork in the Argentinian Chaco, Métraux recalled an occasion when he overheard soldiers camped nearby describe a plan to attack a neighboring indigenous group. As soon as Métraux heard the plan, recalled Clastres, he hatched a plan to warn the unsuspecting indigenous group. "Given that he was undoubtedly dressed in utility clothes similar to military garb," Clastres concluded, "one can understand the Chulupi's misunderstanding."[19]

Alfred Métraux's last musings, his death, and the intellectual tributes it inspired offer a final glimpse into the colonial tensions and contradictions that structured the confrontations with racism that this book has tracked. A striking feature of the stories and tributes Métraux's death brought forth is how they positioned indigenous peoples as portals to the past and as vanishing populations. These stories speak to the persistence of 'vanishing race' theories and their congruence with UNESCO's race campaign. Despite Métraux and UNESCO's efforts to dismantle static race typologies, binary conceptions of temporally distant peoples—either "primitive" or "civilized"—continued to shape Métraux's intellectual milieu. For Métraux's esteemed Americanist colleagues, his death served to remind them of the profound chasms between the indigenous peoples of the Americas they studied and the modern civilization they inhabited as well as the way that Métraux's body of work served the Sisyphean purpose of attempting to reduce this "irreducible distance." To borrow Johannes Fabian's term, they simultaneously praised Métraux's moral and empirical virtues as a scholar while denying the "coevalness" of the people he studied.[20] Yet Métraux's career as a UN technical assistance expert suggests he saw things otherwise. In this context, Métraux began to see the relation between "primitive" and "civilized" through the lens of technological and economic power. Rather than deny the temporal coevalness between "primitive" and "civilized," Métraux conceptualized Western civilization as an irrepressible force that inevitably overwhelmed non-European groups. Yet it was only as he turned sixty and felt death approaching that he seemingly began to confront the civilization in which he so often felt ill at ease. As he faced retirement and considered his dissatisfaction, he began to ask whether reducing the presumed chasm between the West and its Others requires challenging the social structures of the West and how they manage life.

Métraux's death serves as a fitting endpoint for the transnational streams of racial thought this book has tracked. In his final reflections and actions, Métraux reveals an enduring interest in grasping the mechanisms of social contact and change and an unwavering commitment to empirical rigor. These are two epistemic orientations that were shared by many of the key actors and schools of thought this book has tracked: Arthur Ramos and the so-called Nina Rodrigues school; the committee of race experts who drafted the 1950 UNESCO Statement on Race; Te Rangi Hīroa and the many scholars he mentored at the Bishop Museum; Charles Wagley and Eduardo Galvão and their Amazonian interlocutors; the urban sociologists of Rio de Janeiro and São Paulo; and the international cadre of race relations experts who gathered in Honolulu. For this generation of mid-twentieth-century experts, confronting racism was often an empirical matter of grasping the history and social life of local communities as intimately as possible. In their work, lofty theories of social evolution, which presumed a law-like and often unidirectional temporal orientation, were increasingly replaced with conceptual frameworks that oriented researchers to comparatively more situated and relational processes such as accultur-ation, race relations, and economic development. Within this ostensi-bly nonracial epistemic space that was articulated during the interwar period and amplified after World War II, southern American societ-ies such as Brazil and Peru and societies in the Pacific such as New Zealand and Hawai'i became exemplars of racial harmony that North American and European scholars turned to in search of redemptive les-sons that might help them govern racial issues in their own societies.[21] Within the internationalist context of UNESCO, these sociohistorical approaches were quickly incorporated into the emerging international domain of technical assistance that sought to preserve the economic and geopolitical hierarchies that had been created through European imperialism. It no longer became necessary to name race in order to enact and perpetuate racist structures. Like Arthur Ramos's attempts to redeem Nina Rodrigues's work, UNESCO's race experts endeavored to conceal the worst elements of racial thought and render them a distant remnant of a less enlightened past.

Métraux's continued use of terms like "savage" and "civilized" and the ways in which his Americanist colleagues in France identified him with vanishing indigenous tribes suggests that this project of rendering race a remnant was incomplete. In fact, the frame of "salvage ethnog-raphy," which typified his early career and presupposed the inevitable collapse of indigenous societies, continued to inform his work right up until his death. One of Métraux's last contributions before his death was to lend his expertise to the emergent field that Joanna Radin has aptly called "salvage biology," which flourished during the Cold War.[22] In December 1962, Métraux traveled to Geneva, Switzerland, and served as a consultant for the WHO's Scientific Group on Research in Population Genetics of Primitive Groups, whose overarching aim was to urgently document the genetic makeup of "primitive" groups whom they saw as "threatened with imminent cultural disintegration, and, in some instances, loss of physical identity in the face of advancing civilization."[23] In the notes he prepared for this meeting, Métraux described South America as offering a "particularly favorable ground for the study of ethnic isolates" and noted that although the "native population" was rapidly declining outside the Andes, there were still a large number of small groups in the Amazon and Chaco that had "managed to survive due to their isolation."[24] The WHO experts that gathered for this meeting identified Marie-Yvonne Vellard's tribe—the Aché—as one of several "high-priority" groups who were "in grave danger of cultural extinction or decline in numbers" and thus prime targets for genetic research. During the meeting, Alfred Métraux also persuaded the human biologist and Kuru researcher Carleton Gajdusek to travel to Paraguay in order to visit the Aché and conduct research on "child growth and development and human genetics," which he eventually did with help from Pierre Clastres, who was also studying them.[25] Though Métraux spent much of his career studying and documenting, in often painstaking detail, the rich cultural and political lives of indigenous groups, one of his last acts was thus to propel a field that has since played a prominent role crafting a biological conception indigeneity that frames indigenous peoples as storehouses of genetic diversity at risk of vanishing through admixture.[26] For all the effort he expended combating racism, Métraux struggled to challenge what Linda Tuhiwai Smith has called the positional superiority of the West or to give much space to indigenous agency. Yet by this point he was swimming against a new internationalist tide.

Notes

INTRODUCTION: THE REMNANTS OF RACE SCIENCE

1. The members of Marie-Yvonne's tribe refer to themselves as the Aché. The label "Guayaki" is a pejorative one that was used historically by neighboring Tupi-Guarani groups who hunted the Aché and sold the Aché children as slaves to neighboring ranchers.

2. Alfred Métraux, "An Indian Girl with a Lesson for Humanity," *UNESCO Courier* 3, no. 8 (1950): 8.

3. Métraux, "An Indian Girl," 8.

4. Métraux, "An Indian Girl," 8.

5. Métraux, "An Indian Girl," 8.

6. Vincanne Adams, Michelle Murphy, and Adele E. Clarke, "Anticipation: Technoscience, Life, Affect, Temporality," *Subjectivity;* 28, no. 1 (September 2009): 246–65; Aileen Moreton-Robinson, *The White Possessive: Property, Power, and Indigenous Sovereignty* (Minneapolis: University of Minnesota Press, 2015).

7. Perrin Selcer, *The Postwar Origins of the Global Environment: How the United Nations Built Spaceship Earth* (New York: Columbia University Press, 2018), 4.

8. From the UNESCO website: "The Drafting Committee for UNESCO's Constitution was created at its 15th meeting on 10 January 1945. Twelve members of the Committee represented nine countries: Belgium, China, Czechoslovakia,

France, the Netherlands, Norway, the United Kingdom, the United States, and the USSR." Accessed on May 11, 2023: https://tinyurl.com/ycxrhph8.

9. "Constitution of the United Nations Educational, Scientific and Cultural Organization," UNESCO, accessed June 16, 2022, https://www.unesco.org/en/legal-affairs/constitution.

10. I will examine the drafting of the 1950 Statement on Race in closer detail in chapter 2. For other accounts of the drafting of the statement and its significance, see Michelle Brattain, "Race, Racism, and Antiracism: UNESCO and the Politics of Presenting Science to the Postwar Public," *American Historical Review* 112, no. 5 (2007): 1386–413; Jean Gayon, "Do the Biologists Need the Expression 'Human Race'? UNESCO 1950–1951," in *Bioethical and Ethical Issues Surrounding the Trials and Code of Nuremberg: Nuremberg Revisited,* ed. Jacques J. Rozenberg (Lewiston, NY: Edwin Mellen Press, 2003), 23–48; Sonali Thakkar, "The Reeducation of Race: From UNESCO's 1950 Statement on Race to the Postcolonial Critique of Plasticity," *Social Text* 38, no. 2 (June 1, 2020): 73–96; Perrin Selcer, "Beyond the Cephalic Index: Negotiating Politics to Produce UNESCO's Scientific Statements on Race," *Current Anthropology* 53, no. S5 (April 1, 2012): S173–84.

11. Donna J. Haraway, "Remodeling the Human Way of Life: Sherwood Washburn and the New Physical Anthropology, 1950–1980," in *Bones, Bodies, and Behavior: Essays in Behavioral Anthropology,* ed. George Stocking (Madison: University of Wisconsin Press, 1988), 206–60; Robert N. Proctor, "Three Roots of Human Recency: Molecular Anthropology, the Refigured Acheulean, and the UNESCO Response to Auschwitz," *Current Anthropology* 44, no. 2 (2003): 213–39; Staffan Müller-Wille, "Claude Lévi-Strauss on Race, History and Genetics," *BioSocieties* 5 (2010): 330–47.

12. Brattain, "Race, Racism, and Antiracism," 1388.

13. Métraux, "An Indian Girl," 8.

14. Métraux, "An Indian Girl," 8.

15. Michael Adas, "Contested Hegemony: The Great War and the Afro-Asian Assault on the Civilizing Mission Ideology," *Journal of World History* 15, no. 1 (March 2004): 31–63.

16. Ann Laura Stoler, "Racial Histories and Their Regimes of Truth," *Political Power and Social Theory* 11 (1997): 183–206.

17. Stoler, "Racial Histories," 185.

18. Métraux, "An Indian Girl," 8.

19. Métraux, "An Indian Girl," 8.

20. Alana Lentin, "Replacing 'Race,' Historicizing 'Culture' in Multiculturalism," *Patterns of Prejudice* 39, no. 4 (December 1, 2005): 379–96; Kamala Visweswaran, *Un/Common Cultures: Racism and the Rearticulation of Cultural Difference*

(Durham, NC: Duke University Press, 2010); Lee D. Baker, *Anthropology and the Racial Politics of Culture* (Durham, NC: Duke University Press, 2010).

21. Will Kymlicka, *Multiculturalism: Success, Failure, and the Future* (Washington, DC: Migration Policy Institute, 2012), 6.

22. Kymlicka, *Multiculturalism*, 5.

23. Kymlicka, *Multiculturalism*, 6.

24. As Jenny Reardon has argued, this standard narrative was developed in George W. Stocking, *Race, Culture, and Evolution: Essays in the History of Anthropology* (New York: Free Press, 1968); Nancy Stepan, *The Idea of Race in Science: Great Britain, 1800–1960* (Hamden, CT: Archon Books, 1982); Elazar Barkan, *The Retreat of Scientific Racism: Changing Concepts of Race in Britain and the United States Between the World Wars* (Cambridge: Cambridge University Press, 1992).

25. For examples of this interpretation of the UNESCO Statements, see William B. Provine, "Geneticists and the Biology of Race Crossing," *Science* 182, no. 4114 (November 23, 1973): 790–96; Gayon, "Do the Biologists Need the Expression 'Human Race'?," 23–48.

26. My thinking here is indebted to Jenny Reardon's account of the historiography of race science and to Lorraine Daston's work on the biographies of scientific objects. See Jenny Reardon, *Race to the Finish: Identity and Governance in an Age of Genomics* (Princeton, NJ: Princeton University Press, 2005); Lorraine Daston, ed., *Biographies of Scientific Objects* (Chicago: University of Chicago Press, 2000).

27. This campaign is evidenced by an outpouring of antiracist publications and pamphlets during the 1930s and 1940s, such as Julian Huxley and A. C. Haddon's *We Europeans, the Geneticists' Manifesto*, and Ruth Benedict and Gene Weltfish's "The Races of Mankind," which were often concerned with opposing Nazi racial ideologies. In these accounts, the UNESCO Statements on Race figure as the culmination of an intense trans-Atlantic campaign to dismantle Nazi-style racism and as the pinnacle of a decisive shift from rigid hierarchies and biological determinism to relativist and indeterminist conceptions of human variation. See Barkan, *Retreat of Scientific Racism*, 279–340; Stepan, *Idea of Race in Science*; Provine, "Geneticists and the Biology of Race Crossing," 795–96.

28. Reardon, *Race to the Finish*; Lisa Gannett, "Racism and Human Genome Diversity Research: The Ethical Limits of 'Population Thinking,'" *Philosophy of Science* 68, no. 3 (September 2001): S479–92; Lisa Gannett, "Theodosius Dobzhansky and the Genetic Race Concept," *Studies in History and Philosophy of Science Part C: Studies in History and Philosophy of Biological and Biomedical Sciences* 44, no. 3 (September 2013): 250–61; Michael Yudell, *Race Unmasked: Biology*

and Race in the Twentieth Century (New York: Columbia University Press, 2014); Brattain, "Race, Racism, and Antiracism," 1386–1413; Tracy Teslow, *Constructing Race: The Science of Bodies and Cultures in American Anthropology* (New York: Cambridge University Press, 2014); Selcer, "Beyond the Cephalic Index," S173–84; Donna Jeanne Haraway, *Modest-Witness@Second-Millennium.FemaleMan-Meets-OncoMouse: Feminism and Technoscience* (New York: Routledge, 1997); Amade M'charek, *The Human Genome Diversity Project: An Ethnography of Scientific Practice* (Cambridge: Cambridge University Press, 2005); Jenny Bangham, "What Is Race? UNESCO, Mass Communication and Human Genetics in the Early 1950s," *History of the Human Sciences* 28, no. 5 (December 2015): 80–107.

29. See especially Haraway, "Remodeling the Human Way of Life"; Yudell, *Race Unmasked*; Gannett, "Theodosius Dobzhansky and the Genetic Race Concept"; Melinda Gormley, "Scientific Discrimination and the Activist Scientist: L. C. Dunn and the Professionalization of Genetics and Human Genetics in the United States," *Journal of the History of Biology* 42, no. 1 (Spring 2009): 33–72; Veronika Lipphardt, "The Jewish Community of Rome: An Isolated Population? Sampling Procedures and Bio-historical Narratives in Genetic Analysis in the 1950s," *BioSocieties* 5, no. 3 (September 2010): 306–29.

30. Reardon, *Race to the Finish*; Donna Jeanne Haraway, "Race: Universal Donors in a Vampire Culture," in *Modest-Witness@Second-Millennium.FemaleMan-Meets-OncoMouse: Feminism and Technoscience* (New York: Routledge, 1997), 213–67; Sebastián Gil-Riaño, "Redemptive Ancestries: Human Population Genetics, Sex and Antiracism," *Social History of Medicine* 30, no. 2 (May 2017): 448–54.

31. Alondra Nelson, *The Social Life of DNA: Race, Reparations, and Reconciliation After the Genome* (Boston: Beacon, 2016); Kimberly TallBear, *Native American DNA: Tribal Belonging and the False Promise of Genetic Science* (Minneapolis: University of Minnesota Press, 2013); Catherine Nash, *Genetic Geographies: The Trouble with Ancestry* (Minneapolis: University of Minnesota Press, 2015); Peter Wade, Carlos López Beltrán, Eduardo Restrepo, and Ricardo Ventura Santos, *Mestizo Genomics: Race Mixture, Nation, and Science in Latin America* (Durham, NC: Duke University Press, 2014); Catherine Bliss, *Race Decoded: The Genomic Fight for Social Justice* (Stanford, CA: Stanford University Press, 2012).

32. Joanna Radin, "Ethics in Human Biology: A Historical Perspective on Present Challenges," *Annual Review of Anthropology* 47, no. 1 (October 2018): 263, 278.

33. Michelle Murphy, *The Economization of Life* (Durham, NC: Duke University Press, 2017), 135.

34. "Alfred Métraux to Cedric Dover," February 12, 1951, 323.12 A 102, Statement on Race, UNESCO Archives, Paris, France.

35. Rogers Brubaker, *Grounds for Difference* (Cambridge, MA: Harvard University Press, 2015), 51.

36. Brubaker, *Grounds for Difference*, 52.

37. The historical literature on development is now vast. Some useful entry points are Arturo Escobar, *Encountering Development: The Making and Unmaking of the Third World* (Princeton, NJ: Princeton University Press, 2011); Irene L. Gendzier, *Development Against Democracy: Manipulating Political Change in the Third World* (London, Pluto Press, 2011); Michael E Latham, *Modernization as Ideology: American Social Science and "Nation-Building" in the Kennedy Era* (Chapel Hill, N.C.: University of North Carolina Press, 2000); Corinna R. Unger, *International Development: A Postwar History* (London: Bloomsbury, 2018); Daniel Immerwahr, *Thinking Small: The United States and the Lure of Community Development* (Cambridge, MA: Harvard University Press, 2015); Sara Lorenzini, *Global Development: A Cold War History* (Princeton, NJ: Princeton University Press, 2019). More recent studies have emphasized the formative role played by Latin American economists and social scientists in the creation of international development discourse. See Amy C. Offner, *Sorting out the Mixed Economy: The Rise and Fall of Welfare and Developmental States in the Americas* (Princeton, NJ: Princeton University Press, 2019). Margarita Fajardo, *The World That Latin America Created: The United Nations Economic Commission for Latin America in the Development Era* (Cambridge, MA: Harvard University Press, 2021); Christy Thornton, *Revolution in Development: Mexico and the Governance of the Global Economy* (Oakland: University of California Press, 2021).

38. James C. Scott, *Seeing Like a State: How Certain Schemes to Improve the Human Condition Have Failed* (New Haven, CT: Yale University Press, 2008).

39. Helen Tilley, *Africa as a Living Laboratory: Empire, Development, and the Problem of Scientific Knowledge, 1870–1950* (Chicago: University of Chicago Press, 2011), 24.

40. Murphy, *The Economization of Life*, 12.

41. Peter Rivière, "Alfred Métraux: Empiricist and Romanticist," in *Out of the Study and Into the Field*, ed. Robert Parkin and Anne de Sales (New York: Berghahn Books, 2010), 151–70; Charles Wagley, "Alfred Métraux 1902–1963," *American Anthropologist* 66, no. 3 (June 1964): 603–13.

42. Warwick Anderson, *The Cultivation of Whiteness: Science, Health, and Racial Destiny in Australia* (Durham, NC: Duke University Press, 2006); Warwick Anderson, "Hybridity, Race, and Science: The Voyage of the *Zaca*, 1934–1935," *Isis* 103, no. 2 (June 2012): 229–53; Warwick Anderson, "Racial Conceptions in the Global South," *Isis* 105, no. 4 (December 2014): 782–92; Karin Alejandra Rosemblatt, *The Science and Politics of Race in Mexico and the United States,*

1910–1950 (Chapel Hill: University of North Carolina Press, 2018); Julia Rodriguez, "Beyond Prejudice and Pride: The Human Sciences in Nineteenth- and Twentieth-Century Latin America," *Isis* 104, no. 4 (December 2013): 807–17; Julia Rodriguez, "South Atlantic Crossings: Fingerprints, Science, and the State in Turn-of-the-Century Argentina," *American Historical Review* 109, no. 2 (April 2004): 387–416; Wade et al., *Mestizo Genomics*; Marisol de la Cadena, *Indigenous Mestizos: The Politics of Race and Culture in Cuzco, Peru, 1919–1991* (Durham, NC: Duke University Press, 2000); Ricardo Roque, *Headhunting and Colonialism: Anthropology and the Circulation of Human Skulls in the Portuguese Empire, 1870–1930* (Basingstoke, UK: Palgrave Macmillan, 2010); Sarah Walsh, "'One of the Most Uniform Races of the Entire World': Creole Eugenics and the Myth of Chilean Racial Homogeneity," *Journal of the History of Biology* 48, no. 4 (Winter 2015): 613–39; Sarah Walsh, "The Executioner's Shadow: Coerced Sterilization and the Creation of 'Latin' Eugenics in Chile," *History of Science* 60, no. 1 (April 12, 2018); Tony Ballantyne, *Orientalism and Race: Aryanism in the British Empire* (Houndmills, UK: Palgrave, 2002); Marilyn Lake and Henry Reynolds, *Drawing the Global Colour Line: White Men's Countries and the International Challenge of Racial Equality* (Cambridge: Cambridge University Press, 2008); Chris Ballard and Bronwen Douglas, eds., *Foreign Bodies: Oceania and the Science of Race 1750–1940* (Canberra: ANU E Press, 2008); Saul Dubow, *Scientific Racism in Modern South Africa* (Cambridge: Cambridge University Press, 1995); Jerry Dávila, *Diploma of Whiteness: Race and Social Policy in Brazil, 1917–1945* (Durham, NC: Duke University Press, 2003); Nancy Stepan, *"The Hour of Eugenics": Race, Gender, and Nation in Latin America* (Ithaca, NY: Cornell University Press, 1991).

43. Anderson, "Racial Conceptions in the Global South," 784.

44. Nancy P. Appelbaum, Anne S. Macpherson, and Karin Alejandra Rosemblatt, eds., *Race and Nation in Modern Latin America* (Chapel Hill: University of North Carolina Press, 2003).

45. Appelbaum, Macpherson, and Rosemblatt, *Race and Nation in Modern Latin America*.

46. Julia Rodriguez, *Civilizing Argentina: Science, Medicine, and the Modern State* (Chapel Hill: University of North Carolina Press, 2006); Rodriguez, "South Atlantic Crossings"; Alejandra Bronfman, *Measures of Equality: Social Science, Citizenship, and Race in Cuba, 1902–1940* (Chapel Hill: University of North Carolina Press, 2004); Julyan G. Peard, *Race, Place, and Medicine: The Idea of the Tropics in Nineteenth Century Brazilian Medicine* (Durham, NC: Duke University Press, 1999); Rodriguez, "Beyond Prejudice and Pride"; Anadelia A. Romo, *Brazil's Living Museum: Race, Reform, and Tradition in Bahia* (Chapel Hill: University of North Carolina Press, 2010).

47. Stepan, *Hour of Eugenics*; Rodriguez, "Beyond Prejudice and Pride"; Romo, *Brazil's Living Museum*.

48. Dávila, *Diploma of Whiteness*; Romo, *Brazil's Living Museum*; Laura Giraudo and Juan Martin Sanchez, eds., *La ambivalente historia del indigenismo: campo interamericano y trayectorias nacionales, 1940–1970* (Lima: Instituto de Estudios Peruanos, 2011); Rosemblatt, *Science and Politics of Race*.

49. Stepan, *Hour of Eugenics*.

50. Stepan, *Hour of Eugenics*; Marius Turda and Aaron Gillette, *Latin Eugenics in Comparative Perspective* (London: Bloomsbury, 2014); Walsh, "Executioner's Shadow."

51. David Theo Goldberg, *The Racial State* (Malden, MA: Blackwell Publishers, 2002); George W. Stocking, Jr., "Lamarckianism in American Social Science: 1890–1915," *Journal of the History of Ideas* 23, no. 2 (April–June 1962): 239–56; R. W. Connell, "Why Is Classical Theory Classical?" *American Journal of Sociology* 102, no. 6 (May 1997): 1511–64; Warwick Anderson, *Colonial Pathologies: American Tropical Medicine, Race, and Hygiene in the Philippines* (Durham, NC: Duke University Press, 2006).

52. Tilley, *Africa as a Living Laboratory*; Christophe Bonneuil, "Development as Experiment: Science and State Building in Late Colonial and Postcolonial Africa, 1930–1970," *Osiris* 15 (2000): 258–81; Alice L. Conklin, *A Mission to Civilize: The Republican Idea of Empire in France and West Africa, 1895–1930* (Stanford, CA: Stanford University Press, 1997); James Ferguson, *Expectations of Modernity: Myths and Meanings of Urban Life on the Zambian Copperbelt* (Berkeley: University of California Press, 2005); Andrew Zimmerman, *Alabama in Africa: Booker T. Washington, the German Empire, and the Globalization of the New South* (Princeton, NJ: Princeton University Press, 2010); Timothy Mitchell, *Rule of Experts: Egypt, Techno-Politics, Modernity* (Berkeley: University of California Press, 2002); Frederick Cooper, *Decolonization and African Society: The Labor Question in French and British Africa* (Cambridge: Cambridge University Press, 1996).

53. Nils Gilman, *Mandarins of the Future: Modernization Theory in Cold War America* (Baltimore, MD: Johns Hopkins University Press, 2003); Michael E. Latham, *Modernization as Ideology: American Social Science and "Nation-Building" in the Kennedy Era* (Chapel Hill: University of North Carolina Press, 2000); Nick Cullather, "Development? It's History," *Diplomatic History* 24, no. 4 (October 2000): 641–53; Frederick Cooper, *International Development and the Social Sciences: Essays on the History and Politics of Knowledge* (Berkeley: University of California Press, 1997).

54. Thomas McCarthy, *Race, Empire, and the Idea of Human Development* (Cambridge: Cambridge University Press, 2009).

55. David Scott, "Colonial Governmentality," *Social Text*, no. 43 (Autumn 1995): 191–220.

56. Harold Cyril Bibby, "L'Antiracisme Commence Sur Les Bancs de l'école," *UNESCO Courier* 13, no. 10 (1960): 6–12; Juan Comas, "'Scientific' Racism Again?," *Current Anthropology* 2, no. 4 (1961): 303–40.

57. These are projects that Jenny Reardon interprets as antiracist because they involve participatory and collaborative approaches that aim to curtail potential power imbalances between researchers and research subjects. See Jenny Reardon, "The Democratic, Anti-Racist Genome? Technoscience at the Limits of Liberalism," *Science as Culture* 21, no. 1 (March 2012): 25–47.

58. Ghassan Hage, "Recalling Anti-Racism," *Ethnic and Racial Studies* 39, no. 1 (January 2, 2016): 123–33. Hage borrows his notion of recalling antiracism from Bruno Latour's account of modernity. Bruno Latour, "The Recall of Modernity: Anthropological Approaches," *Cultural Studies Review* 13, no. 1 (2007): 11–30.

59. Hage, "Recalling Anti-Racism," 125.

60. My thinking here is also informed by Ash Amin's analysis of the speed by which the hard-won achievements of antiracism can be undone. See Ash Amin, "The Remainders of Race," *Theory, Culture & Society* 27, no. 1 (January 1, 2010): 1–23.

61. Hage, "Recalling Anti-Racism," 126.

1. SUBSTITUTING RACE: ARTHUR RAMOS, BAHIA, AND THE "NINA RODRIGUES SCHOOL"

1. Arthur Ramos, *A Aculturaçao Negra no Brasil* (São Paulo: Companhia Editora Nacional, 1942), 180.

2. Brad Lange, "Importing Freud and Lamarck to the Tropics: Arthur Ramos and the Transformation of Brazilian Racial Thought, 1926–1939," *The Americas* 65, no. 1 (July 2008): 9–34.

3. Thomas Skidmore, *Black into White: Race and Nationality in Brazilian Thought* (New York: Oxford University Press, 1974), 58.

4. Skidmore, *Black into White*, 58.

5. Skidmore, *Black into White*, 173.

6. Vera Lúcia Miranda Faillace, "Apresentação," in *Arquivo Arthur Ramos: Inventário Analítico*, ed. Vera Lúcia Miranda Faillace (Rio de Janeiro: Ediçoes Biblioteca Nacional, 2004), 7–8; Luitgarde Oliveira Cavalcanti Barros, "Introdução," in *Arquivo Arthur Ramos: Inventário Analítico*, ed. Vera Lúcia Miranda Faillace (Rio de Janeiro: Ediçoes Biblioteca Nacional, 2004), 9–16.

7. Julyan G. Peard, *Race, Place and Medicine: The Idea of the Tropics in Nineteenth-Century Brazilian Medicine* (Durham, NC: Duke University Press, 1999).

8. Peard, *Race, Place, and Medicine*.

9. Lange, "Importing Freud and Lamarck to the Tropics," 15.

10. Davila, *Diploma of Whiteness*, 35–37.

11. Arthur Ramos, *A Criança Problema: A Higiene Mental na Escola Primária* (Rio de Janeiro: Livraria Editoria da Cada do Estudante do Brasil, 1939).

12. Peard, *Race, Place, and Medicine*, 101–2.

13. Anadelia Romo, *Brazil's Living Museum: Race, Reform, and Tradition in Bahia* (Chapel Hill: University of North Carolina Press, 2010), 29.

14. The Tropicalistas tended to believe that Bahia's "backwardness" had little to do with race and instead stemmed from a dysfunctional education system and poor sanitation infrastructure. Between 1870 and 1888—the period during which slavery was abolished in Brazil and when classic liberalism flourished—Brazil's intellectual elites widely embraced the Tropicalistas' emphasis on the formative power of the environment and became enamored with positivist conceptions of social evolution that envisioned a progressive path to modern civilization through the application of scientific principles. Such appeals to the formative power of the environment, coupled with Lamarckian conceptions of heredity, offered a compelling solution to the "determinist trap" represented by mainstream European race science. Whereas prominent European and North American race scientists—such as the Harvard naturalist and polygenist Louis Agassiz—interpreted Brazil's intense blurring of racial boundaries as a process of "mongrelization" that doomed it to inferiority, Brazil's abolitionists took comfort in softer conceptions of heredity and believed that miscegenation would gradually whiten and thus uplift the Brazilian population. Thomas Skidmore, "Racial Ideas and Social Policy in Brazil, 1870–1940," in *The Idea of Race in Latin America, 1870–1940*, ed. Richard Graham (Austin: University of Texas Press, 1990), 9.

15. Romo, *Brazil's Living Museum*, 28–46.

16. Romo, *Brazil's Living Museum*, 33.

17. Nina Rodrigues, "Métissage, Dégénérescence, et Crime" (Lyon: A. Storck & Cie, Imprimeurs-Éditeurs, 1899), 1–40.

18. Roberto Motta, "Arthur Ramos, sincretismo e mentalidade pré-lógica em O negro brasileiro" in *Arthur Ramos*, ed. Luitgarde Oliveira Cavalcanti Barros, (Rio de Janeiro: Fundação Miguel de Cervantes, 2011), 30–80.

19. Rodrigues, "Métissage, Dégénérescence, et Crime," 5.

20. Rodrigues, "Métissage, Dégénérescence, et Crime," 6.

21. Rodrigues reasoned that he was justified in making generalizations like these because Serrinha's population developed from a small nucleus of inhabitants, which meant that the town's genealogical relationships were tightly knit and resembled that of a "large family." Yet he could not rely on census data or

other statistics to make these claims and instead based his conclusions about Serrinha's tendency toward degeneration on personal observations that he made during medical examinations and on genealogical information he collected about the town's inhabitants. Accordingly, the data Rodrigues presented as evidence for the degeneration of Serrinha's *mestiço* population consisted in twenty-six case histories detailing the maladies and forms of mental illness of individuals he examined as well as a genealogical table that purportedly showed the "medical history" of a typical family and how it was tending toward degeneration. In his case histories, Rodrigues highlighted the prevalence of *neurasthenia*—a vaguely defined condition that included symptoms such as laziness, fatigue, and emotional disturbance—which he claimed was widespread and afflicted people exercising all sorts of professions. He also highlighted the prevalence of epilepsy and physical deformities as well as general "madness." Rodrigues, "Métissage, Dégénérescence, et Crime," 8.

22. Rodrigues, "Métissage, Dégénérescence, et Crime," 18.

23. Rodrigues, "Métissage, Dégénérescence, et Crime," 21.

24. Rodrigues, "Métissage, Dégénérescence, et Crime," 21.

25. Rodrigues, "Métissage, Dégénérescence, et Crime," 21.

26. *O animismo fetichista* was originally published in Portuguese in 1896 and later translated and published in French. See Raymundo Nina Rodrigues, *L'animisme fétichiste de nègres de Bahia* (Bahia, Brazil: Reis, 1900), 3.

27. Rodrigues, *L'animisme fétichiste*, 131.

28. Rodrigues, *L'animisme fétichiste*, 131.

29. Rodrigues, *L'animisme fétichiste*, 135.

30. Rodrigues, *L'animisme fétichiste*, 135.

31. Arthur Ramos, *Introdução à antropologia brasileira*, 3rd ed. (Rio de Janeiro: Coleção Estudos brasileiros da Livraria-Editora da Casa do Estudante do Brasil, Ser. B, 1961), 11.

32. Micol Seigel, "Beyond Compare: Comparative Method After the Transnational Turn," *Radical History Review* 2005, no. 91 (January 1, 2005): 62–90, 74.

33. Arthur Ramos, *O folclore negro do Brasil: Demopsicologia e psicanalise* (Rio de Janeiro: Livraria-Editora da Casa do Estudante do Brasil, 1954), 8.

34. Ramos, *A Aculturação Negra no Brasila*, 5.

35. Romo, *Brazil's Living Museum*, 48.

36. Romo, *Brazil's Living Museum*, 66.

37. Romo, *Brazil's Living Museum*, 67.

38. Romo, *Brazil's Living Museum*, 68.

39. Ramos, *O folclore negro*, 6.

40. Ramos, *O folclore negro*, 7.

41. Romo, *Brazil's Living Museum*, 71.

42. Romo, *Brazil's Living Museum*, 71.

43. Antonio Sérgio Alfredo Guimarães, "Africanism and Racial Democracy: The Correspondence Between Herskovits and Arthur Ramos (1935 1949)," *EIAL: Estudios Interdisciplinarios de America Latina y El Caribe* 19, no. 1 (2008): 53–79, 57.

44. On Ramos's disputes with rival Brazilian anthropologists such as Heloisa Alberto Torres, see Amurabi Oliveira, "Afro-Brazilian Studies from Psychoanalysis to Cultural Anthropology: An Intellectual Portrait of Arthur Ramos," Bérose, accessed August 24, 2022, https://www.berose.fr/article2327 .html; Mariza Corrêa, "Dona Heloisa e a pesquisa de campo," *Revista de Antropologia* 40 (1997): 11–54.

45. On relations between Landes and Ramos, see Romo, *Brazil's Living Museum*, 130–31; Sally Cooper Cole, *Ruth Landes: A Life in Anthropology* (Lincoln: University of Nebraska Press, 2003), 158–60.

46. The 1935 manifesto was later republished in Arthur Ramos, *Guerra e relações de raça* (Rio de Janeiro: Departamento editorial da União nacional dos estudantes, 1943), 171–74. On Ramos's role in publishing other manifestos against racism, see Oliveira, "Afro-Brazilian Studies," 18.

47. Arthur Ramos, "The Negro in Brazil," *Journal of Negro Education* 10, no. 3 (1941): 515–23, 522.

48. V. F. Calverton, "The Compulsive Basis of Social Thought: As Illustrated by the Varying Doctrines as to the Origins of Marriage and the Family," *American Journal of Sociology* 36, no. 5 (March 1931): 689–734, 718.

49. Calverton, "The Compulsive Basis," 719.

50. Arthur Ramos, *As ciências sociais e os problemas de após-guerra* (Rio de Janeiro: Casa do Estudante do Brasil, 1944), 31.

51. Ramos, *As ciências sociais*, 32.

52. Ramos, *As ciências sociais*, 41.

53. Ramos, *As ciências sociais*, 44.

54. Ramos, *As ciências sociais*, 49.

55. Marcos Chor Maio, "Unesco and the Study of Race Relations in Brazil: Regional or National Issue?," *Latin American Research Review* 36, no. 2 (2001): 121.

56. Chor Maio, "Unesco and the Study of Race Relations in Brazil," 121–22.

57. Klineberg taught at the Universidade de São Paulo between 1945 and 1947. He was recruited to this position by Andre Dreyfus—a former student of the geneticist Theodosius Dobzhansky—and spent two "happy years" living in São Paulo where he helped to create the Departamento de Psicologia. During this period, he corresponded with Arthur Ramos about the possibility of conducting a study on intelligence in the Rio school system. See Otto Klineberg, "Otto Klineberg," in *A History of Psychology in Autobiography*, vol. 6, ed. Gardner

Lindzey (Englewood Cliffs, NJ: Prentice-Hall, 1974), 161–82; Marcos Chor Maio, "A Crítica de Otto Klineberg aos testes de inteligência: O Brasil como laboratório racial," *Varia Historia* 33, no. 61 (January–April 2017): 135–61.

58. The idea for the "Tensions Project" was put forward by participants in UNESCO's General Conference in Mexico City in 1947.

59. Arthur Ramos, "Perspectives sur le Département des Sciences Sociales," Coleçao Arthur Ramos, Biblioteca Nacional, Rio de Janeiro, 1949, http://objdigital .bn.br/objdigital2/acervo_digital/div_manuscritos/mss1306731/mss1306731 .pdf.

60. Ramos, "Perspectives sur le Département."

61. Ramos, "Perspectives sur le Département."

62. Ramos, "Perspectives sur le Département."

63. "Arthur Ramos to Clemente Mariani," August 6, 1949, Coleçao Arthur Ramos, Biblioteca Nacional, Rio de Janeiro, 1949, http://objdigital.bn.br/objdigital2 /acervo_digital/div_manuscritos/mss1304712/mss1304712.pdf.

64. "Arthur Ramos to Clemente Mariani," August 6.

65. "Roger Bastide to Arthur Ramos," July 29, 1949, Coleçao Arthur Ramos, Biblioteca Nacional, Rio de Janeiro, 1949, http://objdigital.bn.br/acervo_digital /div_manuscritos/mss1298949.pdf.

66. "Arthur Ramos to Jorge Kingston," September 14, 1949, Coleçao Arthur Ramos, Biblioteca Nacional, Rio de Janeiro, 1949, http://objdigital.bn.br/objdigital2 /acervo_digital/div_manuscritos/mss1303695/mss1303695.pdf.

67. "Arthur Ramos to Luiz Aguiar Costa-Pinto," September 29, 1949, Coleçao Arthur Ramos, Biblioteca Nacional, Rio de Janeiro, 1949, http://objdigital.bn .br/objdigital2/acervo_digital/div_manuscritos/mss1304863/mss1304863.pdf.

68. "Arthur Ramos to Pedro Calmon," October 13, 1949, Coleçao Arthur Ramos, Biblioteca Nacional, Rio de Janeiro, 1949, http://objdigital.bn.br/objdigital2 /acervo_digital/div_manuscritos/mss1303578/mss1303578.pdf.

69. Arthur Ramos, "The Question of Races and the Democratic World," *International Social Science Bulletin* 1, no 3–4 (1949): 9–20. 9.

70. Ramos, "The Question of Races," 9.

71. The utopia that Ramos describes here is likely based on the the Apapocúva-Guaraní legend of the "Land without Evil," which first appears in the Americanist literature with Curt Nimuendajú in 1914 and which was further studied in the context of the history of Tupi-Guarani migrations by a young Alfred Métraux. See Diego Villar and Isabelle Combès, "La 'Tierra Sin Mal': leyenda de la creación y destrucción de un mito," *Tellus* 13, no 24 (2013), 201–25; Ramos, "The Question of Races," 10.

72. Ramos, "The Question of Races," 10.

73. Ramos, "The Question of Races," 10.

74. Ramos, "The Question of Races," 10.

75. Ramos, "The Question of Races," 13.

76. Ramos, "The Question of Races," 13.

77. Ramos, "The Question of Races," 12.

78. As a definitive example of indirect rule, Ramos referred to the "International Institute of African Languages and Civilizations" and its outstanding leaders— the British colonial administrator Frederick Lugard and the German missionary and linguist Diedrich Hermann Westermann—as the most important of all institutions in applied anthropology whose "experts act as advisers to colonial administrators and aim at showing them that without a knowledge of anthropology, they cannot fulfill their task properly." Ramos, "The Question of Races," 12.

79. Ramos, "The Question of Races," 13.

80. Ramos, "The Question of Races," 13.

81. Staffan Müller-Wille, "Claude Lévi-Strauss on Race, History and Genetics," *BioSocieties* 5 (2010): 330–47.

82. Claude Lévi-Strauss, *Race and History* (Paris: UNESCO, 1952), 6.

83. George W. Stocking, *Race, Culture, and Evolution: Essays in the History of Anthropology* (New York: Free Press, 1968); Nancy Stepan, *The Idea of Race in Science: Great Britain, 1800–1960* (Hamden, CT: Archon Books, 1982); Elazar Barkan, *The Retreat of Scientific Racism: Changing Concepts of Race in Britain and the United States Between the World Wars* (Cambridge: Cambridge University Press, 1992).

84. Nancy Stepan, *"The Hour of Eugenics": Race, Gender, and Nation in Latin America* (Ithaca, NY: Cornell University Press, 1996); Sarah Walsh, *The Religion of Life: Eugenics, Race, and Catholicism in Chile* (Pittsburgh, PA: University of Pittsburgh Press, 2022); Sebastián Gil-Riaño and Sarah Walsh, "Introduction: Race Science in the Latin World," *History of Science* 60, no. 1 (2022): 4–17.

2. RELOCATING RACE SCIENCE AFTER WORLD WAR II: SITUATING THE 1950 UNESCO STATEMENT ON RACE IN THE SOUTHERN HEMISPHERE

1. This was the first of several meetings where UNESCO's 1950 Statement on Race was drafted. The 1949 committee consisted of Ernest Beaglehole, Juan Comas, Luiz Aguiar Costa-Pinto, E. Franklin Frazier, Morris Ginsberg, Humayun Kabir, Claude Lévi-Strauss, and Ashley Montagu. The minutes from the 1949 meetings are summarized in UNESCO, *Summary Report (of the Six Meetings)*, Meeting of Experts on Race Problems, December 29, 1949 (Paris: UNESCO/SS/CONF.1/SR1, 1949).

2. Costa Pinto's tribute appears in the meeting minutes, see UNESCO, *Summary Report (of the Six Meetings)*, 3.

3. Corinna R. Unger, *International Development: A Postwar History* (London: Bloomsbury Academic, 2018).

4. A list of sources that interpret and situate the UNESCO Race Statements within various contexts can be found in notes 10, 29, and 30 of the introduction.

5. UNESCO, *Summary Report (of the Six Meetings)*, 7.

6. UNESCO, Provisional agenda, list of participants, list of documents, Meeting of Experts on Race Problems (Paris: SS/CONF.1/1, SS/CONF.1/4, 1949). https://unesdoc.unesco.org/ark:/48223/pf0000156948?posInSet=8&queryId=N-EXPLORE-32739d2f-2642-4925-9500-4a8a0622f76d

7. Louis Wirth, *Implementation of the Resolution of the Economic and Social Council on the Prevention and Discrimination and the Protection of Minorities*, Committee of Experts on Race Problems (Paris: UNESCO/SS/Conf.1/2, 1949), 3.

8. Wirth, *Implementation of the Resolution*, 2.

9. Wirth, *Implementation of the Resolution*, 2.

10. Wirth, *Implementation of the Resolution*, 3.

11. Wirth, *Implementation of the Resolution*, 4.

12. Wirth, *Implementation of the Resolution*, 4.

13. Wirth, *Implementation of the Resolution*, 4.

14. Wirth, *Implementation of the Resolution*, 4.

15. Wirth, *Implementation of the Resolution*, 1.

16. R. W. Connell, "Why Is Classical Theory Classical?," *American Journal of Sociology* 102, no. 6 (May 1997): 1511.

17. Connell, "Why Is Classical Theory Classical?," 1516.

18. Connell, "Why Is Classical Theory Classical?," 1517.

19. Luiz Aguiar Costa-Pinto, *O Negro no Rio de Janeiro* (São Paulo: Companhia Editora Nacional, 1953), 25.

20. On Costa-Pinto's career and views on race relations, see Marcos Chor Maio, "Uma Polêmica Esquecida: Costa Pinto, Guerreiro Ramos e o Tema Das Relações Raciais," *Dados* 40 (1997): 127–63.

21. For overviews of Comas's career see Miguel León-Portilla, "Juan Comas Camps (1900–1979)," *Hispanic American Historical Review* 60, no. 1 (February 1, 1980): 95–96; Jorge Gómez Izquierdo, "El discurso antirracista de un antropólogo indigenista: Juan Comas Camps," *Desacatos*, no. 4 (2000): 80–102.

22. Laura Giraudo, "Neither 'Scientific' nor 'Colonialist': The Ambiguous Course of Inter-American Indigenismo in the 1940s," *Latin American Perspectives* 39, no. 5 (September 2012): 12–32; Laura Giraudo and Juan Martin Sanchez, *La*

Ambivalente Historia del Indigenismo: Campo Interamericano y Trayectorias Nacionales 1940–1970 (Lima: Instituto de Estudios Peruanos, 2011).

23. Juan Comas, *Racial Myths* (Paris: UNESCO, 1951), 12.

24. Juan Comas, "Cultural Anthropology and Fundamental Education in Latin America," *International Social Science Bulletin* 4, no. 3 (1952): 451–61.

25. Comas, "Cultural Anthropology and Fundamental Education," 451.

26. Casey Walsh, "Eugenic Acculturation: Manuel Gamio, Migration Studies, and the Anthropology of Development in Mexico, 1910–1940," *Latin American Perspectives* 31, no. 5 (September 2004): 118–45.

27. Hobhouse also argued that animals could evolve by overcoming their biological instincts. He called this process "orthogenic evolution" and imagined it as a model for understanding the nature of "race progress" in society. See Chris Renwick, *British Sociology's Lost Biological Roots: A History of Futures Past* (Houndmills, UK: Palgrave Macmillan, 2012), 107–10. For Hobhouse's influence on the development of sociology in Britain, see Stefan Collini, *Liberalism and Sociology: L. T. Hobhouse and Political Argument in England, 1880–1914* (Cambridge: Cambridge University Press, 1979).

28. Morris Ginsberg, "Preface," in Jay Rumney, *Herbert Spencer's Sociology; a Study in the History of Social Theory* (New York: Atherton Press, 1966), 2–3.

29. Ginsberg, "Preface," 3.

30. Leonard Trelawney Hobhouse, Morris Ginsberg, and Gerald Clair William Camden Wheeler, *The Material Culture and Social Institutions of the Simpler Peoples; an Essay in Correlation* (London: Chapman & Hall, 1915), 6.

31. Hobhouse, Ginsberg, and Camden, *Material Culture and Social Institutions*, 6.

32. For example, Ginsberg and Hobhouse classified simpler peoples with categories such as "simple hunters," "higher hunters," "hunter and gatherers," "incipient agriculture," "agriculture pure," and "highest agriculture"; Hobhouse, Ginsberg, and Camden, *Material Culture and Social Institutions*, 6.

33. George Steinmetz, "A Child of the Empire: British Sociology and Colonialism, 1940s–1960s," *Journal of the History of the Behavioral Sciences* 49, no. 4 (September 2013): 353–78.

34. Ernest Beaglehole, *Property; a Study in Social Psychology* (New York: Macmillan, 1932), 19.

35. Beaglehole, *Property*, 20–21.

36. Beaglehole, *Property*, 127.

37. Beaglehole, *Property*, 315.

38. Beaglehole, *Property*, 294.

39. James Ritchie, "What Ever Happened to Cross-Cultural Psychology?," *Bulletin* (New Zealand Psychological Society), no. 92 (1992): 17–21; J. E. Ritchie, "Obituary: Ernest Beaglehole," *Journal of the Polynesian Society* 75, no. 1 (March

1966): 109–19; Harry Lionel Shapiro. *The Anthropometry of Pukapuka, Based upon Data Collected by Ernest and Pearl Beaglehole* (New York: American Museum of Natural History, 1942).

40. Ernest Beaglehole, "Race, Caste, and Class," *Journal of the Polynesian Society* 52, no. 1 (March 1943): 1–11, 6.

41. Beaglehole, "Race, Caste, and Class," 2.

42. Beaglehole, "Race, Caste, and Class," 6.

43. For discussions of this debate in imperial social science and settler colonial anthropology, see L. R. Hiatt, *Arguments About Aborigines: Australia and the Evolution of Social Anthropology* (Cambridge: Cambridge University Press, 1996); Patrick Wolfe, *Settler Colonialism and the Transformation of Anthropology: The Politics and Poetics of an Ethnographic Event* (London: Cassell, 1999).

44. Montagu dedicated this book to Malinowski and to Edward Westermarck, a Finnish social evolutionist. Ashley Montagu, *Coming into Being Among the Australian Aborigines: The Procreative Beliefs of the Australian Aborigines*, 2nd ed. (London: G. Routledge and Sons, 1937; repr. London: Routledge and K. Paul, 1974), xvi. All citations refer to the second edition. In his description of Aboriginal beliefs on kinship, Montagu drew heavily from *The Native Tribes of Central Australia* (1899), a colonial ethnography written by Australian anthropologists Baldwin Spencer and F. J. Gillen as well as from *The Psychology of Primitive People* written by the social psychologist Stanley David Porteus. Balwin Spencer and F. J. Gillen, *The Native Tribes of Central Australia* (London: Macmillan, 1899); Stanley David Porteus, *The Psychology of Primitive People* (New York: Longman, Green, 1931).

45. Montagu, *Coming into Being Among the Australian Aborigines*, 326–27.

46. Montagu, *Coming into Being Among the Australian Aborigines*, 329.

47. My analysis here is influenced by Anne McClintock's concept of anachronistic space. See Anne McClintock, *Imperial Leather: Race, Gender, and Sexuality in the Colonial Contest* (New York: Routledge, 1995).

48. Ashley Montagu, *Man's Most Dangerous Myth: The Fallacy of Race* (New York: Columbia University Press, 1942; repr. Lanham, MD: AltaMira Press, 2001), 259.

49. Pierre Saint-Arnaud and Peter Feldstein, *African American Pioneers of Sociology: A Critical History* (Toronto: University of Toronto Press, 2009), 204–48.

50. See Ernest W. Burgess, "Editor's Preface," in E. Franklin Frazier, *The Negro in the United States* (Chicago: University of Chicago Press, 1940), ix–xvii.

51. On Herskovits and Frazier's debate about African roots and Afro-Brazilian families, see Anadelia A. Romo, *Brazil's Living Museum: Race, Reform, and Tradition in Bahia* (Chapel Hill: University of North Carolina Press, 2010), 124–32.

52. E. Franklin Frazier, *The Negro in the United States* (Chicago: The University of Chicago Press, 1940), 488.

53. Wiktor Stoczkowski, "Racisme, Antiracisme et Cosmologie Lévi-Straussienne," *L'Homme* no. 182 (May 2007): 7–51.

54. Claude Lévi-Strauss, *Race and History* (Paris: UNESCO, 1952), 6.

55. Lévi-Strauss, *Race and History*, 59. For a discussion of how Lévi-Strauss was influenced by Marcel Mauss, see Christopher Johnson, *Claude Lévi-Strauss: The Formative Years* (Cambridge: Cambridge University Press, 2003); Alice L. Conklin, *In the Museum of Man: Race, Anthropology, and Empire in France, 1850–1950* (Ithaca, NY: Cornell University Press, 2013).

56. On the role of genetics in Lévi-Strauss's ideas about historical change, see Staffan Müeller-Wille, "Claude Lévi-Strauss on Race, History and Genetics," *BioSocieties* 5, no. 3 (September 2010): 330–47.

57. See Stoczkowski, "Racisme, Antiracisme, et Cosmologie Lévi-Straussienne"; Müeller-Wille, "Claude Lévi-Strauss on Race, History and Genetics."

58. On the imperial context of grand comparative sociological studies, see Raewyn Connell, *Southern Theory: The Global Dynamics of Knowledge in Social Science* (Cambridge: Polity, 2009); Connell, "Why Is Classical Theory Classical?"

59. UNESCO, *Summary Report (of the Six Meetings)*, 10.

60. UNESCO, *Summary Report (of the Six Meetings)*, 9–11.

61. UNESCO, *Summary Report (of the Six Meetings)*, 8.

62. UNESCO, *Summary Report (of the Six Meetings)*, 10.

63. UNESCO, *Summary Report (of the Six Meetings)*, Fourth Meeting, 1–2.

64. UNESCO, *Summary Report (of the Six Meetings)*, Fourth Meeting, 2.

65. UNESCO, *Summary Report (of the Six Meetings)*, Fourth Meeting, 4.

66. UNESCO, *Summary Report (of the Six Meetings)*, Fourth Meeting, 3.

67. UNESCO, *Summary Report (of the Six Meetings)*, Fourth Meeting, 3.

68. UNESCO, *Summary Report (of the Six Meetings)*, Fourth Meeting, 3.

69. UNESCO, *Summary Report (of the Six Meetings)*, Fourth Meeting, 3.

70. Alfred Métraux, "Race and Civilisation," *UNESCO Courier* 3, no. 6 (1950): 8–9.

71. Charles Wagley, "Alfred Métraux," *American Anthropologist* 66, no. 3 (June 1964): 603–13.

72. Alfred Métraux, "Applied Anthropology in Government: United Nations," in *Anthropology Today: An Encyclopedic Inventory*, ed. A. L. Kroeber (Chicago: University of Chicago Press, 1953), 880–94.

73. "Fallacies of Racism Exposed: UNESCO Publishes Declaration by World Scientists," *UNESCO Courier* 3, no. 6–7 (1950): 1.

74. Joy Rohde, "Gray Matters: Social Scientists, Military Patronage, and Democracy in the Cold War," *Journal of American History* 96, no. 1 (June 2009): 99–122; Mark Solovey, "Project Camelot and the 1960s Epistemological Revolution:

Rethinking the Politics-Patronage-Social Science Nexus," *Social Studies of Science* 31, no. 2 (April 2001): 171–206.

75. Arturo Escobar, *Encountering Development: The Making and Unmaking of the Third World* (Princeton, NJ: Princeton University Press, 1995).

76. James Ferguson, *The Anti-Politics Machine: "Development," Depoliticization, and Bureaucratic Power in Lesotho* (Minneapolis: University of Minnesota Press, 1994).

77. Mary L. Dudziak, *Cold War Civil Rights: Race and the Image of American Democracy.* (Princeton, NJ: Princeton University Press, 2000).

3. VIKINGS OF THE SUNRISE: ALFRED MÉTRAUX, TE RANGI HĪROA AND POLYNESIAN RACIAL RESILIENCE

1. Letter from H. C. Laves to Alfred Métraux, February 7, 1950, Fonds Alfred Métraux, G.C.02.01: Papiers UNESCO 1946–1955, Archives des Ethnologues, Biblothèque d'Anthropologie Sociale, Laboratoire d'Anthropologie Sociale, Collège de France, Paris.

2. Sidney Mintz, "Introduction," in Alfred Métraux, *Voodoo in Haiti* (New York: Schocken, 1972), 2.

3. Charles Wagley, "Alfred Métraux 1902–1963," *American Anthropologist* 66, no. 3 (June 1964): 606.

4. Peter Rivière, "Alfred Métraux: Empiricist and Romanticist," in *Out of the Study and into the Field: Ethnographic Theory and Practice in French Anthropology,* ed. Robert Parkin and Anne de Sales (New York: Berghahn, 2010), 159.

5. Warwick Anderson, "Racial Hybridity, Physical Anthropology, and Human Biology in the Colonial Laboratories of the United States," *Current Anthropology* 53, no. S5 (April 2012): S95–S107.

6. In this phase of his career, Métraux's engagement with race was primarily idiomatic. That is, he used "race" loosely to describe groups bound together by a shared ancestry, language, and culture and as a term that was not yet problematized in European ethnological discourse and not given to exact definition. Métraux's work from this period shows little to no interest in the physiological or biological aspects of the groups he studied (apart from occasional mentions of population statistics) and thus stands in contrast to the work of Argentinian anthropologists that preceded him and were explicitly engaged in anthropometric studies of race and intelligence. On idiomatic uses of race, see Joyce Chaplin's analysis of the "racial idioms" that were crafted by early British settlers in North America prior to the advent of a full-fledged race science: Joyce E. Chaplin, "Natural philosophy and an Early Racial Idiom in North America: Comparing English and Indian Bodies," *William and Mary Quarterly* 54, no. 1

(January 1997): 229–52. On turn-of-the-century Argentinian race science, see Julia Rodríguez, *Civilizing Argentina: Science, Medicine, and the Modern State* (Chapel Hill: University of North Carolina Press, 2006).

7. A vast colonial frontier region, the Gran Chaco is a low lying and semi-arid region that serves as a basin for the Río de la Plata and is today divided among three nation-states: Argentina, Paraguay, and Bolivia. The region is one where indigenous groups from diverse traditions have coexisted since the colonial era, including numerous "Chacoan groups" who belong to the Guaykurú linguistic family, groups of Amazonian origin such as the Arawak and Tupi-Guarani, as well as various mestizo groups who represent blends of different indigenous groups such as the Chiriguano, who represent a mix of Chané and Guarani groups and whose culture draws on both Arawak and Tupi-Guarani traditions. See Isabelle Combès and Kathleen Lowrey, "Slaves Without Masters? Arawakan Dynasties Among the Chiriguano (Bolivian Chaco, Sixteenth to Twentieth Centuries)," *Ethnohistory* 53, no. 4 (Fall 2006): 689–90.

8. Edgardo Krebs, "Alfred Metraux and *The Handbook of South American Indians*: A View from Within," *History of Anthropology Newsletter* 32, no 1 (2005): 3–11.

9. Federico Bossert, "Alfred Métraux y la Utopía del Gran Chaco," *Journal de la Société des Américanistes* 102, no. 2 (2016): 25–44.

10. Rivière, "Alfred Métraux: Empiricist and Romanticist," 151.

11. See Alfred Métraux, *La Civilisation Matérielle des Tribus Tupi-Guarani* (Paris: Librairie Orientaliste Paul Geuthner, 1928), vii–viii.

12. Métraux envisioned that Tucumán would become something like a regional branch of the Gothenburg Museum and sought to replicate exact details of the Swedish museum by hanging the same regional maps and photos found in the museum and insisting on having the same cabinet displays and curtains. See Bossert, "Alfred Métraux y la Utopía del Gran Chaco," 39.

13. Carolyne R. Larson, *Our Indigenous Ancestors: Museum Anthropology and Nation-Making in Argentina, 1862–1943* (University Park: Pennsylvania State University Press, 2015), 124.

14. Bossert, "Alfred Métraux y la Utopía del Gran Chaco," 30.

15. Paul Rivet, "L'institut d'ethnologie de l'Université de Tucumán," *Journal de la Société des Américanistes* 25, no 1 (1933)" 188–89.

16. Bossert, "Alfred Métraux y la Utopía del Gran Chaco," 32.

17. Bossert, "Alfred Métraux y la Utopía del Gran Chaco," 33.

18. Métraux described his encounter as a "horrible spectacle" and claimed that the smallpox was spreading at a terrifying speed that was facilitated by the military commanders who "disposed of all the necessary elements for vaccinating the Indians" but chose to "neglect this simple human responsibility," in Alfred

Métraux, "Etudes d'Ethnographie Toba-Pilagá (Gran Chaco)," *Anthropos* Bd. 32, H. 1./2 (January–April 1937): 171.

19. Alfred Métraux, "La Obra de las Misiones Inglesas en el Chaco," *Journal de la Société des Américanistes* 25, no. 1 (1933): 205–9.

20. According to Federico Bossert, this position was offered to him but never materialized; see Bossert, "Alfred Métraux y la Utopía del Gran Chaco," 35. Edgardo Krebs, on the other hand, claims that Métraux failed to convince authorities to give him this position and suffered a crisis where he considered abandoning anthropology to dedicate himself to the welfare of "the Indians." See Krebs, "Alfred Metraux and *The Handbook of South American Indians*," 6.

21. His wife, Eva Spiro, and children initially lived with him in Tucumán but returned to Europe when his youngest son became ill. Bossert, "Alfred Métraux y la Utopía del Gran Chaco," 33.

22. Alfred Métraux, "Suicide Among the Matako of the Argentine Gran Chaco," *América Indígena* 3, no. 3 (1943): 199–210.

23. Alfred Métraux, *Easter Island: A Stone-Age Civilization of the Pacific*, trans. Michael Bullock (New York: Oxford University Press, 1957), 59.

24. Métraux, *Easter Island*, 60.

25. Alice Te Punga Sommerville, "Māori People in Pacific Spaces," in *Once Were Pacific: Maori Connections to Oceania* (Minneapolis: University of Minnesota Press, 2012), 12.

26. Guillaume de Hevesy, the main proponent of this theory, was a Hungarian linguist based in Paris. His findings were announced in 1932 via a letter that was read during a meeting of the Académie des Inscriptions de Belle-Lettres in Paris and apparently took the Parisian intellectual community by storm. See Christine Laurière, "Fictions d'une mission: Île de Pâques 1934–1935," *L'Homme* 175–176, no. 3 (July–September 2005): 321–43; Steven R. Fischer, *Rongorongo: The Easter Island Script: History, Traditions, Texts* (Oxford: Oxford University Press, 1997), 147.

27. Laurière, "Fictions d'une mission," 321–22.

28. Métraux, *Easter Island*, 10.

29. Laurière, "Fictions d'une mission," 323.

30. Laurière, "Fictions d'une mission," 323.

31. Laurière, "Fictions d'une mission," 325.

32. Laurière, "Fictions d'une mission," 328–29.

33. Métraux, *Easter Island*, 12.

34. Alfred Métraux, *Ethnology of Easter Island* (Honolulu: Bernice P. Bishop Museum, 1940), 4.

35. Alfred Métraux, "Easter Island," in the *Annual Smithsonian Report for 1944* (Washington, DC: U.S. Government Printing Office, 1945), 435–52; 436.

36. Métraux, "Easter Island," 437.

37. Métraux, *Ethnology of Easter Island*, 22.

38. Métraux, "Easter Island," 439.

39. According to Shapiro, the project of classifying the populations of Oceania arose in the writings of eighteenth-century explorers who were stimulated by the "taxonomic work of Linné, Buffon and others" and described two distinct races: "the one more fair, well limbed, athletic, of a fine size, and a kind benevolent temper; the other blacker, the hair just beginning to become woolly and crisp, the body more slender and low, and their temper, if possible more brisk, though somewhat mistrustful." H. L. Shapiro, "The Physical Relationships of the Easter Islanders," in Alfred Métraux, *Ethnology of Easter Island* (Honolulu: Bernice P. Bishop Museum, 1940), 24.

40. Shapiro, "Physical Relationships of the Easter Islanders," 25.

41. Shapiro, "Physical Relationships of the Easter Islanders," 25.

42. Shapiro, "Physical Relationships of the Easter Islanders," 27.

43. For more on this mission, see Warwick Anderson, "Hybridity, Race, and Science: The Voyage of the *Zaca*, 1934–1935," *Isis* 103, no. 2 (June 2012): 229–53.

44. Shapiro, "Physical Relationships of the Easter Islanders," 27.

45. Shapiro, "Physical Relationships of the Easter Islanders," 27.

46. Alfred Métraux, "Easter Island and Melanesia: A Critical Study," *Mankind: The Journal of the Anthropological Society of New South Wales* 2, no. 5 (1938): 97–101.

47. Métraux, "Easter Island and Melanesia," 100.

48. Métraux, "Easter Island and Melanesia," 97–98.

49. Métraux, *Ethnology of Easter Island*, 419–20.

50. Eduardo B. Viveiros de Castro, "Bibliografia Etnológica Básica Tupi-Guarani," *Revista de Antropologia* 27/28 (1984): 7–24, 8.

51. Te Rangi Hiroa (Sir Peter Buck), *An Introduction to Polynesian Anthropology* (Honolulu: Bernice P. Bishop Museum, 1945), 41.

52. Editor, "Polynesian Origins: Results of the Bayard Dominick Expedition," *Journal of the Polynesian Society* 32, no. 4 (128) (December 1923): 250–52.

53. Patrick Vinton Kirch, *On the Road of the Winds: An Archaeological History of the Pacific Islands Before European Contact* (Oakland: University of California Press, 2017), 21.

54. Te Rangi Hīroa, *Introduction to Polynesian Anthropology*, 44.

55. Te Rangi Hīroa, *Introduction to Polynesian Anthropology*, 45.

56. Louis R. Sullivan, "Race Types in Polynesia," *American Anthropologist* 26, no. 1 (1924): 22–26; Louis R. Sullivan, "The Racial Diversity of the Polynesian Peoples," *Journal of the Polynesian Society* 32, no. 2 (126 (1923): 79–84.

57. "Avec Rivet, tout est gaché et perdu" and "des entreprises aussi stupides et vaines que la mission Dakar-Djibouti, mission ethnographie d'un type archaique, quelque chose dans le genre de ce que l'on faisait autour de 1850," Alfred Métraux to Yvonne Oddon, February 20, Box 1 Folder 7, Yvonne Oddon letters from Alfred and Rhoda Bubendey Métraux, Beinecke Rare Book and Manuscript Library, Yale University, New Haven, CT. All quotes from this correspondence in the chapter are my translation.

58. ". . . je me sens beacoup moins préparé que le moindre graduate, mais je fais de mon mieux pour apprendre et completer mon education assez mal fait." Alfred Métraux to Yvonne Oddon, January 16, 1936, Box 1 Folder 7, Yvonne Oddon letters from Alfred and Rhoda Bubendey Métraux, Beinecke Rare Book and Manuscript Library, Yale University, New Haven, CT.

59. "arriver bien à comber mes lacunes d'ici quelques années et à être aussi bien préparé que les types ici." Métraux to Oddon, January 16, 1936.

60. Métraux to Oddon, January 16, 1936.

61. "merveilleuse nature, extremement riche et vigoreuse et doué d'un charme extraordinaire" Métraux to Oddon, January 16, 1936.

62. "avec lui j'apprends plus en ethnographie polynésienne que je ne le ferai en devorant cette énorme littérature océanienne" Métraux to Oddon, January 16, 1936.

63. "intuition profonde de l'ame indigene" Métraux to Oddon, January 16, 1936.

64. Jane L. Carey, "A 'Happy Blending'? Māori Networks, Anthropology and 'Native' Policy in New Zealand, the Pacific and Beyond," in *Indigenous Networks: Mobility, Connections and Exchange*, ed. Jane L. Carey and Jane Lydon (New York: Routledge, 2014), 184–215. In a paper presented at the American Historical Association meeting in 2018, Miranda Johnson also noted how Te Rangi Hīroa and other scholar-activists from the Young Māori Party appropriated settler discourses of racial amalgamation to assert the future promise of the Māori race; see Sebastián Gil-Riaño and Sarah Walsh, Report on HSS Sponsored Panel "The Emergence of Racial Modernities in the Global South," at the American Historical Association Meeting, *History of Science Society Newsletter*, April 2018. https://hssonline.org/resources/publications/newsletter/april-2018-newsletter/report-on-hss-sponsored-panel-the-emergence-of-racial-modernities-in-the-global-south-at-the-2018-american-historical-association-meeting/.

65. J. B. Condliffe, *Te Rangi Hiroa: The Life of Sir Peter Buck* (Christchurch, New Zealand: Whitcombe and Tombs, 1971), 19.

66. Carey, "A 'Happy Blending'?" 189.

67. M. P. K. Sorrenson, "Buck, Peter Henry," in *Dictionary of New Zealand Biography*, first published in 1996, updated May 2002. Te Ara—The Encyclopedia of New Zealand, https://teara.govt.nz/en/biographies/3b54/buck-peter-henry.

68. Sorrenson, "Buck, Peter Henry."
69. John S. Allen "Te Rangi Hiroa's Physical Anthropology," *Journal of the Polynesian Society* 103, no. 1 (March 1994): 11–27; M. P. K. Sorrenson, "Polynesian Corpuscles and Pacific Anthropology: The Home-made Anthropology of Sir Apirana Ngata and Sir Peter Buck," *Journal of the Polynesian Society* 91, no. 1 (March 1982): 7–27.
70. Sorrenson, "Polynesian Corpuscles and Pacific Anthropology."
71. Sorrenson, "Buck, Peter Henry."
72. Allen, "Te Rangi Hiroa's Physical Anthropology," 13.
73. Allen, "Te Rangi Hiroa's Physical Anthropology," 13.
74. For good summaries and analyses of these studies, see Carey, "A 'Happy Blending'?"; Allen, "Te Rangi Hiroa's Physical Anthropology."
75. Te Rangi Hiroa, "Māori Somatology. Racial Averages," *Journal of the Polynesian Society* 31, no. 1 (121) (March 1922): 37–44.
76. Te Rangi Hiroa, "Māori Somatology. Racial Averages. III. (Continued)," *Journal of the Polynesian Society* 31, no. 4 (124) (December 1922): 159–70; 164.
77. Te Rangi Hīroa, "Māori Somatology. Racial Averages. III," 164.
78. For an overview of Te Rangi Hīroa place within anthropology in New Zealand, see: M. P. K. Sorrenson, "Polynesian Corpuscles and Pacific Anthropology: The Home-Made Anthropology of Sir Apirana Ngata and Sir Peter Buck," *Journal of the Polynesian Society* 91, no. 1 (1982): 7–27; for an analysis of the shift from diffusionism to social anthropology and human biology in interwar British anthropology, see Ross L. Jones and Warwick Anderson, "Wandering Anatomists and Itinerant Anthropologists: The Antipodean Sciences of Race in Britain between the Wars," *British Journal for the History of Science* 48, no. 1 (March 2015): 1–16.
79. Sorrenson, "Polynesian Corpuscles and Pacific Anthropology," 11.
80. Sorrenson, "Polynesian Corpuscles and Pacific Anthropology," 11.
81. Kirch, *On the Road of the Winds*, 25.
82. Te Punga Somerville, *Once Were Pacific*.
83. Peter Buck (Te Rangi Hiroa), "The Races of the Pacific," in *Problems of the Pacific: Proceedings of the Second Conference of the Institute of Pacific Relations, Honolulu, Hawaii, July 15 to 29, 1927*, ed. J. B. Condliffe (Chicago: University of Chicago Press, 1928), 232–37; 232.
84. Buck, "Races of the Pacific," 233.
85. Buck, "Races of the Pacific," 233.
86. Te Rangi Hiroa (Sir Peter Henry Buck), *Vikings of the Sunrise* (New York: F. A. Stokes, 1938), 16.
87. Te Rangi Hīroa, *Vikings of the Sunrise*, 16.
88. Te Rangi Hīroa, *Vikings of the Sunrise*, 17.

89. Te Rangi Hiroa (P. H. Buck), "The Passing of the Maori," *Transactions and Proceedings of the Royal Society of New Zealand* 55 (1924): 362–75; 368.

90. Laurière, "Fictions d'une mission," 334.

91. Te Rangi Hiroa, "Passing of the Maori," 364.

92. Te Rangi Hiroa, "Passing of the Maori," 364.

93. Te Rangi Hiroa, "Passing of the Maori," 365.

94. Te Rangi Hiroa, "Passing of the Maori," 366.

95. Te Rangi Hiroa, "Passing of the Maori," 366.

96. Te Rangi Hiroa, "Passing of the Maori," 367.

97. Te Rangi Hiroa, "Passing of the Maori," 368.

98. Te Rangi Hiroa, "Passing of the Maori," 368.

99. Te Rangi Hiroa, "Passing of the Maori," 368.

100. Te Rangi Hiroa, "Passing of the Maori," 369.

101. Te Rangi Hiroa, "Passing of the Maori," 369.

102. Te Rangi Hiroa, "Passing of the Maori," 369–70.

103. Te Rangi Hiroa, "Foreword," in Ernest Beaglehole and Pearl Beaglehole, *Some Modern Maoris* (London: Oxford University Press, 1946), xiii.

104. Te Rangi Hiroa, "Foreword," xiii.

105. Te Rangi Hiroa, "Foreword," xvii; Michael Adas, "Contested Hegemony: The Great War and the Afro-Asian Assault on the Civilizing Mission Ideology," *Journal of World History* 15, no. 1 (March 2004): 31–63.

106. Te Rangi Hiroa, "Foreword," xvii.

107. Alfred Métraux, "Applied Anthropology in Government," in *Anthropology Today: An Encyclopedic Inventory*, ed. A. L. Kroeber (Chicago: University of Chicago Press, 1953), 880–94, 884.

108. Métraux, "Applied Anthropology in Government," 884.

109. Métraux, "Applied Anthropology in Government," 892.

4. A TROPICAL LABORATORY: RACE, EVOLUTION, AND THE DEMISE OF UNESCO'S HYLEAN AMAZON PROJECT

1. "Amazon Meeting Opens in Peru April 30th," *UNESCO Courier* 1, no. 2 (March 1948): 1, 8.

2. "Amazon Meeting Opens," *UNESCO Courier*, 1, 8.

3. "Amazon Meeting Opens," *UNESCO Courier*, 1, 8.

4. "Amazon Meeting Opens," *UNESCO Courier*, 1, 8.

5. Julian Huxley, "Science and the United Nations," *Nature* 156, no. 3967 (November 1945): 553.

6. Perrin Selcer, *Postwar Origins of the Global Environment: How the United Nations Built Spaceship Earth* (New York: Columbia University Press, 2018).

7. Marcos Chor Maio and Magali Romero Sá, "Ciência na periferia: a Unesco, a proposta de criação do Instituto Internacional da Hiléia Amazônica e as origens do Inpa," *História, Ciências, Saúde-Manguinhos* 6, supplement (September 2000): 975–1017.

8. Rodrigo Cesar da Silva Magalhães and Marcos Chor Maio, "Desenvolvimento, ciência e política: o debate sobre a criação do Instituto Internacional da Hiléia Amazônica," *História, Ciências, Saúde-Manguinhos* 14, supplement (December 2007): 169–189; 171.

9. Thomas McCarthy, *Race, Empire, and the Idea of Human Development* (Cambridge: Cambridge University Press, 2009).

10. Jorge Cañizares Esguerra, "New World, New Stars: Patriotic Astrology and the Invention of Indian and Creole Bodies in Colonial Spanish America, 1600–1650," *American Historical Review* 104, no. 1 (1999): 33–68; Rebecca Earle, "'If You Eat Their Food . . .': Diets and Bodies in Early Colonial Spanish America," *American Historical Review* 115, no. 3 (2010): 688–713; Nancy Stepan, *Picturing Tropical Nature* (Ithaca, NY: Cornell University Press, 2001).

11. Warwick Anderson, *The Cultivation of Whiteness: Science, Health, and Racial Destiny in Australia* (Durham, NC: Duke University Press, 2006); Julyan G. Peard, *Race, Place, and Medicine: The Idea of the Tropics in Nineteenth-Century Brazil* (Durham, NC: Duke University Press, 1999).

12. Daniel Clayton and Gavin Bowd, "Geography, Tropicality and Postcolonialism: Anglophone and Francophone Readings of the Work of Pierre Gourou," *L'Espace géographique* 35, no. 3 (July–September 2006): 208–221; 209.

13. For a discussion of the ways in which mid-century sociology of science conceptualized science as an antidote to fascism, see Andrew Zimmerman, *Anthropology and Antihumanism in Imperial Germany* (Chicago: University of Chicago Press, 2010), 114–17.

14. "(Records of the) General Conference, First Session, Held at UNESCO House, Paris from 20 November to 10 December 1946 (Including Resolutions)," in *UNESCO. General Conference. 1 C/Resolutions; UNESCO/C/30* (Paris: UNESCO, 1947), 83.

15. "(Records of the) General Conference," in *UNESCO. General Conference*, 83.

16. "(Records of the) General Conference," in *UNESCO. General Conference*, 169.

17. Brazilian positivists played an important role in the abolition of slavery and in establishing the first republic of Brazil in 1890 and in pushing for a modern secular state. Angela Alonso, "Raízes positivistas do reformismo dos anos 30: o caso Paulo Carneiro," in *Ciência, política e relações internacionais: ensaios sobre Paulo Carneiro*, ed. Marcos Choir Maio (Rio de Janeiro: Editora Fiocruz, 2004), 23–42.

18. Priscila Fraiz, "O acervo da família Carneiro: fonte para o estudo do pensamento e da prática filosófica, política e científica brasileira nos séculos XIX e XX," *História, Ciências, Saúde-Manguinhos* 6, supplement (September 2000): 1125–33.

19. Guaraná is also the active ingredient in several soft drinks and energy drinks. Some of these drinks were first commercialized in the 1920s and 1930s. Magali Romero Sá, "Paulo Carneiro e o curare: em busca do princípio ativo," in *Ciência, política e relações internacionais: ensaios sobre Paulo Carneiro*, ed. Marcos Chor Maio (Rio de Janeiro: Editora Fiocruz, 2004) 43–66; 47.

20. For a detailed account of Carneiro's activities during this period (1931–1937), see Marcos Jungmann Bhering and Marcos Chor Maio, "Entre ciência e política: o positivismo de Paulo Carneiro na Secretaria de Agricultura, Indústria e Comércio de Pernambuco (1935)," *Boletim do Museu Paraense Emílio Goeldi. Ciências Humanas* 5, no. 2 (2010): 435–52.

21. Melissa Teixeira, "Making a Brazilian New Deal: Oliveira Vianna and the Transnational Sources of Brazil's Corporatist Experiment," *Journal of Latin American Studies* 50, no. 3 (January 2018): 613–41.

22. As part of this initiative, Carneiro outlined a research program focused on the toxic and medical properties of plants, which included biochemical studies of curare—a poison used in arrows by several indigenous groups in the Amazon—as well as physiological studies of the effects of plants with caffeine content including coffee, guaraná, and mate. With the help of Cândido Rondon from the newly created Indian Protection Services and with samples from Pernambuco's Botanical Garden, Carneiro began studying the chemical and physiological properties of curare. Yet Carneiro was unable to see these projects come to fruition; see Romero Sá, "Paulo Carneiro e o curare," 49.

23. Chor Maio and Romero Sá, "Ciência na periferia," 975–1017.

24. Patrick Petitjean and Heloisa Maria Bertol Domingues. "Paulo Carneiro: um cientista brasileiro na diplomacia da Unesco (1946–1950)," in *Ciência, Política e relaçoes internacionais: ensaios sobre Paulo Carneiro*, Marcos Chor Maio (org) (Rio de Janeiro: Editora Fiocruz / Ediçoes Unesco, 2004), 195–214.

25. For other discussions of Carneiro's evolutionary standpoint, see Patrick Petitjean and Heloisa M. Bertol Domingues, "A Redescoberta da Amazônia num Projeto da Unesco: o Instituto Internacional da Hiléia Amazônica," *Estudos Históricos* 14, no. 26 (2000): 265–92; Patrick Petitjean and Heloisa Maria Bertol Domingues, "Darwinismo e o projeto da Unesco, do Instituto Internacional da Hiléia Amazônica (1946–1950)," *halshs-00115079* version 1–21 (November 2006): 1–12.

26. Paulo Estevão de Berredo Carneiro, *O Instituto Internacional da Hileia Amazônica: razões e objetivos da sua criação* (Rio de Janeiro: Imprensa Nacional, 1951), 5.

27. Carneiro, *O Instituto Internacional da Hileia Amazônica*, 5.
28. Carneiro, *O Instituto Internacional da Hileia Amazônica*, 5.
29. Carneiro, *O Instituto Internacional da Hileia Amazônica*, 6.
30. Carneiro, *O Instituto Internacional da Hileia Amazônica*, 6.
31. Carneiro, *O Instituto Internacional da Hileia Amazônica*, 6.
32. Carneiro, *O Instituto Internacional da Hileia Amazônica*, 5.
33. Carneiro, *O Instituto Internacional da Hileia Amazônica*, 6.
34. Carneiro, *O Instituto Internacional da Hileia Amazônica*, 12.
35. Carneiro, *O Instituto Internacional da Hileia Amazônica*, 13.
36. Eve Buckley, "Unequal Encounters: Debating Resource Scarcity, Population, and Hunger in the Early Cold War," in *Troubling Encounters in the History of the Human Sciences: Latin America and the United States Empire, 1870s-2000s*, ed. Adam Warren, Julia Rodriguez, and Stephen Casper (Cambridge: Cambridge University Press, in press).
37. Carneiro, *O Instituto Internacional da Hileia Amazônica*, 13.
38. Carneiro, *O Instituto Internacional da Hileia Amazônica*, 15.
39. Glenda Sluga, "UNESCO and the (One) World of Julian Huxley," *Journal of World History* 21 no. 3 (September 2010): 393–418; 404.
40. Sluga, "UNESCO and the (One) World of Julian Huxley," 409–10.
41. Peder Anker, *Imperial Ecology: Environmental Order in the British Empire, 1895–1945* (Cambridge, MA: Harvard University Press, 2001).
42. Anker, *Imperial Ecology*, 206.
43. Julian Huxley, "UNESCO: Its Purpose and Its Philosophy," 1946, 10. UNESCO Archives, Paris Preparatory Commission of the United Nations Educational, Scientific, and Cultural Organization.
44. Julian Huxley, "UNESCO, Its First Year," *The Rotarian* 71, no. 5 (November 1947): 13–15; 14.
45. Huxley, "UNESCO, Its First Year," 14.
46. Thomas Mougey, "Building UNESCO Science from the 'Dark Zone': Joseph Needham, Empire, and the Wartime Reorganization of International Science from China, 1942–6," *History of Science* 59, no. 4 (2021): 461–91.
47. Aant Elzinga, "UNESCO and the Politics of International Co-operation in the Realm of Science," in *Les Sciences coloniales. Figures et institutions*, ed. Patrick Petitjean (Paris: Orstom éditions, 1996): 163–202.
48. Patrick Petitjean, "Blazing the Trail: Needham and UNESCO: Perspectives and Realizations," in *Sixty Years of Science at UNESCO, 1945–2005*, ed. P. Petitjean et al. (Paris: UNESCO, 2006), 43–47.
49. Thomas Bastien Mougey, "Enlightening the Dark Zone: UNESCO, Science and the Technocratic Reordering of the World in the Global South, 1937–1959" (PhD diss., Universitaire Pers Maastricht, 2018), 61.

50. Jürgen Hillig, "Going Global: UNESCO Field Science Offices," in *Sixty Years of Science at UNESCO, 1945–2005*, ed. P. Petitjean et al. (Paris: UNESCO, 2006), 72–75.

51. In 1942, Japanese forces invaded Singapore and took over the Botanical Gardens. During this period, Corner's house in the Botanical Gardens was looted by Australian soldiers, and his wife and son were forced to leave Singapore. Yet Corner remained and protected the Botanical Garden's collections and library and negotiated to have food supplies delivered to the prison camp where British colonial servants were held in deplorable conditions. He also developed the innovative technique of training berok monkeys to collect botanical specimens from tall trees. See David J. Mabberley, "Edred John Henry Corner, C.B.E. 12 January 1906—14 September 1996," *Biographical Memoirs of Fellows of the Royal Society* 45 (November 1999): 77–93.

52. Petitjean and Bertol Domingues, "A Redescoberta da Amazônia num Projeto da Unesco."

53. Chor Maio and Romero Sá, "Ciência na periferia."

54. Chor Maio and Romero Sá, "Ciência na periferia."

55. Chor Maio and Romero Sá, "Ciência na periferia," 989.

56. Chor Maio and Romero Sá, "Ciência na periferia," 989.

57. Alfred Métraux, "Centres for Study of Tropical Life and Resources," Documents sur le project de l'Hylea amazonienne (1946–1948), Fonds Alfred Métraux, FAM.AS.AA.03.01, Archives des ethnologues, Biblothèque d'anthropologie sociale, Laboratoire d'anthropologie sociale, Collège de France, Paris.

58. Alfred Métraux, *Itinéraires* (Paris: Payot, 1978), 193–98.

59. Alfred Métraux, "The Anthropological Collections of the Museu Paraense Emilio Goeldi," Documents sur le project de l'Hylea amazonienne (1946–1948), FAM.AS.AA.03.05, 1.

60. Alfred Métraux, "The Anthropological Collections of the Museu Paraense Emilio Goeldi," 3.

61. "General Information on the Conference (and Scope and Programme of the Proposed International Institute of the Hylean Amazon)," NS/IIHA/1 + Annex I + CORR. Nat. Sci./42 (Paris: UNESCO, February 27, 1948).

62. Fred Soper, "Annex I: Scope and Programme of the Proposed International Institute of the Hylean Amazon," in "General Information on the Conference (and Scope and Programme of the Proposed International Institute of the Hylean Amazon)," NS/IIHA/1 + Annex I + CORR. Nat. Sci./42 (Paris: UNESCO, February 27, 1948), 1.

63. Soper "Annex I," 1.

64. The proposed projects included creating research centers in "pedology [soil science] and physical geography" at Belém, Manáos, Iquitos, and Cayenne;

establishing a reference herbarium and botanical library; creating a "working experimental and botanical garden"; completing a "faunistic census of the region" and establishing a "reference collection at a central location"; studying the "ecology of fresh water fish"; investigating means of "controlling the destructive action of termites and ants"; surveying "the structure and living conditions of a few rural communities"; studying the possibility of deploying mobile "fundamental education" teams consisting of "health experts, ethnologists, agriculturalists and teachers equipped with modern projectors for slides and films, records, wireless, etc."; conducting preparatory studies for the creation of "technical schools specializing in industrial chemistry, agriculture and hygiene"; studying the "causes of the decline of populations of the Hylean Amazon"; investigating means of "protecting the native populations from infectious diseases, with special reference to those they might contract as a result of contact with civilized races"; studying the "folklore, arts and crafts, and linguistic aspects of various indigenous groups"; carrying out "biochemical research concerning edible, medicinal and toxic plants in Hylea"; and contributing to "comparative studies of human physiology in different latitudes and altitudes"; see Soper, "Annex I," 2.

65. "Provisional Programme of Activies of the Unesco Hylean Amazon Project During 1948," IIHA/7/Nat.Sci./52 (Paris: UNESCO, 1948), 1.
66. E. J. H. Corner, "Report on the Process of the Hylean Amazon Project of UNESCO, 1947–48 in South America," NS/IIHA/13 (Paris: UNESCO, November 18, 1948), 7.
67. Corner, "Report on the Process of the Hylean Amazon Project," 8.
68. For the task of leading this survey, Corner hired the Mexican zoologist Candido Bolivar and three other scientists to form part of his team—the Peruvian botanist Rámon Ferreyra, the Ecuadorian anthropologist Anibal Buitrón, and the U.S. geographer Edwin Doran. The team was also joined by two other researchers who were hired by the Peruvian government—Coronel Gerardo Dianderas, the director of the Military Institute of Geography in Lima, and the physical anthropologist Pedro Weiss, who was a professor of pathology from the University of San Marcos in Lima. Corner, "Report on the Process of the Hylean Amazon Project," 10.
69. Candido Bolívar, "Report on the Exploration of the River Huallaga Valley, Peru," NS/IIHA/19 (Paris: UNESCO, January 14, 1949), 3.
70. Corner, "Report on the Process of the Hylean Amazon Project," 2.
71. Corner, "Report on the Process of the Hylean Amazon Project," 2–5.
72. Peter van Dresser, "The Future of the Amazon," *Scientific American* 178, no. 5 (May 1948): 11–15; 11.
73. Van Dresser, "The Future of the Amazon," 11.

74. Van Dresser, "The Future of the Amazon," 13.
75. Van Dresser, "The Future of the Amazon," 13.
76. Van Dresser, "The Future of the Amazon," 13.
77. Van Dresser, "The Future of the Amazon," 14.
78. Mariza Corrêa, "Dona Heloisa e a pesquisa de campo," *Revista de Antropologia* 40 (1997): 11–54.
79. "The Hylean Amazon Project and Anthropology," 1947. Documents sur le project de l'Hylea amazonienne (1946–1948), Fonds Alfred Métraux, FAM. AS.AA.03.03, Archives des ethnologues, Bibliothèque d'anthropologie sociale, Laboratoire d'anthropologie sociale, Collège de France, Paris.
80. "The Hylean Amazon Project and Anthropology."
81. "The Hylean Amazon Project and Anthropology."
82. "The Main Objective of the Anthropological Section of the International Hylean Amazon Institute," n.d. Documents sur le projet de l'Hylea amazonienne (1946–1948), Fonds Alfred Métraux, FAM.AS.AA.03.02, Archives des ethnologues, Bibliothèque d'anthropologie sociale, Laboratoire d'anthropologie sociale, Collège de France, Paris.
83. "The Main Objective of the Anthropological Section," Documents sur le projet de l'Hylea amazonienne (1946–1948).
84. "The Main Objective of the Anthropological Section," Documents sur le projet de l'Hylea amazonienne (1946–1948).
85. Alfred Métraux, "Physical Anthropology in the Hylean Amazon Institute 1947–1948," n.d. Documents sur le project de l'Hylea amazonienne (1946–1948), Fonds Alfred Métraux, FAM.AS.AA.03.04, Archives des ethnologues, Bibliothèque d'anthropologie sociale, Laboratoire d'anthropologie sociale, Collège de France, Paris.
86. Métraux, "Physical Anthropology in the Hylean Amazon Institute 1947–1948."
87. Métraux, "Physical Anthropology in the Hylean Amazon Institute 1947–1948."
88. Métraux, "Physical Anthropology in the Hylean Amazon Institute 1947–1948."
89. Métraux, "Physical Anthropology in the Hylean Amazon Institute 1947–1948."
90. "Regional Studies—The Tropic Division," n.d. Documents sur le project de l'Hylea amazonienne (1946–1948), Fonds Alfred Métraux, FAM.AS.AA.03.09, Archives des ethnologues, Bibliothèque d'anthropologie sociale, Laboratoire d'anthropologie sociale, Collège de France, Paris.
91. "Regional Studies—The Tropic Division," Documents sur le project de l'Hylea amazonienne (1946–1948).
92. "Regional Studies—The Tropic Division," Documents sur le project de l'Hylea amazonienne (1946–1948).
93. "Regional Studies—The Tropic Division," Documents sur le project de l'Hylea amazonienne (1946–1948).

94. Richard Pace, "The legacy of Charles Wagley: An Introduction," *Boletim do Museu Paraense Emílio Goeldi. Ciências Humanas* 9, no. 3 (2014): 597–602.

95. Robert Redfield, Ralph Linton, and Melville J. Herskovits, "Memorandum for the Study of Acculturation," *American Anthropologist* 38, no. 1 (January–March 1936): 149–52.

96. Charles Wagley, *Welcome of Tears: The Tapirapé Indians of Central Brazil* (New York: Oxford University Press, 1977), 4.

97. Wagley, *Welcome of Tears*, 4.

98. Charles Wagley, "Alfred Métraux 1902–1963," *American Anthropologist* 66, no. 3 (June 1964): 605.

99. Wagley, *Welcome of Tears*, 4.

100. Wagley, "Alfred Métraux," 605.

101. Wagley, "Alfred Métraux," 605.

102. Wagley, *Welcome of Tears*, 6.

103. Charles Wagley, "The Effects of Depopulation upon Social Organization as Illustrated by the Tapirape Indians," *Transactions of the New York Academy of Sciences* 3, no. 1, series II (1940): 12–16.

104. Marvin Harris, "Charles Wagley's Contribution to Anthropology," in *Looking Through the Kaleidoscope: Essays in Honor of Charles Wagley. Florida Journal of Anthropology* Special Publication, 1990, no. 6, 1–6; 1.

105. For an expanded version of this argument, see Sebastián Gil-Riaño, "Becoming an Area Expert During the Cold War: Americanism and Lusotropicalismo in the Transnational Career of Anthropologist Charles Wagley, 1939–1971," in *Cold War Social Science: Transnational Entanglements*, ed. M. Solovey and C. Dayé (Cham, Switzerland: Palgrave Macmillan, 2021), 127–59.

106. Charles Wagley, *A Social Survey of an Amazon Community with Recommendations for Future Research* (Paris, UNESCO, 1948), 4.

107. Wagley, *A Social Survey*, 4.

108. Wagley, "A Social Survey, 25.

109. Wagley served as a rapporteur for a 1947 SSRC conference on area studies; see Charles Wagley, *Area Research and Training: A Conference Report on the Study of World Areas* (New York: SSRC, 1948).

110. Seth Garfield, *In Search of the Amazon: Brazil, the United States, and the Nature of a Region* (Durham, NC: Duke University Press, 2014), 9.

111. These agencies include the Ministry of Agriculture's Department of Land and Colonization (Divisão de Terras e Colonização), the Instituto Agronômico do Norte, and the Serviço de Estudos de Grandes Endemias. See Garfield, *In Search of the Amazon*, 22.

112. Garfield, *In Search of the Amazon*.

113. Charles Wagley, *Amazon Town: A Study of Man in the Tropics* (New York: Macmillan, 1953), 1.
114. Wagley, *Amazon Town*, 2.
115. Wagley, *Amazon Town*, 5.
116. Wagley, *Amazon Town*, 5.
117. Wagley, *Amazon Town*, 6.
118. Wagley, *Amazon Town*, 6.
119. Wagley, *Amazon Town*, 7.
120. Wagley, *Amazon Town*, 7.
121. Wagley, *Amazon Town*, 290.
122. Wagley, *Amazon Town*, 289.
123. Wagley, *Amazon Town*, 291.
124. Wagley, *Amazon Town*, 291–92.
125. Mougey, "Enlightening the Dark Zone," 260–63.
126. See Seth Garfield, *Indigenous Struggle at the Heart of Brazil: State Policy, Frontier Expansion, and the Xavante Indians, 1937–1988* (Durham, NC: Duke University Press, 2001).
127. Magalhães and Chor Maio, "Desenvolvimento, ciência e política."
128. Petitjean and Bertol Domingues, "A Redescoberta da Amazônia num Projeto da Unesco."
129. Petitjean and Bertol Domingues, "A Redescoberta da Amazônia num Projeto da Unesco."

5. "PEASANTS WITHOUT LAND": RACE AND INDIGENEITY IN THE ILO'S PUNO—TAMBOPATA PROJECT

1. Alfred Métraux, "'Land Hunger': On the Top of the Andes," *UNESCO Courier* 2 (1955): 4–9; 8.
2. Métraux, "'Land Hunger,'" 6.
3. Métraux, "'Land Hunger,'" 6.
4. Carlos Monge, "Appendix X: Medical and Sanitary Precautions in the Adjustment of Highland Indians to the Lowlands," in *Report of the Joint Field Mission on Indigenous Populations (Andean Indian Mission)*, 1953, 3, Folder 19: Andean Indian Mission, Papers of Pearl and Ernest Beaglehole, JC Beaglehole Room, Victoria University of Wellington Library.
5. For useful analyses of the CPP's politics and methodologies, see Jason Pribilsky, "Development and the 'Indian Problem' in the Cold War Andes: *Indigenismo*, Science, and Modernization in the Making of the Cornell-Peru Project at Vicos," *Diplomatic History* 33, no. 3 (2009): 405–26; Jason Pribilsky, "Developing Selves: Photography, Cold War Science and 'Backwards'

People in the Peruvian Andes, 1951–1966," *Visual Studies* 30, no. 2 (2015): 131–50.

6. Key examples of this work include Michael E. Latham, *Modernization as Ideology: American Social Science and "Nation Building" in the Kennedy Era* (Chapel Hill: University of North Carolina Press, 2000); David C. Engerman et al., eds., *Staging Growth: Modernization, Development, and the Global Cold War* (Amherst: University of Massachusetts Press, 2003); Nils Gilman, *Mandarins of the Future: Modernization Theory in Cold War America* (Baltimore, MD: Johns Hopkins University Press, 2003); and Nick Cullather, *The Hungry World: America's Cold War Battle Against Poverty in Asia* (Cambridge, MA: Harvard University Press, 2013). For synthetic overviews of this literature, see Corinna R. Unger, *International Development: A Postwar History* (London: Bloomsbury Academic, 2019); Sara Lorenzini, *Global Development: A Cold War History* (Princeton, NJ: Princeton University Press, 2019).

7. Nick Cullather, "Development? It's History," *Diplomatic History* 24, no. 4 (2000): 641–53.

8. Joel Isaac, "The Human Sciences in Cold War America," *Historical Journal* 50 (2007): 725–46; David Engerman, "Bernath Lecture: American Knowledge and Global Power," *Diplomatic History* 31, no. 4 (2007): 599–622.

9. One exception is Helen Tilley, *Africa as a Living Laboratory: Empire, Development, and the Problem of Scientific Knowledge, 1870–1950* (Chicago: University of Chicago Press, 2011).

10. Warwick Anderson, "Racial Conceptions in the Global South," *Isis* 105, no. 4 (2014): 782–92.

11. Nancy Stepan, *"The Hour of Eugenics": Race, Gender, and Nation in Latin America* (Ithaca, NY: Cornell University Press, 1996), 194–95.

12. For other useful accounts of eugenics in Latin America, see Karin Alejandra Rosemblatt, "Bodies, Environments, and Race," in *Handbook of the Historiography of Latin American Studies on the Life Sciences and Medicine*, ed. Ana Barahona, Historiographies of Science (Cham, Switzerland: Springer International, 2022), 467–68; Sebastián Gil-Riaño and Sarah Walsh, "Introduction: Race Science in the Latin World," *History of Science* 60, no. 1 (March 1, 2022): 4–17; Sarah Walsh, *The Religion of Life: Eugenics, Race, and Catholicism in Chile* (Pittsburgh, PA: University of Pittsburgh Press, 2022).

13. Marisol de la Cadena, *Indigenous Mestizos: The Politics of Race and Culture in Cuzco, Peru, 1919–1991* (Durham, NC: Duke University Press, 2000), 12–29.

14. de la Cadena, *Indigenous Mestizos*, 15.

15. Andres Rios-Molina, "Racial Degeneration, Mental Hygiene, and the Beginning of Peruvian Psychiatry, 1922–1934," *History of Psychology* 22, no. 3 (August 2019): 225–43; 231.

16. de la Cadena, *Indigenous Mestizos*, 16.
17. Walter Mendoza de Souza and Oscar Martínez, "Las Ideas Eugenésicas en la Creación Del Instituto de Medicina Social," *Anales de La Facultad de Medicina* 60, no. 1 (2014): 55–60; 56.
18. de Souza and Martínez, "Las Ideas Eugenésicas," 59.
19. Stepan, *"Hour of Eugenics,"* 180–87.
20. Marcos Cueto, "Social Medicine and 'Leprosy' in the Peruvian Amazon," *The Americas* 61, no. 1 (2004): 55–80; 61.
21. Marisol de la Cadena, "From Race to Class: Insurgent Intellectuals de Provincia in Peru, 1910–1970," in *Shining and Other Paths*, ed. Steve J. Stern (Durham, NC: Duke University Press, 1998), 22–59; 32.
22. José Carlos Mariátegui, *Siete Ensayos de Interpretación de La Realidad Peruana* (Caracas: Fundación Biblioteca Ayacucho, 2007), 30.
23. Mariátegui, *Siete Ensayos*, 30.
24. Mariátegui, *Siete Ensayos*, 30.
25. de la Cadena, *Indigenous Mestizos*, 24.
26. Marisol de la Cadena, "Are 'Mestizos' Hybrids? The Conceptual Politics of Andean Identities," *Journal of Latin American Studies* 37, no. 2 (2005): 259–84; 275.
27. Osmar Gonzales and Mariana Ortega Breña, "The Instituto Indigenista Peruano: A New Place in the State for the Indigenous Debate," *Latin American Perspectives* 39, no. 5 (2012): 33–44; 39–40.
28. Marcos Cueto, "Social Medicine and 'Leprosy,'" 56.
29. Michael Knipper, "Antropología y 'Crisis de La Medicina': El Patólogo M. Kuczynski-Godard (1890–1967) y Las Poblaciones Nativas en Asia Central y Perú," *Dynamis* 29 (2009): 97–121.
30. At the University of Berlin, Kuczynski-Godard also worked closely with the anthropologist Felix von Luschan; see Knipper, "Antropología y 'Crisis de La Medicina,'" 101.
31. Knipper, "Antropología y 'Crisis de La Medicina,'" 114.
32. Cueto, "Social Medicine and 'Leprosy,'" 78.
33. Maxime H. Kuczynski Godard, *Los Andes Peruanos: Ilave-Ichupampa-Lauramarca-Iguaín Investigaciones Andinas* (Lima: Universidad Nacional Mayor San Marcos, Fondo Editorial, 2004).
34. Maxime H. Kuczynski-Godard, *Estudios Médico-Sociales en Minas de Puno con Anotaciones Sobre las Migraciones Indígenas* (Lima: Ministerio de Salud Pública y Asistencia Social, 1945).
35. Kuczynski-Godard, *Estudios Médico-Sociales*, 19–20.
36. Kuczynski-Godard, *Estudios Médico-Sociales*, 95.
37. Kuczynski-Godard, *Estudios Médico-Sociales*, 97.

38. Maxime H. Kuczynski-Godard et al., *Disección del Indigenismo Peruano: Un Examen Sociologico y Medico-Social* (Lima: Publicaciones del Instituto de Medicina Social, 1948), vi.

39. Kuczynski-Godard et al., *Diseccion del Indigenismo*, vii.

40. Kuczynski-Godard et al., *Diseccion del Indigenismo*, vi.

41. Kuczynski-Godard et al., *Diseccion del Indigenismo*, 4.

42. Kuczynski-Godard et al., *Diseccion del Indigenismo*, 10.

43. Kuczynski-Godard et al., *Diseccion del Indigenismo*, 59.

44. Thomas C. Field, *From Development to Dictatorship: Bolivia and the Alliance for Progress in the Kennedy Era* (Ithaca, NY: Cornell University Press, 2014).

45. Pribilsky, "Development and the 'Indian Problem,'" 412.

46. For an overview of the project, see Jef Rens, "The Andean Programme," *International Labour Review* 84, no. 6 (1961): 423–61.

47. Quoted in J. E. Ritchie, "Obituary: Ernest Beaglehole," *Journal of the Polynesian Society* 75, no. 1 (1966): 109–19; 112.

48. According to Maul, in the early days of its existence the ILO was often referred to as a form of "revolution insurance" presumably because it strove to institutionalize many of the demands made by the international workers movement thereby containing the need for revolution. Daniel Maul, *Human Rights, Development and Decolonization: The International Labour Organization, 1940–70* (New York: Palgrave Macmillan, 2012), 2, 121.

49. Daniel Morrow and Barbara Brookes, "The Politics of Knowledge: Anthropology and Māori Modernity in Mid-Twentieth-Century New Zealand," *History and Anthropology* 24, no. 4 (December 2013): 453–71.

50. Beaglehole himself played a prominent role in promoting this harmonious view of New Zealand. This can be seen in articles he wrote for UNESCO and the ILO where he described a relatively friendly "climate of race relations" that allowed the Māori to more easily adapt to modern life thanks to the "knowledge that he has been considered the social equal of the white man and that the white New Zealander has, by and large, been vitally concerned to change a theoretical equality into a living reality." Ernest Beaglehole, "Maori in New Zealand: A Case Study of Socio-Economic Integration," *International Labour Review* 76, no. 2 (1957): 103–23; 105; Ernest Beaglehole, "Race Relations in the Pacific," *International School Science Bulletin* 2, no. 4 (1950): 489–96.

51. Ernest Beaglehole, *Report of the Joint Field Mission on Indigenous Populations (Andean Indian Mission)*, 1953, 9, Folder 19: Andean Indian Mission, Papers of Pearl and Ernest Beaglehole, JC Beaglehole Room, Victoria University of Wellington Library.

52. Beaglehole, *Report of the Joint Field Mission*, 9.

53. Beaglehole, *Report of the Joint Field Mission*, 36.

54. Beaglehole, *Report of the Joint Field Mission*, 36.

55. Beaglehole, *Report of the Joint Field Mission*, 13.

56. Beaglehole, *Report of the Joint Field Mission*, 29.

57. Beaglehole, *Report of the Joint Field Mission*, 29–30.

58. Gonzales and Breña, "The Instituto Indigenista Peruano," 33–44.

59. Pribilsky, "Development and the 'Indian Problem,'" 412.

60. Monge, "Appendix X," 2.

61. Monge, "Appendix X," 5.

62. Monge, "Appendix X," 6.

63. Monge, "Appendix X," 6.

64. Monge, "Appendix X," 7.

65. Monge, "Appendix X," 8.

66. Monge, "Appendix X," 9.

67. *Actas de la Segunda Conferencia Panamericana de Eugenesia y Homicultura de las Repúblicas Americanas* (Buenos Aires: Imprenta Frascoli y Bindi, 1934), 73–80.

68. Carlos Monge, *Acclimatization in the Andes* (Baltimore, MD: Johns Hopkins University Press, 1948), 66.

69. Monge, *Acclimatization*, 67.

70. Monge, *Acclimatization*, 74.

71. On the influence of U.S. New Deal policies and the Tennessee Valley Authority (TVA) on development thought, see David Ekbladh, *The Great American Mission: Modernization and the Construction of an American World Order* (Princeton, NJ: Princeton University Press, 2011).

72. For an excellent overview of Monge's role in the coca debates, see Adam W. Warren, "Collaboration and Discord in International Debates About Coca Chewing, 1949–1950," *Medicine Anthropology Theory* 5, no. 2 (2018): 35–51.

73. Alfred Métraux, *Itinéraires 1(1935–1953): Carnets de notes et journaux de voyage* (Paris: Payot, 1978), 470.

74. Métraux, *Itinéraires 1*, 471.

75. "Alfred Métraux a Pierre Verger, Lima le 1er Février, 1954" in *Le Pied à l'étrier: Correspondance 12 Mars 1946–5 Avril 1963* (Paris: Jean-Michel Place, 1994), 189–90.

76. Alfred Métraux, "Las Migraciones Internas de los Indios Aymara en el Peru Contemporaneo," in *Estudios Antropologicos Publicados en Homenaje al Doctor Manuel Gamio* (Mexico City: Sociedad Mexicana de Antropologia, 1956), 391.

77. Métraux, "Las Migraciones Internas," 392.

78. Métraux, "Las Migraciones Internas," 393.

79. Métraux, "Las Migraciones Internas," 397.

80. Métraux, "Las Migraciones Internas," 397.

81. Métraux, "Las Migraciones Internas," 405.

82. Métraux, "Las Migraciones Internas," 406.

83. Alfred Métraux, *Informe del Sr Métraux Sobre la Emigracion Interna y Externa de los Indios Aymaras y Acerca de la Colonizacion en el Valle del Tambopata* (Geneva: ILO, 1954), 16.

84. Alfred Métraux, "The Social and Economic Structure of the Indian Communities of the Andean Region," *International Labour Review* 79 (1959): 225–43.

85. Métraux, "Social and Economic Structure," 231.

86. Métraux, "Social and Economic Structure," 239.

87. Jason Guthrie, "The International Labor Organization and the Social Politics of Development, 1938–1969" (PhD diss., University of Maryland, 2015), 154.

88. Pribilsky, "Development and the 'Indian Problem,'" 418.

89. Pribilsky, "Development and the 'Indian Problem,'" 418.

90. Guthrie, "International Labor Organization," 154.

91. Raúl Necochea López, "Demographic Knowledge and Nation-Building: The Peruvian Census of 1940," *Berichte Zur Wissenschaftsgeschichte* 33, no. 3 (2010): 280–96; 287.

92. Necochea-López, "Demographic Knowledge and Nation-Building," 287.

93. Alberto Arca Parró, "Problemas Demográficos, 'Proyecto Puno-Tambopata,'" *La Prensa* (Lima), Tuesday, April 5, 1955, Dossier: "Projet Andin 1956" (1928–1956), fr/cdf/las/FAM.AS.AA.08, Fonds Alfred Métraux, Bibliothèque d'Anthropologie Sociale, College de France, Paris, France.

94. Arca Parró, "Problemas Demográficos."

95. The organizers of the symposium gave Monge a prominent place in the program. Monge gave the opening paper as well as the symposium's closing remarks.

96. Carlos Monge, "Clausura del Simposio," in *Migración e Integración en el Perú*, ed. Henry F. Dobyns and Mario C. Vázquez (Lima: Editorial Estudios Andinos, 1963), 187.

6. A BRAZILIAN RACIAL DILEMMA: MODERNIZATION AND UNESCO'S RACE RELATIONS STUDIES IN BRAZIL

1. Alfred Métraux, "An Inquiry into Race Relations in Brazil," *UNESCO Courier* 5, no. 8–9 (August–September, 1952): 6.

2. Métraux, "Inquiry into Race Relations in Brazil," 6.

3. On U.S. racial liberalism and race relations inquiry, see Leah N. Gordon, *From Power to Prejudice: The Rise of Racial Individualism in Midcentury America* (Chicago: University of Chicago Press, 2015); Walter A. Jackson, *Gunnar Myrdal and America's Conscience: Social Engineering and Racial Liberalism, 1938–1987* (Chapel Hill: University of North Carolina, 1994); Ruth Feldstein, *Motherhood*

in *Black and White: Race and Sex in American Liberalism, 1930–1965* (Ithaca, NY: Cornell University Press, 2018); Franz Samelson, "From 'Race Psychology' to 'Studies in Prejudice': Some Observations on the Thematic Reversal in Social Psychology," *Journal of the History of the Behavioral Sciences* 14, no. 3 (1978): 265–78.

4. Gunnar Myrdal, *American Dilemma: The Negro Problem & Modern Democracy* (New York: Harper & Brothers., 1944), xliii.

5. The literature on the history of Chicago sociology is vast. Some useful entry points are: Henry Yu, *Thinking Orientals: Migration, Contact, and Exoticism in Modern America* (New York: Oxford University Press, 2001); Andrew Abbott, *Department and Discipline: Chicago Sociology at One Hundred* (Chicago: University of Chicago Press, 2017); Davarian L. Baldwin, "Black Belts and Ivory Towers: The Place of Race in U.S. Social Thought, 1892–1948," *Critical Sociology* 30, no. 2 (March 1, 2004): 397–450.

6. Florestan Fernandes, *The Negro in Brazilian Society*, trans. Jacqueline D. Skiles (New York: Columbia University Press, 1969), xv.

7. On dependency theory in Latin American social science, see Margarita Fajardo, *The World That Latin America Created: The United Nations Economic Commission for Latin America in the Development Era*, (Cambridge, MA: Harvard University Press, 2022); Karin Alejandra Rosemblatt, "Modernization, Dependency, and the Global in Mexican Critiques of Anthropology," *Journal of Global History* 9, no. 1 (2014): 94–121; Jeremy Adelman and Margarita Fajardo, "Between Capitalism and Democracy: A Study in the Political Economy of Ideas in Latin America, 1968–1980," *Latin American Research Review* 51, no. 3 (2016): 3–22.

8. Gordon, *From Power to Prejudice*, 2.

9. Hadley Cantril, "Psychology Working for Peace," *American Psychologist* 4, no. 3 (1949): 69–73; 70.

10. For accounts of the SSD's early years and the intellectual, political, and organizational challenges it faced, see Peter Lengyel, *International Social Science, the UNESCO Experience* (New Brunswick, NJ: Transaction Books, 1986); Perrin Selcer, "The View from Everywhere: Disciplining Diversity in Post–World War II International Social Science," *Journal of the History of the Behavioral Sciences* 45, no. 4 (2009): 309–29; Teresa Tomás Rangil, "Citizen, Academic, Expert, or International Worker? Juggling with Identities at UNESCO's Social Science Department, 1946–1955," *Science in Context* 26, no. 1 (2013): 61–91; Per Wisselgren, "From Utopian One-Worldism to Geopolitical Intergovernmentalism: UNESCO's Department of Social Sciences as an International Boundary Organization, 1946–1955," *Serendipities: Journal for the Sociology and History of the Social Sciences* 2, no. 2 (2017): 148–82.

11. E. P. Hollander, "Obituary: Otto Klineberg (1899–1992)," *American Psychologist* 48, no. 8 (1993): 909–10; 909.

12. For instance, his research showed that children from the Yakima tribe in Washington State who took intelligence tests made very few errors and were generally less concerned with speed and more concerned with accuracy when compared with white children.

13. Otto Klineberg, *Race Differences* (New York: Harper, 1935); Otto Klineberg, *Negro Intelligence and Selective Migration* (New York, Columbia University Press, 1935; Otto Klineberg, *The Race Question in Modern Science: Race and Psychology*, UNESCO Publication (Paris: UNESCO, 1951).

14. Otto Klineberg, "The Place of Psychology in UNESCO's Social Science Program," *Transactions of the New York Academy of Sciences* 18 (1956): 456–61; 457.

15. Otto Klineberg, *Tensions Affecting International Understanding; a Survey of Research*, Bulletin, Social Science Research Council (New York: Social Science Research Council, 1950), 93.

16. Klineberg, *Tensions Affecting International Understanding*, 125.

17. Klineberg argued that the "informational approach" had been particularly effective in school courses concerned with race relations and would continue to be so as long as it struck a balance in describing the "common elements in all cultures, without neglecting information concerning cultural differences." Echoing claims that would figure in the UNESCO Statements on Race, Klineberg also stressed that the "informational approach" to attitude change should place special emphasis on information concerning "the absence of biologically fixed differences in culture" and should also emphasize the "range of individual variations within a particular group." Klineberg, *Tensions Affecting International Understanding*, 155.

18. On Klineberg's activities in Brazil, see Marcos Chor Maio, "A Crítica de Otto Klineberg aos testes de inteligência. O Brasil como laboratório racial," *Varia Historia* 33 (April 2017): 135–61.

19. Quoted in Chor Maio, "A Crítica de Otto Klineberg," 149.

20. Klineberg, *Tensions Affecting International Understanding*, 192–193.

21. Hadley Cantril, ed., *Tensions That Cause Wars: Common Statement and Individual Papers by a Group of Social Scientists Brought Together by UNESCO* (Urbana: University of Illinois Press, 1950).

22. Marcos Chor Maio, "Gilberto Freyre and the UNESCO Research Project on Race Relations in Brazil," in *Luso-Tropicalism and Its Discontents: The Making and Unmaking of Racial Exceptionalism*, ed. Warwick Anderson, Ricardo Roque, and Ricardo Ventura Santos (New York: Berghahn Books, 2018), 116.

23. Otto Klineberg, "International Tensions a Challenge to the Sciences of Man," *The Lancet* 254, no. 6584 (November 5, 1949): 851–54; 854.

24. See Anadelia A. Romo, *Brazil's Living Museum: Race, Reform, and Tradition in Bahia* (Chapel Hill: University of North Carolina Press, 2010), 115–19.

25. Donald Pierson, *Negroes in Brazil: A Study of Race Contact at Bahia* (Chicago: University of Chicago Press, 1942), vii.

26. During his visit to Brazil, Park met many of Brazil's leading social scientists including Arthur Ramos with whom he would maintain an active correspondence. Having established these connections, Park encouraged U.S. graduate students including Pierson and the Columbia anthropologist Ruth Landes to conduct research in Brazil. When Park traveled to Brazil a second time in 1937, Donald Pierson was already two years into his doctoral research in Salvador, Bahia, where he had been living with his wife, Helen Pierson, since 1935. During this second trip, Park met with Arthur Ramos in Rio and toured Salvador, Bahia, with the Piersons. During this visit, Park and Pierson spent long hours walking through the city of Salvador making observations and interviewing people and theorizing about the nature of Bahian society. Also during this visit, Pierson and Park observed Candomblé rituals. Lícia do Prado Valladares, "A Visita Do Robert Park Ao Brasil, o 'Homem Marginal' e a Bahia Como Laboratório," *Caderno CRH* 23 (2010): 35–49.

27. Pierson, *Negroes in Brazil*, viii.

28. Pierson, *Negroes in Brazil*, xv–xvi.

29. Robert Park, "Introduction," in Pierson, *Negroes in Brazil*, xix.

30. See Andrew Zimmerman, *Alabama in Africa: Booker T. Washington, the German Empire, and the Globalization of the New South* (Princeton, NJ: Princeton University Press, 2010), 219–27.

31. Charles Johnson, a black sociologist and former student of Park, was the one who invited him to this position. See Valladares, "A Visita Do Robert Park," 36.

32. Sally Cooper Cole, *Ruth Landes: A Life in Anthropology* (Lincoln: University of Nebraska Press, 2003), 150.

33. For overviews of the history of the Chicago school of sociology and the rise of the "caste school of race relations," see Oliver C. Cox, "The Modern Caste School of Race Relations," *Social Forces* 21, no. 2 (1942): 218–26; John P. Jackson, *Social Scientists for Social Justice: Making the Case Against Segregation, Critical America* (New York: New York University Press, 2001), 17–42; Alice O'Connor, *Poverty Knowledge: Social Science, Social Policy, and the Poor in Twentieth-Century U.S. History* (Princeton, NJ: Princeton University Press, 2001), 74–99; Pierre Saint-Arnaud, *African American Pioneers of Sociology: A Critical History*, trans. Peter Feldstein (Toronto: University of Toronto Press, 2009); Kamala Visweswaran, *Un/Common Cultures: Racism and the Rearticulation of Cultural Difference* (Durham, NC: Duke University Press, 2010); Henry

Yu, *Thinking Orientals: Migration, Contact, and Exoticism in Modern America* (New York: Oxford University Press, 2001); Davarian L. Baldwin, "Black Belts and Ivory Towers: The Place of Race in U.S. Social Thought, 1892–1948," *Critical Sociology* 30, no. 2 (2004); Michael Banton, "Race Relations," in *A Companion to Racial and Ethnic Studies*, ed. David Theo Goldberg and John Solomos (Malden, MA: Blackwell, 2002), xiii. For an analysis of the contributions of American and German sociology to imperial projects in the Global South and to the elaboration of an "imperial racism" in the early twentieth century, see Zimmerman, *Alabama in Africa*.

34. Daniel Immerwahr, "Caste or Colony? Indianizing Race in the United States," *Modern Intellectual History* 4, no. 2 (August 2007): 275–301.

35. Warner's article informed subsequent studies such as John Dollard's *Caste and Class in a Southern Town* (New Haven, CT: Yale University Press, 1937); Allison Davis and John Dollard's *Children of Bondage* (Washington DC: American Council of Education, 1940), and Allison Davis, Burleigh B. Gardner, and Mary R. Gardner, *Deep South: A Social Anthropological Study of Caste and Class*, ed. W. Lloyd Warner (Chicago, IL: University of Chicago Press, 1941).

36. W. Lloyd Warner, "American Caste and Class," *American Journal of Sociology* 42, no. 2 (1936): 234–37; 234.

37. Pierson, *Negroes in Brazil*, 348.

38. Pierson, *Negroes in Brazil*, 348.

39. Christophe Brochier, "De Chicago à São Paulo: Donald Pierson et la sociologie des relations raciales au Brésil," *Revue d'Histoire des Sciences Humaines* 25, no. 2 (2011): 293–324.

40. Brochier, "De Chicago à São Paulo," 296–97.

41. Alfred Métraux to Melville Herskovits, January 29, 1951, in *Race Question and Protection of Minorities, Pt. II from 1/VII/50 to 31/XII/51*, Dossier 323.1, Box 165, UNESCO Archives, Paris, France.

42. Records of the General Conference of Unesco, Fifth Session, Florence, 1950: Resolutions; 5 C/Resolutions, https://unesdoc.unesco.org/ark:/48223/pf0000114589, 40.

43. Alfred Métraux to Alva Myrdal, January 26, 1951, in *Race Question and Protection of Minorities, Pt. II from 1/VII/50 to 31/XII/51*, Dossier 323.1, Box 165, UNESCO Archives, Paris, France.

44. Métraux to Myrdal, January 12, 1951, in *Race Question and Protection of Minorities, Pt. II from 1/VII/50 to 31/XII/51*, Dossier 323.1, Box 165, UNESCO Archives, Paris, France.

45. Marcos Chor Maio, "UNESCO and the Study of Race Relations in Brazil: Regional or National Issue?," *Latin American Research Review* 36, no. 2 (2001): 118–36, 124.

46. Alfred Métraux au Director général, January 22, 1951, "Mission au Brésil 16 Nov–20 Dec 1950)," in *Race Question and Protection of Minorities, Pt. II from 1/VII/50 to 31/XII/51*, Dossier 323.1, Box 165, UNESCO Archives, Paris, France.

47. Métraux, "Mission au Brésil."

48. Métraux, "Mission au Brésil."

49. Métraux, "Mission au Brésil."

50. Never one to pass up an ethnographic opportunity, Métraux during this visit to Bahia also traveled to a remote Sertão village named Mirandela whose inhabitants confirmed the existence of the Kiriri—an indigenous group thought to have become extinct; see Alfred Métraux, "Une Nouvelle Langue Tapuya de La Région de Bahia (Brésil)," *Journal de La Société Des Américanistes* 40 (1951): 51–58.

51. Bastide joined the Universidade de São Paulo in 1938 when he took over the chair of sociology vacated by Claude Lévi-Strauss. See Stefania Capone, "Transatlantic Dialogue: Roger Bastide and the African American Religions," *Journal of Religion in Africa* 37, no. 3 (January 1, 2007): 336–70,340. According to Christophe Brochier, Pierson took over from the American anthropologist Robert Elliot Lowrie, see Brochier, "De Chicago a São Paulo," 296.

52. The committee was composed of Roger Bastide, Oracy Nogueira, Mario Wagner, Octavio Costa da Eduardo, and a "psychologist" to be appointed later.

53. Métraux, "Mission au Brésil."

54. Métraux, "Mission au Brésil." The quote in the original French reads: "Des conflits et des tensions sont en train de naître, à la suite du rapide développement industriel de Sao Paulo. Cette ville géante nous offre une occasion unique de connaître les facteurs susceptibles de sucsiter des antagonismes raciaux qui, autrefois, étaient à l'état latent ou manquaient de virulence. Je n'ignore pas qu'en organisant une étude à Sao Paulo nous risquons de parvenir à des conclusions qui ne répondront pas aux espoirs de ceux qui ont présenté et voté la résolution citée plus haut, mais ce serait trahir l'esprit scientifique qui doit animer notre enquête que d'écarter les problèmes nouveaux pour nous en tenir à un état de choses heureux mais périmé. L'enquête de Bahia ne peut offrir de la question raciale au Brésil qu'une image incomplète."

55. As Ian Merkel has shown, Freyre's interpretation of race mixing was very warmly received by French scholars in the 1940s before the UNESCO studies; see Ian Merkel, "Brazilian Race Relations, French Social Scientists, and African Decolonization: A Transatlantic History of the Idea of Miscegenation," *Modern Intellectual History* 17, no. 3 (September 2020): 801–32.

56. Romo, *Brazil's Living Museum.*

57. Charles Wagley, "Serendipity in Bahia: The History of Research Cooperation," *Universitas* 6–7 (1970): 29–41; 30.

58. Wagley, "Serendipity in Bahia," 31.

59. The Kubitschek government's state-led program was called the "Targets Plan," which targeted six key areas of growth: energy, transportation, food, basic industries, education, and the integration of the Brazilian interior into the national economy, which was to be achieved by the construction of Brasilia as a new state capital. On Kubitschek's plans and development debates in Brazil in the 1950s, see Rafael R. Ioris, "'Fifty Years in Five'and What's in It for Us? Development Promotion, Populism, Industrial Workers and Carestia in 1950s Brazil," *Journal of Latin American Studies* 44, no. 2 (2012): 261–84.

60. On Teixeira's role in Rio's educational system, see Jerry Dávila, *Diploma of Whiteness: Race and Social Policy in Brazil, 1917–1945* (Durham, NC: Duke University Press, 2003), 34–35.

61. Romo, *Brazil's Living Museum*, 133.

62. Romo, *Brazil's Living Museum*, p. 136–42.

63. Romo, *Brazil's Living Museum*, 142.

64. Charles Wagley, ed., *Race and Class in Rural Brazil*, 2nd ed., Race and Society (New York: UNESCO, 1963), 36.

65. Wagley, *Race and Class*, 57–58.

66. Wagley, *Race and Class*, 98–99.

67. Wagley, *Race and Class*, 98–99.

68. Wagley, *Race and Class*, 7–8.

69. Wagley, *Race and Class*, 9.

70. Wagley, *Race and Class*, 9.

71. Wagley, ed., *Race and Class*, 14.

72. Wagley, *Race and Class*, 154–55.

73. Wagley, *Race and Class*, 155.

74. Darién J. Davis, *Avoiding the Dark: Race and the Forging of National Culture in Modern Brazil*, Routledge Revivals (Abingdon, UK: Routledge, 2018).

75. Florestan Fernandes wrote the first draft of the essay, which was then read and revised by Roger Bastide after discussion between both.

76. For a useful overview of the São Paulo and UNESCO studies and how their methods differed from those of U.S. race relations research, see Roger Bastide, "Race Relations in Brazil," *International Social Science Bulletin*, 9, no. 4 (1957): 495–512.

77. The quote in the original Portuguese reads, "Nesse sentido, parece que o preconceito racial tende a desenvolverse como conseqüência natural do contacto inter-mitente ou contínuo de pessoas ou grupos de pessoas pertencentes a "raças" diversas(8), sempre que condições de desigualdade econômica e social contrastam marcas raciais com discrepâncias notórias quanto às ocupações, as riquezas, ao nível de vida, à posição social e à educação." Roger Bastide and

Florestan Fernandes, "O Preconceito Racial em São Paulo: Projeto do Estudo," in *Race Question and Protection of Minorities, Pt. II from 1/VII/50 to 31/XII/51*, 85, Dossier 323.1, Box 165, UNESCO Archives, Paris, France.

78. Bastide and Fernandes, "O Preconceito Racial," 88.

79. The quote in the original Portuguese reads, "A antiga ideologica racial não entrou em colapso immediatio nem perdue a função que possuia na ordem social escravocrata." Bastide and Fernandes, "O Preconceito Racial," 88.

80. Bastide and Fernandes, "O Preconceito Racial," 88.

81. Florestan Fernandes, "The Negro in Brazilian Society: Twenty-Five Years Later," in *Brazil, Anthropological Perspectives: Essays in Honor of Charles Wagley* (New York: Columbia University Press, 1979), 96–114; 96.

82. Roger Bastide and Florestan Fernandes, *Brancos e negros em São Paulo: ensaio sociológico sôbre aspectos da formação, manifestações atuais e efeitos do preconceito de côr na sociedade paulistana*, 2nd ed. (São Paulo: Companhia Editora Nacional, 1959), 348.

83. Bastide and Fernandes, *Brancos e negros*, xiii.

84. Oracy Nogueira, "Skin Color and Social Class," *VIBRANT—Vibrant Virtual Brazilian Anthropology* 5, no. 1 (2008): i–xxv; xii.

85. He also dedicated the book that came out of his UNESCO-sponsored study—*O Negro no Rio de Janeiro: relações de raças numa sociedade em mudança* (The Negro of Rio de Janeiro: race relations in a changing society)—to Ramos and Rodrigues.

86. Luiz Aguiar Costa-Pinto, *O Negro no Rio de Janeiro: relações de raças numa sociedade em mudança* (São Paulo: Companhia Editora Nacional, 1953), 33.

87. Costa-Pinto, *O Negro no Rio*, 311.

88. Costa-Pinto, *O Negro no Rio*, 24.

89. Costa-Pinto, *O Negro no Rio*, 327. Costa-Pinto borrowed the term from the sociologist Renzo Sereno's study of Puerto Rico. See Renzo Sereno, "Cryptomelanism: A Study of Color Relations and Personal Insecurity in Puerto Rico," *Psychiatry* 10, no. 3 (1947): 261–69.

90. Costa-Pinto, *O Negro no Rio*, 327.

91. Costa-Pinto, *O Negro no Rio*, 264.

92. Costa-Pinto, *O Negro no Rio*, 268.

93. Costa-Pinto, *O Negro no Rio*, 282.

94. Costa-Pinto, *O Negro no Rio*, 286.

95. The quote in the original Portuguese reads: "Vista assim—e assim é a dialética das coisas—a idéia da negritude, antes de mais nada, é o florescimento na cabeça de uma elite negra de uma se- mente que lá foi plantada pelas atitudes dos brancos. Noutras palavras, do mesmo modo que se pode aqui mais uma vez repetir que não há um problema do negro—pois o problema é o

branco que tem sôbre o negro falsas idéias e age de acôrdo com essas idéias falsas—também se poderia ·dizer, inversamente, que a idéia da negritude não é negra—é branca, é o reflexo invertido, na cabeça de negros, da idéia que os brancos fazem sôbre êle, é o resultado da tomada de consciência (também em têrmos falsos, diga-se de passagem) da resistência que o branco faz à ascenção social do negro. É, em suma, um racismo às avesas." Costa-Pinto, *O Negro no Rio*, 333.

96. Costa-Pinto, *O Negro no Rio*, 334.
97. Luiz de Aguiar Costa-Pinto, "Rio de Janeiro: Melting-Pot of Peoples," *UNESCO Courier* 5, no. 8–9 (1952): 10.
98. Fajardo, *World That Latin America Created*.
99. Jerry Dávila, *Hotel Trópico: Brazil and the Challenge of African Decolonization, 1950–1980* (Durham, NC: Duke University Press, 2010).
100. Barbara Freitag, "Florestan Fernandes Revisited," *Estudos Avançados* 19 (December 2005): 229–43.

7. A WHITE WORLD PERSPECTIVE AND THE COLLAPSE OF GLOBAL RACE RELATIONS INQUIRY

1. Richard Wright, *The Color Curtain: A Report on the Bandung Conference* (Cleveland, OH: World Publishing, 1956), 13.
2. Wright, *Color Curtain*, 14.
3. On Wright's connections to Chicago sociology, see Gabriel N. Mendes, *Under the Strain of Color: Harlem's Lafargue Clinic and the Promise of an Antiracist Psychiatry* (Ithaca, NY: Cornell University Press, 2015).
4. For an excellent analysis of Chicago sociologists' intellectual and political outlook, see Henry Yu, *Thinking Orientals: Migration, Contact, and Exoticism in Modern America* (New York: Oxford University Press, 2001).
5. Robert Vitalis makes a provocative case that Wright's romantic vision of Bandung as a gathering of all the darker and nonaligned nations and peoples has inflected much of the subsequent Bandung historiography at the expense of nuanced knowledge of the event; see Robert Vitalis, "The Midnight Ride of Kwame Nkrumah and Other Fables of Bandung (Ban-Doong)," *Humanity: An International Journal of Human Rights, Humanitarianism, and Development* 4, no. 2 (2013): 261–88.
6. Andrew Lind, ed., *Race Relations in World Perspective; Papers Read at the Conference on Race Relations in World Perspective, Honolulu, 1954* (Honolulu: University of Hawaii Press, 1955), ix.
7. Adom Getachew, *Worldmaking After Empire: The Rise and Fall of Self-Determination* (Princeton, NJ: Princeton University Press, 2019), 25.

8. Perrin Selcer, "The View from Everywhere: Disciplining Diversity in Post–World War II International Social Science," *Journal of the History of the Behavioral Sciences* 45, no. 4 (2009): 309–29; 314.

9. Andrew Lind, "Prospectus: Race Relations in World Perspective Conference," 3, General Correspondence: ZK Matthews Papers (DCAS ACC 101), University of South Africa Archives, Pretoria.

10. Lind, "Prospectus," 3.

11. Maile Arvin, *Possessing Polynesians: The Science of Settler Colonial Whiteness in Hawai'i and Oceania* (Durham, NC: Duke University Press, 2019), 98.

12. Arvin, *Possessing Polynesians*, 114–22.

13. "Opening Session, June 28, 1954," 2, Conference on Race Relations in World Perspective, J. B. Condliffe Papers, BANC MSS C-B 901, The Bancroft Library, University of California, Berkeley.

14. Rebecca Lemov, "Anthropology's Most Documented Man, Ca. 1947: A Prefiguration of Big Data from the Big Social Science Era," *Osiris* 32, no. 1 (September 2017): 21–42.

15. "Opening Session, June 28, 1954," 2, Conference on Race Relations in World Perspective.

16. Robert Park also played an important role in adapting the ecological model of succession for the purpose of studying social change in humans. See Robert E. Park, "Succession, an Ecological Concept," *American Sociological Review* 1, no. 2 (1936): 171–79.

17. Christine Leah Manganaro, "Assimilating Hawai'i: Racial Science in a Colonial 'Laboratory,' 1919–1939" (PhD diss., University of Minnesota, 2012), 262.

18. Manganaro, "Assimilating Hawai'i," 260–71.

19. Lind, "Prospectus," 1.

20. Lind, "Prospectus," 1.

21. Lind, *Race Relations in World Perspective*, ix.

22. Andrew Lind, "The Conference on Race Relations in World Perspective," *Social Process* 19 (1955): 13–19; 14.

23. "Handbook of General Information for Conference Participants," Conference on Race Relations in World Perspective, J. B. Condliffe Papers, BANC MSS C-B 901, The Bancroft Library, University of California, Berkeley.

24. Masuoka immigrated to Hawai'i from Japan when he was thirteen and went on to complete a PhD in sociology under E. B. Reuter. He then spent most of his career in the Department of Social Science at Fisk University—a historically black university. See Greg Robinson, "Jitsuichi Masuoka," in *Densho Encyclopedia*, accessed October 24, 2022, https://encyclopedia.densho.org/Jitsuichi%20Masuoka.

25. "Second Plenary Session—June 30, 1954," 1, Conference on Race Relations in World Perspective, J. B. Condliffe Papers, BANC MSS C-B 901, The Bancroft Library, University of California, Berkeley.

26. "Second Plenary Session—June 30, 1954," 1–2, Conference on Race Relations in World Perspective.

27. "Second Plenary Session—June 30, 1954," 2, Conference on Race Relations in World Perspective.

28. "Second Plenary Session—June 30, 1954," 3, Conference on Race Relations in World Perspective.

29. "Second Plenary Session—June 30, 1954," 3, Conference on Race Relations in World Perspective.

30. The society's founder and charismatic leader, Gopal Krishna Gokhale, was one of Mahatma Gandhi's mentors. Gokhale visited Gandhi during his tumultuous years in South Africa and was the first to invite him back to India and join the battle against colonialism. See Elena Valdameri, *Indian Liberalism Between Nation and Empire: The Political Life of Gopal Krishna Gokhale* (London: Routledge, 2022).

31. "P. Kodanda Rao," in "Handbook of General Information for Conference Participants," Conference on Race Relations in World Perspective, J. B. Condliffe Papers, BANC MSS C-B 901, Bancroft Library, University of California, Berkeley.

32. "Second Plenary Session—June 30, 1954," 3, Conference on Race Relations in World Perspective.

33. "Third Plenary Session—July 1, 1954," 3, Conference on Race Relations in World Perspective, J. B. Condliffe Papers, BANC MSS C-B 901, The Bancroft Library, University of California, Berkeley.

34. Hourani played an important role in institutionalizing Middle East studies at Oxford University; see Roger Owen, "Albert Hourani and the Making of Modern Middle East Studies in the English-Speaking World: A Personal Memoir," in *Arabic Thought Beyond the Liberal Age: Towards an Intellectual History of the Nahda*, ed. Jens Hanssen and Max Weiss (Cambridge: Cambridge University Press, 2016), 41–61.

35. "Fourth Plenary Session—July 2, 1954," 1, Conference on Race Relations in World Perspective, J. B. Condliffe Papers, BANC MSS C-B 901, The Bancroft Library, University of California, Berkeley.

36. Albert Hourani, "The Concept of Race Relations: Thoughts After a Conference," *International Social Science Bulletin* 7, no. 2 (1955): 335–40; 335.

37. Hourani, "Concept of Race Relations," 336.

38. Hourani, "Concept of Race Relations," 335.

39. "Suggestions for Further Research," Conference on Race Relations in World Perspective, J. B. Condliffe Papers, BANC MSS C-B 901, The Bancroft Library, University of California, Berkeley.

40. "Twenty-third Plenary Session—July 23, 1954," 1, Conference on Race Relations in World Perspective, J. B. Condliffe Papers, BANC MSS C-B 901, The Bancroft Library, University of California, Berkeley.

41. "Twenty-third Plenary Session—July 23, 1954," 2, Conference on Race Relations in World Perspective.

42. "Twenty-third Plenary Session—July 23, 1954," 2, Conference on Race Relations in World Perspective.

43. Andrew Lind, "The Conference on Race Relations in World Perspective," *Social Process* 19 (1955): 13–19; 16.

44. Lind, "Conference on Race Relations," 18.

45. "Proposal for an International Organization in the Field of Race Relations," Conference on Race Relations in World Perspective, J. B. Condliffe Papers, BANC MSS C-B 901, The Bancroft Library, University of California, Berkeley.

46. "Twenty-third Plenary Session—July 23, 1954," 2, Conference on Race Relations in World Perspective.

47. After the conference, Conant went on to become a policy expert specializing in the geopolitics of oil and energy. Melvin Conant, *Race Issues on the World Scene: A Report on the Conference on Race Relations in World Perspective, Honolulu, 1954* (Honolulu, Hawai'i: University of Hawai'i Press, 1955), 142.

48. Lind, "Conference on Race Relations," 18.

49. Pierre Saint-Arnaud, *African American Pioneers of Sociology: A Critical History*, trans. Peter Feldstein (Toronto: University of Toronto Press, 2009), 217.

50. Saint-Arnaud, *African American Pioneers*, 217.

51. See M. F. Keen, *Stalking Sociologists: J. Edgar Hoover's FBI Surveillance of American Sociology* (New Brunswick, NJ: Transaction, 2004); Saint-Arnaud, *African American Pioneers of Sociology*.

52. Kevin Gaines, "E. Franklin Frazier's Revenge: Anticolonialism, Nonalignment, and Black Intellectuals' Critiques of Western Culture," *American Literary History* 17, no. 3 (2005): 506–29.

53. E. Franklin Frazier to Alfred Métraux, October 5, 1954, 323.12 Race Relations, UNESCO Archives, Paris.

54. Alfred Métraux to E. Franklin Frazier, October 22 1954, 323.12 Race Relations, UNESCO Archives, Paris.

55. Alfred Métraux to E. Franklin Frazier, December 10, 1954, 323.12 Race Relations, UNESCO Archives, Paris.

56. Alfred Métraux to Alva Myrdal, December 20, 1954, 323.12 Race Relations, UNESCO Archives, Paris.

57. Métraux to Myrdal, December 20, 1954.

58. Métraux to Myrdal, December 20, 1954.

59. E. Franklin Frazier and William O. Brown to Alfred Métraux, January 12, 1955, 323.12 Race Relations, UNESCO Archives, Paris.

60. Frazier and Brown to Métraux, January 12, 1955.

61. Frazier and Brown to Métraux, January 12, 1955.

62. J. A. Barnes to Alfred Métraux, January 28, 1955, 323.12 Race Relations, UNESCO Archives, Paris.

63. Alfred Métraux to J. A. Barnes, February 3, 1955, 323.12 Race Relations, UNESCO Archives, Paris.

64. Métraux to Barnes, February 3, 1955.

65. Métraux to Barnes, February 3, 1955.

66. Alfred Métraux to E. Franklin Frazier, February 3, 1955, 323.12 Race Relations, UNESCO Archives, Paris.

67. Métraux to Frazier, February 3, 1955.

68. Métraux to Frazier, February 3, 1955.

69. Métraux to Frazier, February 3, 1955.

70. Métraux to Frazier, February 3, 1955.

71. Métraux to Frazier, February 3, 1955.

72. Georges Balandier to E. Franklin Frazier, February 18, 1955, 323.12 Race Relations, UNESCO Archives, Paris.

73. E. Franklin Frazier to Alva Myrdal, May 31, 1955, 323.12 Race Relations, UNESCO Archives, Paris.

74. Frazier to Myrdal, May 31, 1955.

75. Frazier to Myrdal, May 31, 1955.

76. Frazier to Myrdal, May 31, 1955.

77. Frazier to Myrdal, May 31, 1955.

78. Circular letter from International Society for the Scientific Study of Race Relations to All Members, Folder: International Society for the Scientific Study of Race Relations, 1956–1957, Horace Mann Bond Papers (MS 411), Special Collections and University Archives, University of Massachusetts Amherst Libraries.

79. Alfred Métraux to T. H. Marshall, December 19, 1956, SSMemo/ME/hjl, 323.12 Race Relations, UNESCO Archives, Paris.

80. Métraux to Marshall, December 19, 1956.

81. "Foreword: Recent Research on Race Relations," *International Social Science Bulletin* 10, no. 3 (1958): 343–44.

82. Anthony H. Richmond, "Britain," *International Social Science Bulletin* 10, no. 3 (1958): 344–71; 370.

83. Barbara E. Ward, "East Africa," *International Social Science Bulletin* 10, no. 3 (1958): 372–86; 372.

84. Ward, "East Africa," 373.

85. "Foreword: Recent Research on Race Relations—II," *International Social Science Journal* 13, no. 2 (1961): 175–76; 175.

86. "Foreword: Recent Research on Race Relations—II," *International Social Science Journal*, 175.

87. "Foreword: Recent Research on Race Relations—II," *International Social Science Journal*, 175.

88. Yu, *Thinking Orientals*, 43.

89. Eduardo Bonilla-Silva, "Rethinking Racism: Toward a Structural Interpretation," *American Sociological Review* 62, no. 3 (June 1997): 465.

CONCLUSION: "RACISM CONTINUES TO HAUNT THE WORLD"

1. Alfred Métraux, "Does Life End at Sixty?," *UNESCO Courier*, April 1963, 20–23; 20.

2. Ronald Niezen, *The Origins of Indigenism: Human Rights and the Politics of Identity* (Berkeley: University of California Press, 2003); Marisol de la Cadena and Orin Starn, eds., *Indigenous Experience Today* (London: Taylor & Francis, 2007); Miranda C. L. Johnson, *The Land Is Our History: Indigeneity, Law, and the Settler State* (New York: Oxford University Press, 2016).

3. "UNESCO's Action in the Field of Race Relations and the Statements on Race; Note by the Secretariat," Paris, March 2 1964, 3, https://unesdoc.unesco.org/ark:/48223/pf0000157691?posInSet=1&queryId=001af11b-6b73-4df3-9053-e44db9e4350e, SS/Race/1WS/0364.42/SS, 4.

4. For instance, the scientific director for the 1964 meeting was the Belgian human biologist Jean Hiernaux, who had previously served as a university administrator in the Belgian Congo and in Rwanda. The 1964 committee also included (among others) the French medical anthropologist and Caribbeanist Jean Benoist, the Brazilian geneticist Francisco Salzano, the hematologist Yaya Kane from the Senegalese Pasteur Institute, the "sociometrist" Ramakrishna Mukherjee from the Indian Statistical Institute, and the physical anthropologist Adelaida Díaz de Ungría from Venezuela. In addition to Hiernaux, who participated in the 1964 and 1967 meetings, notable participants in the 1967 meeting include the Trinidadian sociologist Lloyd Brathwaite, the French Africanist Georges Balandier, the U.S. sociologist Leonard Broom, who had studied the Japanese interment camps and made several trips as a Fulbright scholar to the Australian National University, and Muddathir Abdel Rahim, a political scientist from the University of Khartoum in Sudan. See "Biography: Jean Hiernaux," Musée royal de l'Afrique centrale, Tervuren, Fonds Jean Hiernaux,

https://archives.africamuseum.be/agents/people/167; For a list of the participants in the 1967 meeting, see *Final Report: Meeting of Experts on Race and
Racial Prejudice*, Paris, December 14, 1967, 6, https://unesdoc.unesco.org/ark:/
48223/pf0000186092?posInSet=9&queryId=802c5867-0133-4428-9182-21842c
90c74b, SHC/CS/122/8.

5. *Final Report: Meeting of Experts on Race and Racial Prejudice*, Paris, December
14 1967, 2.
6. *Final Report: Meeting of Experts on Race and Racial Prejudice*, Paris, December
14 1967, 2.
7. *Final Report: Meeting of Experts on Race and Racial Prejudice*, Paris, December
14 1967, 21.
8. *Final Report: Meeting of Experts on Race and Racial Prejudice*, Paris, December
14 1967, 22.
9. *Final Report: Meeting of Experts on Race and Racial Prejudice*, Paris, December
14 1967, 22.
10. *Final Report: Meeting of Experts on Race and Racial Prejudice*, Paris, December
14 1967, 2.
11. *Final Report: Meeting of Experts on Race and Racial Prejudice*, Paris, December
14 1967, 2.
12. Notable pieces of legislation against racism leading up to the 1978 declaration
include: the ILO's Convention Against Discrimination in Respect to Employment and Occupation (1958), UNESCO's Convention Against Discrimination
in Education (1960), the UN Declaration on the Elimination of All Forms of
Racial Discrimination (1963), and the UN International Convention on the
Elimination of All of Racial Discrimination (1965). See Natan Lerner, "New
Concepts in the UNESCO Declaration on Race and Racial Prejudice," *Human
Rights Quarterly* 3, no. 1 (February 1981): 48.
13. *Declaration on Race and Racial Prejudice: Adopted by the General Conference
of Unesco at Its Twentieth Session, Paris, 27 November 1978* (Paris: Unesco,
1979), 12.
14. *Declaration on Race and Racial Prejudice*, 10.
15. *Declaration on Race and Racial Prejudice*, 13.
16. Métraux, "Does Life End at Sixty?," 23.
17. "Alfred Métraux a Pierre Verger," Paris, le 5 avril 1963, in *Le pied à l'étrier: correspondance*, 12 mars 1946–5 avril 1963, ed. Jean-Pierre le Boulier (Paris: Cahiers
de Gradhiva, 1994), 306.
18. Claude Levi-Strauss and R. d'Harcourt, "ALFRED MÉTRAUX (1902–1963),"
Journal de La Société Des Américanistes 52 (1963): 301–11.
19. Pierre Clastres, "Hommage à Alfred Métraux," in *Présence d'Alfred Métraux*
(Paris: Acéphale-Les amis de Georges Bataille, 1992), 31–34; 32.

20. Johannes Fabian, *Time and the Other: How Anthropology Makes Its Object* (New York: Columbia University Press, 1983).

21. A similar point is made about Mexico and the United States in Karin Alejandra Rosemblatt, *The Science and Politics of Race in Mexico and the United States, 1910–1950* (Chapel Hill: University of North Carolina Press, 2018).

22. Joanna Radin, *Life on Ice: A History of New Uses for Cold Blood* (Chicago: University of Chicago Press, 2017).

23. *Research in Population Genetics of Primitive Groups: Report of a WHO Scientific Group* (Geneva: World Health Organization, 1964), 4.

24. "Note sur qulques groupes indigénes de l'Amerique du sud relativement isolés et primitifs," 28 novembre 1962, Fonds Alfred Métraux, FAM.AS.AA.15.14: Documents de l'Organisation Mondiale de la Santé (O.M.S), Archives des Ethnologues, Biblothèque d'Anthropologie Sociale, Laboratoire d'Anthropologie Sociale, Collège de France, Paris.

25. "Carleton Gajdusek to Pierre Clastres, July 26, 1963," in *Paraguayan Indian Expeditions to the Guayaki and Chako Indians, August 25, 1963 to September 28, 1963*, ed. Daniel Carleton Gajdusek (Bethesda, MD: National Institute of Neurological Diseases and Blindness, National Institutes of Health, 1963), 76.

26. Kim TallBear, "Genomic Articulations of Indigeneity," *Social Studies of Science* 43, no. 4 (2013): 509–33.

Index

Page numbers in *italics* refer to illustrations or tables.

Printed and bound by CPI Group (UK) Ltd, Croydon, CR0 4YY

16/04/2025

14658572-0001